A Biologist's Guide to Artificial Intelligence

A Biologist's Guide to Artificial Intelligence

Building the Foundations of Artificial Intelligence and Machine Learning for Achieving Advancements in Life Sciences

Edited by

Ambreen Hamadani

National Institute of Technology, Srinagar, Jammu and Kashmir, India

Nazir A. Ganai

Sher-e-Kashmir University of Agricultural Sciences and Technology, Kashmir, Jammu and Kashmir, India

Henna Hamadani

Faculty of Veterinary Sciences and Animal Husbandry, Sher-e-Kashmir University of Agricultural Sciences and Technology, Kashmir, Jammu and Kashmir, India

J. Bashir

National Institute of Technology, Srinagar, Jammu and Kashmir, India

ACADEMIC PRESS

An imprint of Elsevier

ELSEVIER

Academic Press is an imprint of Elsevier
125 London Wall, London EC2Y 5AS, United Kingdom
525 B Street, Suite 1650, San Diego, CA 92101, United States
50 Hampshire Street, 5th Floor, Cambridge, MA 02139, United States

Notices
Knowledge and best practice in this field are constantly changing. As new research and experience broaden our understanding, changes in research methods, professional practices, or medical treatment may become necessary.

Practitioners and researchers must always rely on their own experience and knowledge in evaluating and using any information, methods, compounds, or experiments described herein. In using such information or methods they should be mindful of their own safety and the safety of others, including parties for whom they have a professional responsibility.

To the fullest extent of the law, neither the Publisher nor the authors, contributors, or editors, assume any liability for any injury and/or damage to persons or property as a matter of products liability, negligence or otherwise, or from any use or operation of any methods, products, instructions, or ideas contained in the material herein.

ISBN: 978-0-443-24001-0

For information on all Academic Press publications visit our website at https://www.elsevier.com/books-and-journals

Publisher: Stacy Masucci
Acquisitions Editor: Linda Bushman
Editorial Project Manager: Billie Jean Fernandez
Production Project Manager: Swapna Srinivasan
Cover Designer: Matthew Limbert

Typeset by TNQ Technologies

Working together
to grow libraries in
developing countries

www.elsevier.com • www.bookaid.org

Contents

CHAPTER 4   Decoding life: Genetics, bioinformatics, and artificial intelligence 47

Parvaze A. Sofi, Sajad Majeed Zargar, Ambreen Hamadani, Sadiah Shafi, Aaqif Zaffar, Ishrat Riyaz, Deepak Bijarniya and P.V. Vara Prasad

CHAPTER 10   Advancing precision agriculture through artificial intelligence: Exploring the future of cultivation 151

Rohitashw Kumar, Muneeza Farooq and Mahrukh Qureshi

Contributors

Syed Aadam Ahmad
Department of Information Technology, Cluster University Srinagar, Srinagar, Jammu and Kashmir, India

Sami Alshmrany
Department of Physics, Goverment Degree College Sumbal, Sumbal, Jammu and Kashmir, India

Syed Immamul Ansarullah
Department of Computer Applications, Government Degree College Sumbal, Sumbal, Jammu and Kashmir, India

Alwi Bamhdi
Department of Computer Science, Umm AlQura University, Mecca, Saudi Arabia

Nelofar Banday
Division of Floriculture and Landscape Architecture, SKUAST Kashmir, Srinagar, Jammu and Kashmir, India; Sher-e-Kashmir University of Agricultural Sciences and Technology of Kashmir, Srinagar, Jammu and Kashmir, India

Umar Bashir
National Institute of Technology, Srinagar, Jammu and Kashmir, India

Deepak Bijarniya
SKUAST Kashmir, Srinagar, India

Mahvish Khurshid Bijli
National Institute of Technology, Srinagar, Jammu and Kashmir, India

Megdalia Bromhal
University of North Carolina Wilmington, Computer Science Department, Wilmington, NC, United States

Gulustan Dogan
University of North Carolina Wilmington, Computer Science Department, Wilmington, NC, United States

Muneeza Farooq
College of Agricultural Engineering and Technology, Sher-e-Kashmir University of Agricultural Sciences and Technology of Kashmir, Srinagar, Jammu and Kashmir, India

Ufaq Fayaz
Division of Food Science and Technology, Sher-e-Kashmir University of Agriculture Sciences and Technology of Kashmir, Srinagar, Jammu and Kashmir, India

Arfat Firdous
Faculty of Computer and Information Systems, Islamic University of Madinah, Madinah, Saudi Arabia

Nazir Ahmad Ganai
Sher-e-Kashmir University of Agricultural Sciences and Technology of Kashmir, Srinagar, Jammu and Kashmir, India

Ambreen Hamadani
National Institute of Technology, Srinagar, Jammu and Kashmir, India

Henna Hamadani
Sher-e-Kashmir University of Agricultural Sciences and Technology of Kashmir, Srinagar, Jammu and Kashmir, India

Shamsul Hauq
National Institute of Technology, Srinagar, Jammu and Kashmir, India

Syed Zameer Hussain
Division of Food Science and Technology, Sher-e-Kashmir University of Agriculture Sciences and Technology of Kashmir, Srinagar, Jammu and Kashmir, India

Mehreen Khaleel
Wildlife Research and Conservation Foundation, Srinagar, Jammu and Kashmir, India

Asra Khanam
University of Kashmir, Srinagar, India

Burhan Khursheed
Department of Electronics and Communication, NIT, Srinagar, India

Mudasir Manzoor Kirmani
Department of Computer Science, Division of Social Science, FoFy, SKUAST-Kashmir, Srinagar, Jammu and Kashmir, India

Rohitashw Kumar
College of Agricultural Engineering and Technology, Sher-e-Kashmir University of Agricultural Sciences and Technology of Kashmir, Srinagar, Jammu and Kashmir, India

Aqsa Ashraf Makhdomi
National Institute of Technology, Srinagar, Jammu and Kashmir, India

Faheem Syeed Masoodi
University of Kashmir, Srinagar, India

SukhDev Mishra
Department of Biostatistics, ICMR-National Institute of Occupational Health, Ahmedabad, Gujarat, India

Qazi Hammad Mueen
Department of Biological Sciences, Middle East Technical University, Üniversiteler Mahallesi, Çankaya, Ankara, Turkey

Naureen Murtaza
Department of Environmental Sciences, Faculty of Engineering and Technology, Jamia Millia Islamia, New Delhi, India

Uzmat Ul Nisa
National Institute of Technology, Srinagar, Jammu and Kashmir, India

O.J. Ogieriakhi
Department of Mathematical and Physical Sciences, College of Basic and Applied Sciences, Glorious Vision University, Ogwa, Edo State, Nigeria

O.H. Onyijen
Department of Mathematical and Physical Sciences, College of Basic and Applied Sciences, Glorious Vision University, Ogwa, Edo State, Nigeria

S. Oyelola
Department of Mathematical and Physical Sciences, College of Basic and Applied Sciences, Glorious Vision University, Ogwa, Edo State, Nigeria

Tahiya Qadri
Division of Food Science and Technology, Sher-e-Kashmir University of Agriculture Sciences and Technology of Kashmir, Srinagar, Jammu and Kashmir, India

Syed Fatima Qadri
Wildlife Research and Conservation Foundation, Srinagar, Jammu and Kashmir, India

Shazeena Qaiser
Government Dental College, Srinagar, Jammu and Kashmir, India

Mahrukh Qureshi
College of Agricultural Engineering and Technology, Sher-e-Kashmir University of Agricultural Sciences and Technology of Kashmir, Srinagar, Jammu and Kashmir, India

Rukia Rahman
University of Kashmir, Srinagar, Jammu and Kashmir, India

Ishrat Riyaz
SKUAST Kashmir, Srinagar, India

Shabia Shabir
Islamic University of Science and Technology (IUST), Awantipora, Jammu and Kashmir, India

Sadiah Shafi
SKUAST Kashmir, Srinagar, India

Immad A. Shah
ICMR-National Institute of Occupational Health, Ahmedabad, Gujarat, India

Parvaze A. Sofi
SKUAST Kashmir, Srinagar, India

Doorva Vaidya
University of North Carolina Wilmington, Computer Science Department, Wilmington, NC, United States

P.V. Vara Prasad
Kansas State University, Manhattan, KS, United States

Tabinda Wani
Division of Floriculture and Landscape Architecture, SKUAST Kashmir, Srinagar, Jammu and Kashmir, India

Nazrana Rafique Wani
Division of Food Science and Technology, Sher-e-Kashmir University of Agriculture Sciences and Technology of Kashmir, Srinagar, Jammu and Kashmir, India

Arielle Yoo
University of California, Davis, CA, United States

Abida Yousuf
National Institute of Technology, Srinagar, Jammu and Kashmir, India

Aaqif Zaffar
SKUAST Kashmir, Srinagar, India

Aaqib Zahoor
National Institute of Technology, Srinagar, Jammu and Kashmir, India

Sajad Majeed Zargar
SKUAST Kashmir, Srinagar, India

Exploring artificial intelligence through a biologist's lens

Shabia Shabir[1] and Ambreen Hamadani[2]
[1]*Islamic University of Science and Technology (IUST), Awantipora, Jammu and Kashmir, India;*
[2]*National Institute of Technology, Srinagar, Jammu and Kashmir, India*

Introduction

Artificial intelligence (AI) is a simulation technique that can create intelligent agents, usually with the learning capability of using perception in the form of data or sensors. Machine learning, being an area within AI, focuses on the design and development of algorithms to induce the learning capability in AI agents. Statistics is the technique that is extensively being used for analytical surveys, especially in Data Science. However, the main difference between statistical concepts and AI is that the former works by assuming a situation (hypothesis) and then testing its validity through some tests. In contrast, AI can create/predict a situation based on the data being provided (Krenn et al., 2022). Data mining is a prominent area of AI that focuses on extracting valuable insights, hidden patterns, or unknown knowledge from large datasets, databases, or data warehouses. It draws inspiration from various fields such as machine learning, AI, and statistics to develop techniques and algorithms for effective data analysis. By employing data mining methods, researchers and analysts can uncover meaningful relationships, trends, and patterns within complex and vast datasets that may not be apparent through traditional methods of analysis. These extracted insights can then be utilized for decision-making, predictive modeling, anomaly detection, and other valuable applications in diverse industries and domains (López, 2005). While machine learning, AI, and statistics share common goals, they employ different approaches to achieve them. Each field brings its unique methodologies and techniques to table. AI finds application in a wide range of areas, including robotics, strategic planning and scheduling, manufacturing, and maintenance. In robotics, AI techniques are used to develop intelligent systems that can perceive, reason, and act in the physical world. Strategic planning and scheduling benefit from AI algorithms to optimize decision-making processes and resource allocation. Manufacturing utilizes AI for automation, quality control, and process optimization. Maintenance processes can be enhanced through predictive analytics and machine learning algorithms to detect potential failures and schedule maintenance activities proactively. Collaboration and knowledge sharing among researchers from different fields are crucial. By exchanging their insights and experiences, experts from various domains can contribute to the development of new technologies and approaches. This interdisciplinary collaboration can lead to enhanced understanding and the extraction

A Biologist's Guide to Artificial Intelligence. https://doi.org/10.1016/B978-0-443-24001-0.00001-4

of hidden knowledge within their respective fields, ultimately fostering innovation and advancement in AI applications. Fig. 1.1 gives a better understanding of the relationship between statistics and AI.

Due to the increase in volume/size and complexity of biological data, there became a necessity to introduce AI techniques in the field to create various prospective and predictive models. This would enhance the diagnostic system in the medical field and help in error-free analysis (Na, 2020). Various techniques in AI have been introduced that are well suited to the type of medical data available such as gene expression/sequence, protein/molecular structure, or images. Even the surveys collected from patients could be helpful in creating various analytical models for future perception.

Before we go through the various AI techniques used extensively in biological science, we must have a good knowledge of the various concepts often used in AI. Some of them are listed below:

1. **Dataset:** The dataset is formed after the data are collected from heterogeneous sources through the process of extract, transform, and load (ETL) or ELT. This shall process the data in a standardized form suitable for analysis (Hassoun et al., 2021). This is usually in the form of rows and columns, wherein the columns/attributes represent the features or attributes of a particular row or selection. Each row consists of inputs $(X_1, X_2, X_3, ..., X_n)$ and output as (Y).
2. **Feature selection:** Various features are considered to be irrelevant concerning the type of analysis we need to perform. Although the whole dataset may be meaningful data, however, we need to select some of the features that are most relevant and can strongly predict the output of the AI model (Setua et al., 2017). This is done through the process of feature selection using various techniques like filter or wrapper methods (Li et al., 2019).
3. **Training and testing dataset:** The dataset is divided into two sub-datasets, that is, training dataset is used when we want to train the model, wherein the labeled output is provided with an input set. This labeled output indicates the class to which the record belongs, whereas the testing dataset is without the output that is used to assess the performance of the model, that is, how well

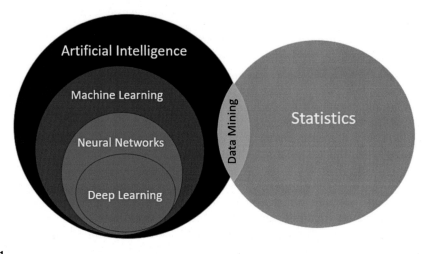

FIGURE 1.1

Artificial intelligence (AI) in relation to statistics and other fields.

a particular learned model predicts (Géron, 2019). The model learns from training by adjusting its internal parameters or weights based on the patterns and relationships it discovers. The size and diversity of the dataset determine the model's ability to generalize and make accurate predictions on unseen data.

4. **Machine learning:** It refers to the process by which an AI system acquires knowledge or improves its performance through experience or data, thus enabling the system to make predictions, recognize patterns, solve problems, and perform various tasks. Learning in AI is of two types - supervised and unsupervised learning (Ang et al., 2016).

5. **Supervised learning:** In this learning, a labeled dataset is being provided with defined output as a training set and the test dataset, which is without the output is being used to determine the performance of the model. Some of the models that work on this principle include classification models like KNN, trees, random forest, logistic regression, NaiveBayes, SVM, etc.

6. **Unsupervised learning:** In this learning, an unlabeled dataset with no defined output is provided to make the model learn about the various similarities in the data. Some of the models that work on this principle include various clustering models like K-means, PCA, etc, and association analysis such as Apriori, FP-growth, and Hidden Markow models.

7. **Cost function:** In machine learning, there is a measure to assess the discrepancy between the model's predictions and the actual output, known as the error or loss function. This function quantifies the extent to which the model deviates from accurately capturing the relationship between the input and output data. While accuracy metrics indicate how well the model is performing overall, they do not provide specific insights on how to enhance its performance. To address this, a corrective or optimization function is utilized to determine when the model is most accurate or when the error is minimized. This function aids in the computation of adjustments that can be made to improve the model's predictions. By iteratively optimizing the model based on the error function, machine learning algorithms can refine their performance and enhance their accuracy (Lim and Kang, 2011). An example illustrating the concept of a cost function can be seen in a scenario where a robot is assigned the task of lifting and stacking boxes, potentially encountering obstacles in the process. In this case, the cost function is utilized to determine the minimum value of the root mean square error (RMSE). The RMSE is calculated by taking the square root of the average of the squared differences between the predicted values and the actual values. By employing the cost function and minimizing the RMSE, the robot can optimize its movements and actions to minimize the discrepancy between its predicted outcomes and the actual outcomes. This allows the robot to improve its accuracy and efficiency in completing the given task, navigating around obstacles, and achieving the desired stacking of boxes.

8. **Overfitting and underfitting:** Overfitting occurs when an AI model attempts to learn from data that may contain noise and tries to fit each data point precisely on a curve. This phenomenon arises when the model lacks flexibility, leading to poor predictions on new data points. Consequently, the model may reject every new data point during prediction. Overfitting is often associated with models that have high variance, and it can be triggered by factors such as unclean data or an insufficient training dataset. On the other hand, underfitting refers to a situation where the model fails to learn the relationships between variables in the data or makes accurate predictions or classifications on new data points. Since the model does not fully grasp the underlying patterns, it accepts every new data point during prediction, regardless of its accuracy.

Underfitting typically occurs when there is unclean data or when the model has high bias, implying that it oversimplifies the relationships between variables. In summary, overfitting and underfitting represent two extremes in model performance. Overfitting occurs when the model learns the training data too well, while underfitting arises when the model fails to capture the underlying patterns in the data effectively. Both situations can be problematic and can be mitigated by using appropriate techniques such as regularization, cross-validation, or increasing the complexity of the model.

Machine learning algorithms—the foundations of AI

The first challenging task when applying AI is to select the appropriate model based on which certain predictions and patterns would be identified. The task is challenging because the better the model, the better would be the decision. Various steps are involved in the selection of the model:

1. **Defining a task:** This involves specifying a problem and objective that needs to be resolved by the AI model. This would give a clear insight into what the purpose of an AI system should be. The problem statement is being provided at this stage.
2. **Obtaining dataset:** Once the problem is known, sufficient and relevant data should be made available for implementation to avoid overfitting and underfitting issues. The dataset may include the relevant features or attributes that need to be focused on for analytical purposes. In the case of classification, labeled target output should be made available.
3. **Designing test and training set:** Once the dataset is available, we need to create subsets of the dataset to determine the records that need to be utilized for training purposes and the rest for testing purposes. The more the size and diversity of the training set, the better the ability of the model to provide accurate predictions on unseen data. Further, the test set estimates how well the model is likely to perform on unseen or real-world data (Bishop and Nasrabadi, 2006).
4. **Selection of the model:** This involves choosing the most suitable algorithm or architecture that can effectively solve the problem at hand. The selection depends on various factors, including the nature of the data, the complexity of the task, the data available, the desired performance metrics, etc. Certain criteria can help out in the process of selection (Greener et al., 2022): (a) If the data are sufficient to be provided to the model with a fixed number of labeled features, then we can predict the class or value by either going with traditional classification algorithms like multilayer perception, SVM, random forest, etc., with labeled data or else go with clustering algorithm in case of unlabeled data. (b) If the data are more than sufficient with a large number of unlabeled features and there is a connection between various entities in the data, then a graph convolutional network (GCN) can be used (Ye et al., 2023). (c) If, in the large dataset, no connections are being identified, and the data are spatial or image, then the choice could be 2D/3D convolutional network or else if data are sequential instead of spatial, then the choice could be recurrent neural network (RNN) or 1D convolutional network (Esteva et al., 2017).
5. **Training and testing:** Once the model is selected, we need to train it using a training dataset, and it can tune the model through various iterations by adjusting its internal parameters or weights based on the patterns and relationships it discovers. The test set would determine the accuracy of the model in the case of the labeled training dataset.

Integrating AI with biological science

With the increase in size, complexity, and type of biological data, it became a necessity to introduce some computational methodology to improve the diagnostic and prediction system. AI has helped and affected various fields of biology like genome analytics where genomic data are analyzed using various AI algorithms, and further can be used for protein structure and genes. It may help in identifying the genetic variants associated with diseases and help in drug development. Apart from that medical diagnostics is another field, wherein AI analyze medical tests in the form of images (MRI/X-rays) for disease detection and classification. AI helps in the protection of the ecosystem by analyzing environmental behavior and changes. It helps in resource management and conservation. Various AI algorithms listed below have effectively helped in various fields of biological science.

Logistic regression

Logistic regression is a statistical model used for binary classification tasks, where the goal is to determine the probability that an input belongs to one of two classes. It is a popular and widely used algorithm in machine learning and statistics. A logistic regression model is based on a logistic function, also known as a sigmoid function, which maps any real-valued number to a value between 0 and 1. In logistic regression, the input features are combined linearly using weights, and the results are passed through a sigmoid function to obtain a predicted probability. In mathematics, a logistic regression model can be represented as:

$P(y = 1|x) = \text{sigmoid}(w^T x + b)$. where: $P(y = 1|x)$ is the probability that the output variable y equals 1 given the input characteristic x. "w" is the weight vector associated with the input features, and "b" is the negative term.

In training, logistic regression models are fitted to the training data, which are labeled and are having fixed number of features (Kleinbaum et al., 2002). Logistic regression has many advantages, including simplicity, interpretability, and efficiency, and has a good benchmark. It can handle both numerical and categorical input features and can be extended to handle multi-class classification problems. However, logistic regression assumes a linear relationship between input characteristics and the log-oddness of the outcome. If the relationship is not linear, that is, complex feature relationship, feature engineering, or more complex models may be required. This model also overfits if the number of features is large.

An application of AI can be observed in the field of protein-variant effect prediction, where researchers in biology have introduced a method called DeMaSk. This method aims to predict the impact of missense mutations in proteins by leveraging deep mutation scanning (DMS) databases and sequence homologs. DeMaSk exhibits state-of-the-art performance in accurately predicting the effect of amino acid substitutions and can be readily applied to protein sequences of any type. This approach holds promise for enhancing our understanding of the functional consequences of mutations in proteins and can have implications in various biological and biomedical research domains (Munro and Singh, 2020). Apart from that, its application is found in chemical/biochemical reaction kinetics where the researchers have presented solutions for detecting chemical kinetic and analogous models, which involve the systems where the concentrations of intermediates are not complex in terms of analysis (Haario and Taavitsainen, 1998).

Support vector machine

Support vector machines (SVMs) are versatile machine learning algorithms employed for classification and regression tasks. They excel in scenarios with intricate decision boundaries and high-dimensional data. SVMs are particularly suitable for labeled data and a fixed number of features. The fundamental concept of SVMs involves identifying an optimal hyperplane that effectively separates data points belonging to different classes. In binary classification, the objective is to construct a hyperplane that maximizes the margin between the two classes. The margin represents the perpendicular distance between the hyperplane and the nearest data points of each class, known as support vectors. By finding the hyperplane with the largest margin, SVMs can robustly classify new data points based on their position about this decision boundary. SVMs have demonstrated effectiveness in various domains and offer a valuable tool for tackling complex classification problems and handling high-dimensional datasets. The various steps involved in this algorithm include data preparation, feature mapping, hyperplane optimization, margin maximization, regularization, and tuning. After the SVM model is trained, it can predict the class of the test input data or data points by checking which side of the hyperplane they lie on. SVM performs linear and nonlinear classification and regression; however, it is difficult to scale up the model for large datasets.

Application of this model has been found in protein function prediction, wherein Gene Ontology terms have been assigned to human protein chains for better performance (Cozzetto et al., 2016) and transmembrane-protein topology prediction, which integrates both signal peptide and re-entrant (Nugent and Jones, 2009).

Random forest is a machine learning algorithm that belongs to the family of ensemble methods. It performs predictions by connecting multiple decision trees and is widely used in classification and post-hoc applications. Random forest allows you to measure the importance of features based on their contribution to the accuracy of the sample. It can handle high-dimensional data and has been successfully used in fields such as finance, health care, image recognition, and text analysis. The chance of overfitting is reduced, and better generalization for unseen data is provided. It is known for its ability to handle complex data sets and deliver accurate predictions. However, the number of trees, depth per tree, and other hyperparameters must be carefully tuned to optimize the performance for the particular problem.

Application of random forest is found in the development of protein−ligand interaction functions (Wang and Zhang, 2017) and in the prediction of disease-associated genome mutations, wherein nonsynonymous single nucleotide polymorphisms (nsSNPs) prevalent in genomes are closely related to inherited diseases (Bao et al., 2005).

Gradient boosting

This model is used for regression and classification tasks. This model is built by ensembling various weak machine-learning models like decision trees. Various implementations of gradient boosting are available, with XGBoost, LightGBM, and CatBoost being popular libraries. These services streamline the curriculum and provide additional resources to increase productivity and productivity. It should be noted that gradient boosting can require careful tuning of hyperparameters and can be computationally intensive compared to some other algorithms. However, due to its high interpretability, strong predictive capabilities, and less sensitivity to feature scaling, it is widely used in various industries such as finance, healthcare, web search, and recommendation systems.

The application of this model lies in gene expression profiling, wherein gene expression values were predicted based on XGBoost (Li et al., 2019).

Clustering

Clusters are used in AI and machine learning to group similar data points based on their characteristics or patterns. This is an unsupervised learning method, which means it does not need any labeled data for training. The purpose of clustering is to identify underlying structures or patterns in a data set, where data points in one cluster are more similar than those in another cluster. Clustering can provide insight into the underlying distribution of data ho, identify natural clusters, or help explore and understand data. Certain cluster validation metrics can assess the performance of the clustering process and is suitable for low dimensional data. However, it is difficult to scale up for large datasets, and there may be some contradictory outputs due to the inclusion of noisy datasets.

Application of clustering can be found in differential gene expression analysis, wherein similarly expressed genes are identified across the patient under study (Altman and Krzywinski, 2017). Apart from this, its application is found in protein structure prediction, wherein a strategy named "SPICKER" has been presented to identify near-native folds. This is done by clustering protein structures generated during computer simulations (Zhang and Skolnick, 2004).

Genetic algorithm

The genetic algorithm is an inspiration from Charles Darwin's principle of natural evolution. It is an optimization algorithm and is often used in AI and machine learning to solve complex optimization problems. The basic idea behind genetic algorithms is to mimic the evolutionary process of organisms to find the best solution to the problem. The algorithm contains a population of preferred solutions, usually represented as sequences of values, chromosomes, or individuals. These individuals undergo genetic operations such as selection, crossover, and mutation that can produce new offspring with improved fitness. Genetic algorithms have been successfully applied to various problems, including parameter optimization, feature selection, scheduling, vehicle routing, and many more, but their performance can be affected by factors such as choice of genetic operators, population number, and termination criteria affect. Often, these processes need to be optimized to achieve the best results. Application of genetic algorithm is found in eliminating multiple sequence alignment process by integrating it with the data mining approach and overcoming the complexity of motif discovery (Baloglu and Kaya, 2006).

Fuzzy logic

Fuzzy logic is a mathematical framework for reasoning and decision-making under uncertainty and ambiguity. It provides a way to deal with ambiguity or ambiguity, resulting in flexible and nuanced decision-making processes. In traditional binary theory, statements are either true or false. In real-world situations, however, there are usually multiple members of a truth or group. Fuzzy logic extends binary logic by introducing the concept of fuzzy sets that generate the number of members for elements based on their number of members (Khan and Quadri, 2017). Fuzzy thinking is not a substitute for classical thinking but rather a complementary tool that deals with situations of uncertainty and ambiguity. It provides a framework for making decisions based on incomplete or ambiguous information, allowing for greater flexibility and robustness in solving complex real-world problems. Application of Fuzzy logic is found in bioinformatics and biomedical engineering (Bordon et al., 2015; Xu, 2008).

Neural network/multilayer perceptron

A multilayer perceptron (MLP) is a popular architecture within the field of artificial neural networks (ANNs). It comprises interconnected artificial neurons or perceptrons and serves as a fundamental framework in the domain of deep learning. MLPs are widely employed for various tasks, including classification, regression, and pattern recognition. Their ability to model complex relationships and handle large datasets makes them a valuable tool in many applications within the field of machine learning.

Each node in an MLP receives information from the previous stage, computes that information, and passes the result to the next stage. The layers between the input and output layers are called hidden layers, and they enable the network to see a complex representation of the input data. The main building block of MLP is the perceptron, which is a mathematical model of a biological neuron. A perceptron takes the weighted inputs, sums them, applies the activation function, and produces the output. The activation function introduces nonlinearity into the network, allowing the identification of nonlinear relationships in the data. MLPs have been successfully used for a wide range of tasks, including classification, regression, and pattern recognition. However, they have some limitations, such as the need for large amounts of labeled training data and the tendency to fit too complex data. A variety of strategies can be used to reduce these issues, including regular attendance, dropouts, and early delays. MLP can fit fixed-size datasets with fewer layers like CNN, which makes it easy and faster to get trained. However, it has the disadvantage of getting overfitted easily, and a large number of parameters are to be managed. Interpretability is not that easy. Application of this model has been found in protein secondary structure detection and analysis (Buchan and Jones, 2019), disease diagnostics (Setua et al., 2017) and computational drug toxicity prediction (Mayr et al., 2016).

Convolutional neural network

In recent times, advanced neural network architectures such as convolutional neural networks (CNNs) and RNNs have gained significant popularity in the field of deep learning. However, the MLP remains a fundamental concept that aids in understanding neural networks.

CNNs, specifically, have shown remarkable effectiveness in analyzing visual data like images and videos. Inspired by the human visual system, CNNs have achieved great success in computer vision tasks. A key component of a CNN is the convolutional layer, where a small matrix called a filter or kernel is convolved with the input image. This operation calculates element-wise products and sums at each position, resulting in a feature map that captures local patterns and structures within the image. By utilizing multiple filters, the network can learn to recognize various objects at different levels of abstraction. CNNs excel at processing spatial input data arranged in a grid-like format, making them suitable for tasks with varying image sizes.

Overall, CNNs have revolutionized the field of computer vision and continue to be a pivotal tool for image analysis, object detection, and other related tasks. CNNs have revolutionized computer vision, achieving   state-of-the-art   performance   in   image   classification,   object   recognition,   logical segmentation, and image generation. They are also used in other areas beyond vision, such as in natural language processing (NLP) and speech and pattern recognition. However, it is harder to train deeper architectures, thus making predictions complex. Application of CNN is found in protein residue contact and distance prediction predicts the 3D shape of a protein from its amino acid sequence (Senior et al., 2020) and medical image recognition in case of skin cancer (Esteva et al., 2017).

Recurrent neural network

Recurrent neural networks are a type of ANN designed to use the concept of recursive connections to process sequential data. Unlike feedforward neural networks, which process data strictly sequentially, RNNs have response links that maintain information and pass from one ladder to another. The key feature of RNN is that it can capture dependencies and patterns in a sequence of data by holding a hidden state that summarizes the findings so far At each step in the sequence, RNN takes input updates and hidden state depending on the current input and previous hidden state. It works well with variable sized sequential data (for example, biological sequences or time-series data) can model complex temporal relationships. RNNs have been successfully applied in various sequential or time series tasks in many areas of biology, such as NLP, speech recognition, machine translation, sentiment analysis, signature recognition, and validation in patterns and generating sequences of arbitrary length because they can dynamically customize hidden state based on the input context. A limitation of traditional RNNs is the long training time and high computing memory. Apart from this, it faces difficulty in capturing long-term dependencies due to frequent decay, where the slope decreases sharply with long-term propagation To consume on top of this, several RNNs have been developed, such as long-term password memory (LSTM) and gated repetitive unit (GRU) networks, which incorporate gating mechanisms, thus allowing the network to selectively update and recall information over long sequences. Application of RNN is found in protein engineering (Alley et al., 2019; Choi, 2015) and DNA sequencing (Quang and Xie, 2016).

Graph convolutional network

GNN stands for graph neural network, which is a type of neural network designed to work with graph-structured data. Graphs have nodes (also called vertices) and edges that connect pairs of nodes. The model is specifically designed to learn and process information from the nodes and edges of a graph, enabling them to model and analyze complex relationships and dependencies in graph data. Input data are variably sized and characterized by the connection between various types of entities like spatial. The main idea behind GNN is to propagate information in a graph by recursively updating hidden representations based on each node's neighboring nodes. This process is usually done in multiple layers, where each layer collects information from neighboring nodes and adds to the current node representation. It can be used for various graph-related tasks, such as node classification, link prediction, graph classification, and recommendation processing. Attempts are made in situations where the graph structure contains valuable information to be used for prediction or analysis. Significant attention has been gained by the model in recent years due to its ability to model and reason about complex relational data. It has been successfully applied to various domains, including social network analysis, knowledge graph reasoning, drug discovery, recommendation systems, and traffic prediction, among others. GNNs possess high computing memory for large densely connected graphs and provide a powerful framework for learning and making predictions on data with inherent graph structures and most relevant associations. However, the model is hard to train deeper architectures.

Application of GNN is found in predicting drug properties for modeling polypharmacy side effects (Zitnik et al., 2018) and in interpreting molecular structures for antibiotic discovery (Gainza et al., 2020; Stokes et al., 2020) and knowledge extraction in modeling relational data (Schlichtkrull et al., 2017).

Research challenges

While the adoption of AI in research is gaining momentum, there are still challenges involved in research, development, and technology dissemination. These will be discussed in detail in the later chapters. We give a brief overview in this chapter.

AI research in biology presents both research and implementation challenges. From a research perspective, one challenge is the interpretation and explainability of AI models. Understanding the reasoning behind AI predictions or decisions in biological contexts is essential for researchers to gain insights and validate the results. Another research challenge lies in the availability and quality of biological data. Obtaining diverse and well-curated data for training AI models can be difficult due to the complexity and heterogeneity of biological systems. Additionally, integrating multiomics data and developing effective algorithms for data fusion and analysis pose research challenges.

On the implementation side, scalability and efficiency are crucial challenges. Applying AI techniques to large-scale biological datasets requires scalable algorithms and computing infrastructure to process and analyze the data efficiently. Moreover, implementing AI models in real-world biological applications often requires addressing practical issues such as data privacy and security, regulatory compliance, and interoperability with existing systems and workflows. Successful implementation also necessitates collaboration between AI researchers, biologists, and clinicians, bridging the gap between these domains and ensuring effective knowledge transfer. By addressing these research and implementation challenges, AI in biology can lead to transformative advancements in understanding biological systems, driving healthcare innovations, and facilitating personalized medicine.

Conclusion

When dealing with biological data, it is advisable to initially explore traditional machine learning algorithms before delving into deep learning methods. Traditional machine learning techniques can be highly effective and efficient, particularly in cases where the dataset is small, contains a limited number of features, or lacks structured relationships among the features. On the other hand, deep learning is most advantageous when data are abundant; each data point consists of multiple objects, and there are clear relationships among the features, such as adjacent pixels in images or biological data like DNA, RNA, protein sequences, and microscopy images.

While deep learning holds immense potential for analyzing biological data, it is crucial to consider factors such as data availability, feature count, and the structured relationships between features. Traditional machine learning approaches offer the benefits of rapid development and practical applicability, making them valuable options to consider. Moreover, the field of AI can draw inspiration from biology in the development of new algorithm architectures. Biological systems have evolved to handle and interpret complex data effectively. Studying biological systems can inspire AI researchers to design novel algorithms that can enhance the capabilities of AI systems. Overall, fostering cross-disciplinary collaboration among biologists, computer scientists, and engineers can pave the way for the next generation of AI in biology. By combining their expertise and integrating principles from biology, researchers can address challenges related to data, theory, model development, and other obstacles encountered in AI. This collaborative approach has the potential to drive advancements in both AI and biology, leading to exciting breakthroughs and discoveries at the intersection of these disciplines.

References

Alley, E.C., Khimulya, G., Biswas, S., AlQuraishi, M., Church, G.M., 2019. Unified rational protein engineering with sequence-based deep representation learning. Nature Methods 16 (12), 1315–1322. https://doi.org/10.1038/s41592-019-0598-1.

Altman, N., Krzywinski, M., 2017. Points of significance: clustering. Nature Methods 14 (6), 545–546. https://doi.org/10.1038/nmeth.4299.

Ang, J.C., Mirzal, A., Haron, H., Hamed, H.N.A., 2016. Supervised, unsupervised, and semi-supervised feature selection: a review on gene selection. IEEE/ACM Transactions on Computational Biology and Bioinformatics 13 (5), 971–989. https://doi.org/10.1109/tcbb.2015.2478454.

Baloglu, U.B., Kaya, M., 2006. Top-down motif discovery in biological sequence datasets by genetic algorithm. In: Proceedings - 2006 International Conference on Hybrid Information Technology, ICHIT 2006, vol. 2, pp. 103–107. https://doi.org/10.1109/ICHIT.2006.253597.

Bao, L., Zhou, M., Cui, Y., 2005. nsSNPAnalyzer: identifying disease-associated nonsynonymous single nucleotide polymorphisms. Nucleic Acids Research 33 (2), W480–W482. https://doi.org/10.1093/nar/gki372.

Bishop, C.M., Nasrabadi, N.M., 2006. Pattern Recognition and Machine Learning. Springer, 2006.

Bordon, J., Moskon, M., Zimic, N., Mraz, M., 2015. Fuzzy logic as a computational tool for quantitative modelling of biological systems with uncertain kinetic data. IEEE/ACM Transactions on Computational Biology and Bioinformatics 12 (5), 1199–1205. https://doi.org/10.1109/tcbb.2015.2424424.

Buchan, D.W.A., Jones, D.T., 2019. The PSIPRED protein analysis workbench: 20 years on. Nucleic Acids Research 47 (1), W402–W407. https://doi.org/10.1093/nar/gkz297.

Choi, E., 2015. Doctor AI: Predicting Clinical Events via Recurrent Neural Networks. https://doi.org/10.48550/arxiv.1511.05942.

Cozzetto, D., Minneci, F., Currant, H., Jones, D.T., 2016. FFPred 3: feature-based function prediction for all gene ontology domains. Scientific Reports 6 (1). https://doi.org/10.1038/srep31865.

Esteva, A., Kuprel, B., Novoa, R.A., Ko, J., Swetter, S.M., Blau, H.M., Thrun, S., 2017. Dermatologist-level classification of skin cancer with deep neural networks. Nature 542 (7639), 115–118. https://doi.org/10.1038/nature21056.

Gainza, P., Sverrisson, F., Monti, F., Rodolà, E., Boscaini, D., Bronstein, M.M., Correia, B.E., 2020. Deciphering interaction fingerprints from protein molecular surfaces using geometric deep learning. Nature Methods 17 (2), 184–192. https://doi.org/10.1038/s41592-019-0666-6.

Géron, A., 2019. On Machine Learning with Scikit-Learn, Keras, and TensorFlow.

Greener, J.G., Kandathil, S.M., Moffat, L., Jones, D.T., 2022. A guide to machine learning for biologists. Nature Reviews Molecular Cell Biology 23 (1), 40–55. https://doi.org/10.1038/s41580-021-00407-0.

Haario, H., Taavitsainen, V.M., 1998. Combining soft and hard modelling in chemical kinetic models. Chemometrics and Intelligent Laboratory Systems 44 (1–2), 77–98. https://doi.org/10.1016/S0169-7439(98)00166-X.

Hassoun, S., Jefferson, F., Shi, X., Stucky, B., Wang, J., Rosa, E., 2021. Artificial intelligence for biology. Integrative and Comparative Biology 61 (6), 2267–2275. https://doi.org/10.1093/icb/icab188.

Khan, S.S., Quadri, S.M.K., 2017. Structure identification and IO space partitioning in a nonlinear fuzzy system for prediction of patient survival after surgery. International Journal of Intelligent Computing and Cybernetics 10 (2), 166−182. https://doi.org/10.1108/IJICC-06-2016-0021.

Kleinbaum, D.G., Klein, M., Pryor, E.R., 2002. Logistic Regression: A Self-Learning Text. Springer.

Krenn, M., Pollice, R., Guo, S.Y., Aldeghi, M., Cervera-Lierta, A., Friederich, P., dos Passos Gomes, G., Häse, F., Jinich, A., Nigam, A.K., Yao, Z., Aspuru-Guzik, A., 2022. On scientific understanding with artificial intelligence. Nature Reviews Physics 4 (12), 761−769. https://doi.org/10.1038/s42254-022-00518-3.

Li, W., Yin, Y., Quan, X., Zhang, H., 2019. Gene expression value prediction based on XGBoost algorithm. Frontiers in Genetics 10. https://doi.org/10.3389/fgene.2019.01077.

Lim, Y., Kang, S., 2011. Development of link cost function using neural network concept in sensor network. KSII Transactions on Internet and Information Systems 5 (1), 141−156. https://doi.org/10.3837/tiis.2011.01.008.

López, B., 2005. Artificial Intelligence Research and Development. IOS Press.

Mayr, A., Klambauer, G., Unterthiner, T., Hochreiter, S., 2016. DeepTox: toxicity prediction using deep learning. Frontiers in Environmental Science 3. https://doi.org/10.3389/fenvs.2015.00080.

Munro, D., Singh, M., 2020. DeMaSk: a deep mutational scanning substitution matrix and its use for variant impact prediction. Bioinformatics 36 (22−23), 5322−5329. https://doi.org/10.1093/bioinformatics/btaa1030.

Na, D., 2020. User guides for biologists to learn computational methods. Journal of Microbiology 58 (3), 173−175. https://doi.org/10.1007/s12275-020-9723-1.

Nugent, T., Jones, D.T., 2009. Transmembrane protein topology prediction using support vector machines. BMC Bioinformatics 10. https://doi.org/10.1186/1471-2105-10-159.

Quang, D., Xie, X., 2016. DanQ: a hybrid convolutional and recurrent deep neural network for quantifying the function of DNA sequences. Nucleic Acids Research 44 (11), e107. https://doi.org/10.1093/nar/gkw226.

Schlichtkrull, M., Kipf, T.N., Bloem, P., Van Den Berg, R., Titov, I., Welling, M., 2017. Modeling relational data with graph convolutional networks. arXiv. https://arxiv.org.

Senior, A.W., Evans, R., Jumper, J., Kirkpatrick, J., Sifre, L., Green, T., Qin, C., Žídek, A., Nelson, A.W.R., Bridgland, A., Penedones, H., Petersen, S., Simonyan, K., Crossan, S., Kohli, P., Jones, D.T., Silver, D., Kavukcuoglu, K., Hassabis, D., 2020. Improved protein structure prediction using potentials from deep learning. Nature 577 (7792), 706−710. https://doi.org/10.1038/s41586-019-1923-7.

Setua, S., Khan, S., Yallapu, M.M., Behrman, S.W., Sikander, M., Khan, S.S., Jaggi, M., Chauhan, S.C., 2017. Restitution of tumor suppressor MicroRNA-145 using magnetic nanoformulation for pancreatic cancer therapy. Journal of Gastrointestinal Surgery 21 (1), 94−105. https://doi.org/10.1007/s11605-016-3222-z.

Stokes, J.M., Yang, K., Swanson, K., Jin, W., Cubillos-Ruiz, A., Donghia, N.M., MacNair, C.R., French, S., Carfrae, L.A., Bloom-Ackermann, Z., Tran, V.M., Chiappino-Pepe, A., Badran, A.H., Andrews, I.W., Chory, E.J., Church, G.M., Brown, E.D., Jaakkola, T.S., Barzilay, R., Collins, J.J., 2020. A deep learning approach to antibiotic discovery. Cell 181 (2), 475−483. https://doi.org/10.1016/j.cell.2020.04.001.

Wang, C., Zhang, Y., 2017. Improving scoring-docking-screening powers of protein−ligand scoring functions using random forest. Journal of Computational Chemistry 38 (3), 169−177. https://doi.org/10.1002/jcc.24667.

Xu, D., 2008. Applications of Fuzzy Logic in Bioinformatics. Imperial College Press.

Ye, Z., Qu, Y., Liang, Z., Wang, M., Liu, Q., 2023. Explainable fMRI -based brain decoding via spatial temporal-pyramid graph convolutional network. Human Brain Mapping 44 (7), 2921−2935. https://doi.org/10.1002/hbm.26255.

Zhang, Y., Skolnick, J., 2004. SPICKER: a clustering approach to identify near-native protein folds. Journal of Computational Chemistry 25 (6), 865−871. https://doi.org/10.1002/jcc.20011.

Zitnik, M., Agrawal, M., Leskovec, J., 2018. Modeling polypharmacy side effects with graph convolutional networks. Bioinformatics 34 (13), i457−i466. https://doi.org/10.1093/bioinformatics/bty294.

The synergy of AI and biology: A transformative partnership

Mahvish Khurshid Bijli[1,a], Uzmat Ul Nisa[1,a], Aqsa Ashraf Makhdomi[1,a] and Henna Hamadani[2,a]

[1]*National Institute of Technology, Srinagar, Jammu and Kashmir, India;* [2]*Sher-e-Kashmir University of Agricultural Sciences and Technology of Kashmir, Srinagar, Jammu and Kashmir, India*

Introduction

Artificial intelligence (AI) and biology, two seemingly distinct disciplines, have converged in a groundbreaking intersection that is revolutionizing the way we understand and explore the complexities of life. AI, with its ability to process vast amounts of data, recognize patterns, and learn from experience, has found a natural home in the realm of biology, where intricate biological systems and processes present challenges and opportunities that can benefit from intelligent computational approaches (Bhardwaj et al., 2022). This marriage between AI and biology holds immense promise, opening up new frontiers in healthcare, genomics, drug discovery, personalized medicine, and beyond. The integration of AI and biology allows us to delve deeper into the intricate workings of living organisms, unlocking hidden insights and unraveling the mysteries of life (Dutta et al., 2022). AI algorithms can analyze large-scale biological datasets, such as genomics, proteomics, and metabolomics, with remarkable speed and accuracy. By deciphering complex biological patterns and relationships, AI empowers scientists and researchers to make sense of the intricate mechanisms that underpin health and disease (Richards et al., 2022).

In the realm of healthcare, AI has the potential to revolutionize diagnostics, prognosis, and treatment (Bohr and Memarzadeh, 2020). Machine learning (ML) algorithms can analyze medical images, such as magnetic resonance imaging (MRI) and computed tomography (CT) scans, with exceptional precision, aiding in the early detection of diseases and providing personalized treatment recommendations. AI-driven decision support systems can help physicians interpret patient data, predict disease progression, and optimize treatment plans, leading to more targeted and effective interventions (Comito et al., 2020).

Genomics, the study of an organism's complete set of DNA, has experienced a paradigm shift with the integration of AI. ML algorithms can shift through vast genomic datasets, identifying genetic variations, uncovering disease associations, and enabling the development of personalized medicine. AI algorithms can predict the risk of developing certain diseases, identify genetic markers that

---

[a]Authors have equal contribution.

A Biologist's Guide to Artificial Intelligence. https://doi.org/10.1016/B978-0-443-24001-0.00002-6

13

influence drug response, and facilitate the design of tailored therapies for individuals based on their unique genetic makeup (Libbrecht and Noble, 2015).

In the realm of drug discovery, AI has emerged as a game-changer. Traditional drug development is a time-consuming and costly process. However, AI algorithms can expedite the identification and optimization of potential drug candidates by predicting their efficacy, safety, and potential side effects. By analyzing massive databases of molecular structures and biological interactions, AI algorithms can guide researchers toward promising drug targets and accelerate the discovery of novel therapies (Chan et al., 2019).

Moreover, AI in biology extends beyond these applications. AI algorithms can aid in ecological modeling (Schuwirth et al., 2019), bioinformatics (Li et al., 2020), protein folding predictions (Noé et al., 2020), and the analysis of biological networks (Muzio et al., 2021), opening up new avenues for understanding the intricate web of life. The synergy between AI and biology is not only enhancing our knowledge of biological systems but also pushing the boundaries of scientific discovery and innovation.

As AI continues to advance and our understanding of biology deepens, the possibilities for this exciting integration are boundless. The collaboration between AI and biology holds the potential to revolutionize healthcare, reshape our approach to medicine, and provide profound insights into the fundamental nature of life itself (Yu et al., 2018). It is an extraordinary partnership that has evolved over history to touch every sphere of life (Fig. 2.1).

It is paving the way for a future where technology and biology intertwine, augmenting our capabilities and propelling us toward new frontiers of scientific exploration and improved human well-being.

The transformative power of AI in biology

The AI era has arrived in biology, marking a transformative shift in scientific exploration. This paradigm shift is driven by remarkable advancements in hardware technology, including sensors, Internet of Things (IoT) devices, and computing devices. These technologies enable the collection of vast amounts of data from various biological sources.

In conjunction with hardware advancements, software technology has played a pivotal role in the proliferation of AI in biology. Advanced algorithms and techniques have been developed to interpret, process, and extract meaningful insights from abundant biological data. These algorithms have the capability to analyze complex patterns, identify correlations, and unveil previously hidden relationships within the data.

The integration of hardware and software advancements has propelled the popularity of AI in biology. Researchers and scientists are leveraging these tools to conduct cutting-edge research, accelerate discoveries, and unravel the mysteries of life. AI in biology holds immense potential for improving healthcare, advancing drug discovery, understanding complex biological processes, and contributing to genomics, proteomics, and personalized medicine.

As the field continues to evolve, we can expect further innovations and breakthroughs as AI continues to shape the future of biological research and its applications. The collaborative efforts of biologists, data scientists, and technologists are fueling this exciting era of AI in biology, bringing us closer to unraveling the complexities of life and unlocking new frontiers in scientific exploration.

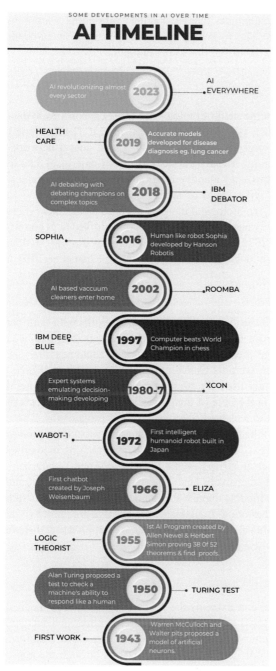

FIGURE 2.1

Artificial intelligence (AI): a brief timeline some noteworthy innovations in AI.

This section focuses on elucidating several AI techniques that have gained substantial traction in the field of biology, spanning diverse applications. The following subsections provide an in-depth exploration of these techniques and their significant contributions to the domain.

Machine learning

Machine learning has gained significant attention in the field of biology due to its capabilities in data mining, prediction, and analysis. ML algorithms, such as decision trees (Song and Lu, 2015), support vector machines (Pisner and Schnyer, 2019), random forests (Rigatti, 2017), and neural networks (Lawrence, 1993), are extensively employed in biology. ML algorithms learn from biological data to classify samples, predict outcomes, discover patterns, and identify relationships between variables. The applications of ML in biology encompass a wide range of areas from disease prediction to ecological conservation (Zaghloul and Achari, 2022) to biological imagery analysis using UAVs (Gao et al., 2020) and satellite images of plant fields (Kislov and Korznikov, 2020). It is utilized in genomics (Whalen et al., 2022), proteomics (Desaire et al., 2022), drug discovery (Vamathevan et al., 2019), disease diagnosis (Hamadani, 2023), and personalized medicine (Ozer et al., 2020; Khan et al., 2020).

Deep learning

Deep learning (DL) (Kelleher, 2019), a subset of ML, utilizes artificial neural networks with multiple layers to extract complex patterns and features from large-scale biological datasets. DL has been highly effective in image analysis, natural language processing (NLP), genomic sequence analysis, protein structure prediction, and medical image interpretation. Convolutional neural networks (CNNs) and recurrent neural networks (RNNs) are commonly used DL architectures in biology. CNNs and RNNs have emerged as particularly effective tools in the field of image analysis within the context of various applications, such as ecological conservation, medical diagnostics, and prognostics. These advanced neural network architectures have been applied to diverse image datasets, including camera-based images for ecological monitoring (McClure et al., 2020), as well as medical imaging modalities (Suzuki, 2017; Klang, 2018) like X-rays (Jaiswal et al., 2019), MRIs (Lundervold and Lundervold, 2019), and CT scans (Yang et al., 2021), to comprehensively analyze a patient's health status and predict their future conditions. However, these DL models mainly act as black-box models, wherein we can get accurate results but they lack a sense of reasoning. Explainable models that are more logical and visualizable are required in biology to get insights about the particular case or application that we are dealing with (Ryo et al., 2021).

Natural language processing

Natural language processing significantly enhances and expands the scope of standardized medical data by extracting valuable insights from unstructured sources such as clinical notes and medical publications (Spasic and Nenadic, 2020). The primary objective of NLP methods is to transform unstructured text into structured and machine-readable data, enabling its analysis through the application of ML approaches. NLP, as a fundamental component of human−computer interaction, facilitates the efficient processing of extensive scientific data while effectively filtering out inappropriate spam content.

In the domain of healthcare, NLP plays a pivotal role in isolating and managing complex data. By incorporating AI techniques, NLP enables the real-time acquisition of critical data from reliable sources, empowering healthcare providers to make informed decisions. Leveraging NLP, virtual healthcare assistants employ models activated by medical terminology, allowing seamless interactions and facilitating the retrieval of valuable insights. This seamless integration of AI and NLP enables the healthcare sector to collect crucial data from trustworthy sources in real time, ultimately improving patient care and outcomes (Wen et al., 2019).

Reinforcement learning

Reinforcement learning (RL) algorithms are instrumental in optimizing various aspects of biology, including biological experiments, drug discovery, and treatment plans. RL agents acquire knowledge by interacting with their environment and receiving feedback in the form of rewards or penalties (Sutton and Barto, 2018). In the context of biology, RL is applied to optimize critical areas such as drug combinations, therapeutic interventions, and experimental conditions. In drug discovery, RL plays a vital role in designing optimal therapeutic interventions (Serrano et al., 2018). RL agents explore and learn from the vast space of potential interventions, identifying promising candidates and refining their strategies based on the observed outcomes. This iterative process helps in the discovery of novel therapies and treatment regimens. Furthermore, RL algorithms aid in determining optimal experimental conditions. By leveraging RL, researchers can systematically explore experimental parameter spaces and identify the ideal conditions that maximize the desired outcomes (Neftci and Averbeck, 2019). This approach enables efficient experimentation and accelerates the discovery of new insights in various biological domains.

Data integration and fusion

AI techniques are employed to integrate and fuse diverse biological datasets from multiple sources, such as physical data, genomics, transcriptomics, proteomics, and clinical data. By integrating these datasets, AI can uncover hidden relationships, identify biomarkers, and enable comprehensive analyses of biological systems. In the study of complex phenomena such as diseases, it is recognized that no single data type can fully capture the underlying complexity and factors involved. To address this challenge, integrative methods that combine data from multiple technologies have become crucial statistical and computational approaches (Stahlschmidt et al., 2022). These methods aim to provide a comprehensive and relevant systems view by integrating diverse data sources.

The development of effective models is a key challenge in implementing these integrative approaches. The goal is to create models that can effectively capture the intricate relationships and interactions between various data types, ultimately providing a holistic understanding of the phenomenon under study (Kang et al., 2022). These models need to account for the heterogeneity, variability, and interdependencies present in the integrated data.

Genetic algorithms

Genetic algorithms (GAs) (Goldberg, 2013) are optimization algorithms inspired by the principles of natural selection and evolutionary processes. In the context of biology, GAs are utilized to optimize complex biological systems, including protein folding (Lin et al., 2019), metabolic pathway optimization, and drug design (Spiegel and Durrant, 2020).

GAs offer a powerful approach to finding optimal solutions within large and complex search spaces. By mimicking natural selection, GAs iteratively generate and evaluate candidate solutions, selecting the most promising ones based on a fitness criterion. These selected solutions then undergo genetic operations such as crossover and mutation to create new candidate solutions. Through successive generations, GAs progressively converge toward better solutions, resembling the process of biological evolution.

The need for AI in biology

While biology can certainly exist and make progress without AI, integrating AI into the field brings numerous advantages and advancements that are difficult to overlook. Here are some reasons why AI has become increasingly indispensable in biology.

1. **Handling complex and big data:** With the advent of high-throughput technologies, biology has witnessed an explosion of data (Hamadani et al., 2023b). AI excels at processing and analyzing vast amounts of complex biological data, including genomic sequences, proteomic profiles, and medical records. Without AI, the sheer volume and complexity of biological data would be challenging to manage and interpret effectively.
2. **Efficient data analysis:** AI algorithms enable efficient and high-throughput analysis of biological data, uncovering patterns, correlations, and insights that may not be readily apparent through traditional methods (Hamadani and Ganai, 2022). This accelerates scientific discoveries, enabling researchers to make breakthroughs in understanding diseases, genetic mechanisms, and ecological systems.
3. **Prediction and modeling:** AI's ability to build predictive models is invaluable in biology. By learning from existing data, AI can make predictions about disease risks, treatment outcomes, and drug–target interactions. These predictive models enhance decision-making in precision medicine, drug discovery, and ecological research, leading to more targeted and effective interventions (Hamadani et al., 2022a).
4. **Automation of laborious tasks:** AI technologies can automate repetitive and labor-intensive tasks in biology, such as data processing, image analysis, and laboratory experiments. By freeing up researchers' time from these mundane tasks, AI allows them to focus on higher-level analysis, critical thinking, and creative problem-solving (Hamadani and Khan, 2015).
5. **Uncovering complex relationships:** Biology involves studying intricate biological systems with multifaceted relationships and interactions. AI algorithms can uncover hidden patterns and relationships within biological networks, gene regulatory circuits, and ecological systems. This provides a deeper understanding of the underlying mechanisms and helps reveal emergent properties that are challenging to detect with traditional approaches (Hamadani et al., 2022a).
6. **Accelerating drug discovery and development:** AI has the potential to significantly accelerate the drug discovery process by analyzing large chemical libraries, predicting drug–target interactions, and optimizing lead compounds. This efficiency in identifying potential therapeutic candidates saves time and resources and reduces the cost of drug development, leading to the faster availability of life-saving treatments.

7. **Personalized medicine and healthcare:** AI enables personalized approaches to medicine by analyzing individual patient data and tailoring treatments based on specific characteristics. This precision medicine approach improves patient outcomes, reduces adverse effects, and maximizes the efficiency of healthcare interventions.

While biology can exist without AI, integrating AI technologies enhances the field's capabilities, accelerates scientific progress, and enables novel approaches to understanding biological systems, developing therapies, and addressing complex challenges in healthcare and the environment. AI has the potential to unlock new frontiers in biology, making it an indispensable tool for researchers and practitioners in the field.

Some applications

The applications of AI in biology are wide-ranging and have the potential to transform various aspects of the field. All the applications of AI in biology are hard, to sum up in a single book chapter since it is seeping into every sphere of life and finding applications everywhere. Some major applications of AI in biology are stated in the sections that follow.

Healthcare

The integration of AI into drug delivery systems holds immense potential for revolutionizing healthcare. AI-driven formulation design, targeted drug delivery, real-time monitoring, predictive modeling, intelligent devices, and improved patient adherence are transforming the landscape of drug administration. As AI continues to evolve, its synergy with drug delivery systems will drive advancements, leading to personalized, precise, and effective treatments, ultimately improving the quality of life for patients worldwide. AI can accelerate the drug discovery process by analyzing vast amounts of biological and chemical data. AI algorithms can identify potential drug targets, optimize lead compounds, and predict drug efficacy and safety. This enables researchers to prioritize and design drug candidates with higher success rates, ultimately speeding up the development of new therapies for pathological as well as metabolic conditions like obesity (Mansoor et al., 2021). Drug delivery systems play a crucial role in ensuring optimal therapeutic outcomes and minimizing side effects in medical treatments. The integration of AI has brought remarkable advancements to this field, enabling personalized and precise drug administration (Hassanzadeh et al., 2019). By harnessing AI's capabilities in data analysis, modeling, and decision-making, drug delivery systems are being revolutionized, leading to improved treatment efficacy, patient adherence, and overall healthcare outcomes (Dawoodbhoy et al., 2021).

Precision medicine and personalized treatment

Precision medicine, also referred to as personalized medicine or individualized medicine, is a healthcare approach that takes into consideration the individual's unique characteristics, like their genetic makeup, environment, and lifestyle, and provides medical decisions, treatments, and interventions specifically for that individual. The goal of personalized medicine is to move away from treating everyone the same way and instead provide care that is specifically suited to each person's unique needs. It focuses on tailoring treatments to the individual so that they are more targeted and effective on them (Sitapati et al., 2017). Precision medicine involves several steps that work together in a cyclical and iterative process.

Data collection: In this step, comprehensive information about an individual's unique characteristics, such as their genes, medical history, environment, lifestyle, and other relevant factors is gathered. AI tools can be used to collect and aggregate patient data from various sources, such as electronic health records, wearable devices, and genetic databases. By automating data extraction and cleaning, AI ensures accurate and comprehensive datasets for analysis in further steps.

Molecular profiling: In this step, the individual's molecular profile, including their genetic variations, gene expression patterns, and biomarkers, is analyzed to identify the specific molecular characteristics related to their disease or condition. Through the use of ML algorithms, AI can analyze complex molecular data and identify relevant biomarkers, genetic variations, and molecular signatures associated with specific diseases or treatment responses, which aid in comprehensive molecular profiling.

Computational Analysis: After molecular profiling, computational analysis is employed to interpret the vast amount of data generated from the profiling process. This analysis involves the application of various computational techniques, algorithms, and bioinformatics tools to identify patterns, correlations, and potential therapeutic targets within the molecular data (Hamadani et al., 2023b). Computational analysis with AI techniques can result in the efficient handling of complex and high-dimensional data and provide insights into patterns, correlations, and potential therapeutic targets. Through ML and data mining, AI enhances the interpretation and analysis of the extensive information derived from molecular profiling.

Generate treatment plan: Based on the insights gained from molecular profiling and computational analysis, a personalized treatment plan is generated for the patient. This plan takes into account the patient's unique molecular characteristics, disease stage, potential therapeutic targets, and other relevant factors. The goal is to develop a treatment strategy that is specifically tailored to the individual patient, maximizing the chances of a successful outcome. AI can help in the generation of treatment plans. By integrating patient-specific data, such as genomic information, medical history, and clinical guidelines, AI algorithms can offer evidence-based recommendations. These AI-driven recommendations empower healthcare professionals to make informed decisions regarding treatment options, dosages, and potential drug interactions, ultimately leading to more personalized and effective treatment plans.

Treatment testing: Once the treatment plan is formulated, it is important to conduct testing to evaluate its effectiveness. This testing involves administering the chosen treatment to the patient and closely monitoring their response. The patient's response to the treatment is assessed through various methods, including imaging, laboratory tests, and clinical evaluation. This step helps determine whether the treatment is achieving the desired results and allows for adjustments or modifications to the treatment plan if needed. AI plays a critical role in treatment testing by monitoring and analyzing patient responses. Real-time patient data, including imaging results, laboratory tests, and electronic health records, can be analyzed by AI algorithms. This analysis enables the detection of subtle changes, prediction of treatment response, and identification of potential adverse events. By leveraging AI, healthcare providers can promptly intervene, make personalized adjustments to treatment plans, and improve patient outcomes.

Enhanced formulation design: AI algorithms analyze vast datasets of drug properties, recipients, and patient-specific factors to design optimized drug formulations. ML algorithms identify complex patterns and relationships, enabling the development of formulations with improved solubility, stability, and bioavailability. AI assists in tailoring the composition of drug carriers and delivery systems, resulting in enhanced drug efficacy and targeted delivery (Hassanzadeh et al., 2019).

Targeted drug delivery: AI plays a crucial role in developing targeted drug delivery systems. By leveraging patient-specific data such as genetic information, biomarker profiles, and medical imaging, AI algorithms assist in identifying the optimal target sites for drug delivery. This allows for the precise localization of therapeutic agents to desired tissues, cells, or organs, maximizing efficacy while minimizing off-target effects (Gao et al., 2023).

Real-time monitoring and control: AI algorithms process real-time patient data, including physiological parameters and drug concentration levels, to enable continuous monitoring and control of drug delivery processes. By analyzing these data, AI algorithms provide feedback and make adjustments to drug delivery parameters, ensuring precise dosing and adaptability to individual patient needs. Real-time monitoring enhances therapeutic response and minimizes the risk of under or overdosing (Saraswat et al., 2022).

Predictive modeling and optimization: Utilizing ML and computational modeling, AI predicts drug release profiles from various delivery systems. This enables the optimization of drug release kinetics, helping to develop controlled-release systems and personalized dosing strategies. By simulating drug behavior and interactions within the body, AI-driven models facilitate the design of optimal drug delivery systems for specific drugs and patient populations (Kumar and Ram, 2021).

Intelligent drug delivery devices: AI contributes to the development of intelligent drug delivery devices. By analyzing patient data, environmental conditions, and physiological responses, AI algorithms optimize drug delivery parameters in real time. This leads to the development of smart devices capable of adjusting drug dosage, timing, or delivery based on individual patient needs and conditions, ultimately improving treatment outcomes (Tan et al., 2022).

Improved patient adherence: AI-based technologies facilitate the monitoring and support of patient adherence to medication regimens. By analyzing data from wearable devices, smart packaging, or patient-reported information, AI algorithms track medication usage patterns. AI-powered interventions, reminders, and personalized feedback enhance patient adherence, resulting in better treatment outcomes and disease management (Harrer et al., 2019).

Decision support systems: AI-powered decision support systems can assist healthcare professionals, veterinarians, and breeders (Hamadani et al., 2019) in selecting the most appropriate drug delivery systems and dosing regimens for individual patients as well as animals (Hamadani and Ganai, 2022). By integrating patient-specific data, treatment guidelines, and clinical knowledge, these systems can provide personalized recommendations, improving treatment outcomes and reducing medication errors (Braun et al., 2021).

Genomics and genetic research

AI plays a crucial role in analyzing and interpreting genomic data. It can assist in genome sequencing, genome annotation, and variant calling, providing insights into the genetic basis of diseases and traits. AI techniques also aid in studying gene expression patterns, gene regulatory networks, and the impact of genetic variations on cellular processes.

Genomic data analysis: AI techniques such as ML and DL have been applied to analyze large-scale genomic data. These algorithms can identify patterns, extract meaningful insights, and make predictions. AI has helped in identifying disease-causing genetic variations, understanding gene expression patterns, and classifying different types of genetic mutations.

Genomic sequencing and assembly: AI algorithms have improved the efficiency and accuracy of genomic sequencing and assembly processes. By leveraging ML techniques, researchers can reduce errors, optimize sequencing protocols, and enhance the quality of assembled genomes.

Variant calling and interpretation: Identifying genetic variants from sequencing data is a complex task. AI algorithms have been developed to accurately detect and classify genetic variants. These algorithms use pattern recognition and statistical modeling to distinguish between disease-causing variants and benign ones, aiding in genetic diagnosis and personalized medicine.

Disease diagnosis and risk prediction: AI algorithms can analyze a patient's genomic data, medical history, and other relevant factors to predict disease risk and aid in early diagnosis. By integrating genomic information with clinical data, AI can provide more accurate and personalized risk assessments for various conditions, including cancer, cardiovascular diseases, and genetic disorders.

Gene editing and CRISPR/Cas9: AI algorithms have been used to improve the precision and efficiency of gene editing techniques like CRISPR/Cas9. AI models can design and guide RNAs, predict off-target effects, and optimize gene editing protocols, making gene editing safer and more precise.

Image analysis and medical imaging

AI algorithms excel in analyzing complex biological images, such as histopathology slides, medical scans, and microscopy images. AI-based image analysis can aid in diagnosing diseases, identifying cellular structures, detecting abnormalities, and monitoring treatment responses. It enhances the efficiency and accuracy of image interpretation, benefiting fields like pathology, radiology, and cancer research. AI has made remarkable advancements in the field of image analysis and medical imaging, providing new tools and techniques that enhance the accuracy, speed, and efficiency of medical diagnostics. Here are some ways AI is impacting image analysis and medical imaging:

Image segmentation: AI algorithms, particularly DL models like CNNs, have greatly improved image segmentation tasks. These algorithms can accurately delineate structures and identify regions of interest within medical images, such as tumors, organs, or blood vessels. Image segmentation assists in quantifying and analyzing anatomical structures, aiding in diagnosis and treatment planning.

Image classification and detection: AI models are trained to classify medical images into different categories or detect specific abnormalities. For example, AI algorithms can identify cancerous lesions in mammograms, detect lung nodules in chest X-rays or CT scans, and diagnose diabetic retinopathy in retinal images. AI-based classification and detection systems help radiologists and other healthcare professionals in making more accurate and timely diagnoses. Image analysis is also used in the classification and detection of other data as well.

Computer-aided diagnosis (CAD): AI is used to develop computer-aided diagnosis systems that assist radiologists and clinicians in interpreting medical images. By analyzing large datasets and leveraging ML techniques, CAD systems can provide second opinions, flag potential abnormalities, and improve diagnostic accuracy. CAD systems have been developed for various modalities, including mammography, MRI, CT, and ultrasound.

Image reconstruction and enhancement: AI algorithms can reconstruct and enhance medical images to improve their quality and aid in diagnosis. For example, AI techniques such as super-resolution and denoising algorithms can enhance the resolution and reduce noise in images, leading to clearer and more detailed representations of anatomical structures.

Image registration and fusion: AI can align and fuse multiple medical images taken at different times or using different modalities. Image registration techniques help in tracking disease progression, monitoring treatment response, and providing more comprehensive information for diagnosis. AI-based fusion methods combine complementary information from different imaging modalities to generate a more comprehensive and informative representation of the patient's condition.

Radiomics and quantitative imaging: AI plays a crucial role in extracting quantitative features from medical images, known as radiomics. AI algorithms analyze these features and identify correlations with clinical outcomes, aiding in prognosis, treatment response prediction, and personalized medicine. Radiomics-based AI models can extract intricate patterns and texture features that may not be apparent to the human eye.

Real-time image analysis: AI algorithms can process medical images in real time, enabling rapid analysis and decision-making during critical procedures. This capability is particularly useful in interventional radiology, where AI can assist in guiding procedures, targeting lesions, and optimizing treatment outcomes. The integration of AI into image analysis and medical imaging holds great promise for improving diagnostics, treatment planning, and patient outcomes. As AI algorithms continue to evolve and more comprehensive datasets become available, further advancements in this field are expected in this field.

Biological network analysis

AI techniques enable the analysis of large-scale biological networks, such as protein–protein interaction networks and gene regulatory networks. By uncovering network properties, identifying key nodes or modules, and predicting interactions, AI helps in understanding biological processes, disease mechanisms, and drug targets. AI has become instrumental in advancing biological network analysis, which involves the study of complex interactions and relationships within biological systems. The contributions of AI to this field are listed below.

Network reconstruction: AI algorithms can analyze large-scale biological datasets, such as gene expression data or protein-protein interaction data, to reconstruct biological networks. By applying ML techniques, AI can identify patterns, dependencies, and regulatory relationships among biological entities, leading to the construction of comprehensive and accurate network models.

Network visualization and exploration: AI enables the visualization and exploration of complex biological networks. AI-powered visualization tools can represent network structures in an intuitive and informative manner, allowing researchers to identify key nodes, clusters, and connectivity patterns. Such visualizations aid in the interpretation and understanding of biological networks.

Network analysis and prediction: AI algorithms can analyze biological networks to extract meaningful insights and make predictions. For example, AI techniques like graph mining and network embedding can uncover important network motifs, functional modules, and topological properties. AI can also predict missing edges or connections in the network, helping researchers discover new biological interactions or predict the behavior of unobserved entities.

Functional annotation and pathway analysis: AI assists in functional annotation by associating biological entities with their putative functions or roles within a network. By integrating multiple data sources and leveraging ML methods, AI algorithms can predict the functions of genes, proteins, or other biological components based on their network context. AI also enables pathway analysis by identifying enriched pathways and signaling cascades within a network, providing insights into biological processes and disease mechanisms.

Drug target identification and drug repurposing: AI can analyze biological networks to identify potential drug targets and repurpose existing drugs for new indications. By considering network properties, AI algorithms can predict how perturbing specific nodes or edges within the network may affect disease-related processes. This information helps in identifying potential drug targets or repurposing existing drugs to target specific components of the network.

Network-based disease analysis: AI facilitates the analysis of disease-related networks to understand the molecular mechanisms underlying diseases. AI algorithms can integrate genomic, transcriptomic, and proteomic data to construct disease-specific networks and identify dysregulated pathways or critical nodes associated with the disease. This knowledge can aid in uncovering disease biomarkers, identifying therapeutic targets, and designing personalized treatment strategies.

Network medicine and systems pharmacology: AI contributes to the field of network medicine by integrating biological networks with patient-specific data to develop personalized treatment approaches. AI algorithms can analyze patient-specific networks, identify network perturbations, and predict how these perturbations can be therapeutically targeted. This approach enables the design of precision medicine strategies tailored to individual patients' network profiles. AI's ability to handle large-scale datasets, detect complex patterns, and learn from diverse data sources makes it an invaluable tool for biological network analysis. It enhances our understanding of the intricate relationships within biological systems and offers new avenues for therapeutic interventions and precision medicine.

Ecology and conservation

AI can assist in monitoring and managing ecosystems by analyzing ecological data. It aids in species identification, habitat mapping, and biodiversity assessment. AI algorithms can process large-scale environmental data, such as satellite imagery and sensor data, to understand ecosystem dynamics, predict species distributions, and support conservation efforts. AI has emerged as a powerful tool in ecology and conservation efforts, aiding in the study, monitoring, and management of ecosystems and endangered species. A few ways in which AI is changing the field are listed below.

Species identification and monitoring: AI algorithms can analyze images, audio recordings, or sensor data to automatically identify and track species in their natural habitats. By training ML models on large datasets, AI can recognize species based on their visual or acoustic characteristics. This technology helps in monitoring wildlife populations, identifying endangered species, and detecting illegal activities like poaching or logging.

Habitat monitoring and land cover classification: AI algorithms can analyze satellite imagery or aerial photographs to monitor changes in habitats and classify land cover types. By combining remote sensing data with AI techniques, researchers can track deforestation, land degradation, and habitat fragmentation. This information is crucial for understanding ecosystem health, predicting biodiversity patterns, and guiding conservation planning.

Predictive modeling and species distribution: AI algorithms can analyze environmental data, such as climate variables, topography, and vegetation indices, to predict species distribution patterns. By training models on species occurrence records, AI can identify suitable habitats, predict range shifts due to climate change, and assess the potential impacts of habitat loss or degradation. This information aids in conservation prioritization and management decisions. Models are used to study the effects of various factors on animals and humans (Hamadani et al., 2023a).

Ecological data analysis: AI techniques, including ML and data mining, can analyze large ecological datasets to extract meaningful patterns and insights. AI algorithms can identify ecological relationships, detect community dynamics, and assess ecosystem functioning. This knowledge contributes to a better understanding of ecological processes and supports evidence-based conservation practices.

Conservation planning and decision support: AI-based tools can assist in conservation planning by providing decision support systems. These tools consider multiple variables, such as species distributions, habitat quality, connectivity, and human impacts, to identify priority areas for conservation action. AI algorithms can optimize conservation strategies, considering trade-offs between conflicting objectives, resource limitations, and uncertainties in the data.

Wildlife and animal behavior analysis: AI algorithms can analyze animal behavior data, such as movement patterns, foraging behavior, or social interactions, to gain insights into ecological processes. AI enables the automated processing of large-scale tracking data, helping researchers understand species' responses to environmental changes, habitat preferences, and migration patterns. This information is valuable for conservation management and designing protected areas.

Antipoaching and wildlife protection: AI-based systems can aid in antipoaching efforts by detecting and alerting authorities to potential threats in real time. For example, AI algorithms can analyze camera trap images or acoustic data to identify illegal activities, such as poaching, logging, or wildlife trafficking. These systems enable more effective monitoring and enforcement, contributing to the protection of endangered species. AI has the potential to revolutionize ecology and conservation by enabling more efficient data analysis, real-time monitoring, and evidence-based decision-making. As AI continues to evolve and integrate with other technologies like remote sensing and sensor networks, its role in supporting ecological research and conservation efforts will become increasingly critical.

Synthetic biology and bio-engineering
AI-driven design and optimization tools facilitate the engineering of biological systems. AI algorithms can aid in gene synthesis, DNA assembly, and metabolic pathway engineering. They optimize designs based on desired properties and help researchers create novel biological entities and biosynthetic pathways. AI is playing a significant role in advancing synthetic biology and bioengineering, revolutionizing the design and development of biological systems. Here are some ways AI is contributing to this field:

DNA sequence design: AI algorithms can generate optimized DNA sequences for specific purposes, such as gene expression, protein engineering, or metabolic pathway design. By analyzing large databases of genetic sequences and leveraging ML techniques, AI can predict the performance and functionality of DNA sequences, leading to the design of more efficient and reliable genetic constructs.

Genetic circuit design: AI enables the design of complex genetic circuits with specific behaviors and functions. AI algorithms can model and simulate the behavior of genetic circuits, allowing researchers to predict and optimize circuit performance before implementation. This technology facilitates the construction of synthetic biological systems for applications like biosensors, biofuel production, or bioremediation.

Metabolic engineering: AI assists in metabolic pathway engineering by optimizing the design and regulation of biochemical pathways in organisms. AI algorithms can analyze metabolic networks, predict enzyme activities, and suggest genetic modifications to improve production yields or enhance metabolic efficiency. This technology accelerates the development of microbial cell factories for the production of pharmaceuticals, chemicals, and biofuels.

Protein design and engineering: AI algorithms are used to design and engineer proteins with desired properties. By modeling protein structures and simulating protein folding, AI can predict the structure—function relationship and optimize protein sequences for specific applications. This technology aids in the development of novel enzymes, therapeutics, and biomaterials.

Bioprocess optimization: AI helps optimize bioprocesses, such as fermentation or cell culture, for the production of biological compounds. AI algorithms can analyze process variables, sensor data, and historical records to predict optimal process conditions, control strategies, and quality attributes. This technology improves process efficiency, reduces costs, and ensures consistent product quality.

Data integration and analysis: AI algorithms can integrate and analyze large-scale biological data from diverse sources, such as genomics, proteomics, and metabolomics. AI techniques, including ML and data mining, enable the extraction of meaningful insights, identification of biological patterns, and prediction of biological behavior. This information aids in understanding complex biological systems and guiding bioengineering efforts.

Biodesign automation: AI facilitates the automation of the biodesign process by developing software tools and platforms that assist in designing, modeling, and simulating biological systems. These tools allow researchers to generate and test numerous design variants, enabling rapid prototyping and iteration in bioengineering projects. AI-based biodesign automation improves efficiency and accelerates the development of novel biological systems. The integration of AI and bioengineering is transforming the field of synthetic biology, enabling the design and construction of sophisticated biological systems with diverse applications. As AI continues to advance, it will play an increasingly vital role in bioengineering, allowing for the development of novel biological solutions to address a wide range of societal and environmental challenges.

Robot-assisted surgery using artificial intelligence

Robotic-assisted surgery combined with AI has ushered in a new era in surgical interventions. This innovative approach leverages the precision and dexterity of robotics and the intelligent decision-making capabilities of AI algorithms. By integrating these technologies, robotic-assisted surgery with AI enhances surgical accuracy, improves patient outcomes, and expands the possibilities of complex procedures (Gumbs and Gayet, 2022). This note explores the transformative potential of robotic-assisted surgery using AI and highlights its key benefits.

Enhanced precision and dexterity: Robotic-assisted surgery offers superior precision and dexterity compared to conventional surgical techniques. The robotic arms, guided by AI algorithms, can execute movements with exceptional steadiness and range of motion, surpassing human capabilities. This allows surgeons to perform intricate maneuvers with enhanced accuracy, minimizing the risk of errors and enabling precise tissue manipulation (Omisore et al., 2022).

Real-time visualization and imaging: AI algorithms integrated with robotic systems provide real-time visualization and advanced imaging capabilities. Computer vision algorithms enhance image quality, reduce noise, and assist in real-time image analysis. Surgeons can benefit from detailed 3D views, improved depth perception, and augmented reality overlays, allowing them to better visualize anatomical structures and make informed decisions during the procedure (Gorpas et al., 2019).

Intelligent surgical assistance: AI algorithms act as intelligent surgical assistants, offering real-time guidance and decision support. By analyzing patient-specific data, medical records, and intra-operative information, AI algorithms provide insights and recommendations to surgeons. They can identify critical structures, suggest optimal instrument placement, and alert surgeons to potential risks or complications. This assistance enhances surgical precision, reduces variability, and promotes safer outcomes (Rasouli et al., 2021).

Automation of repetitive tasks: Robotic-assisted surgery with AI enables the automation of repetitive tasks, freeing up surgeons to focus on critical decision-making. AI algorithms can automate precise and repetitive actions such as suturing, tissue manipulation, or instrument retraction (Attanasio et al., 2021). This reduces fatigue and enhances efficiency in the operating room, ultimately leading to improved patient care and surgical outcomes.

Continuous real-time monitoring: AI algorithms continuously monitor various parameters during surgery, including vital signs, tissue properties, and instrument movements. By analyzing these data in real-time, AI algorithms can detect subtle changes, assess tissue viability, and alert surgeons to potential complications (Gorpas et al., 2019). This real-time monitoring empowers surgeons to make timely interventions and adapt their surgical approach, optimizing patient safety and outcomes.

Postoperative analysis and learning: Robotic-assisted surgery with AI facilitates postoperative analysis and learning. AI algorithms can analyze large datasets of surgical procedures, outcomes, and patient characteristics to identify patterns and trends (Aminsharifi et al., 2020). This data-driven analysis enables the refinement of surgical techniques, the development of evidence-based best practices, and the optimization of future procedures. Surgeons can benefit from shared knowledge and collective experience, leading to continuous improvement in surgical interventions.

These advantages are true for both veterinary and medical surgeries. Furthermore, provided below is an elaborate explanation of the functioning of AI-powered robotic-assisted surgery (Ahmad et al., 2023).

1. **Setup and calibration:** The surgical team sets up the robotic surgical system, which typically includes a console, robotic arms, and surgical instruments. The system is calibrated to ensure accurate movements and feedback (Roberti et al., 2020). AI software and algorithms are installed, providing the intelligence and decision-making capabilities required for surgery.
2. **Preoperative planning and imaging:** Prior to surgery, the surgeon reviews the patient's medical imaging, such as CT scans or MRI, to assess the anatomy and plan the procedure. AI algorithms analyze the imaging data, helping identify important structures, plan the optimal surgical approach, and predict potential outcomes (Antonelli et al., 2019). This AI-driven preoperative planning enhances surgical precision and efficiency.
3. **Console operation:** The surgeon sits at the robotic console, which provides a 3D view of the surgical field and controls the robotic arms. The console is equipped with hand controls, foot pedals, and visual feedback systems (Larkins et al., 2023). AI algorithms assist in real-time tracking of the surgeon's hand movements, translating them into precise robotic actions.
4. **Robotic arm and instrumentation:** The robotic arms, guided by AI algorithms, mimic the surgeon's movements with enhanced precision and stability. Surgical instruments are attached to the robotic arms using docking mechanisms (Zhang et al., 2021). The AI software ensures seamless communication and synchronization between the robotic arms, instruments, and the surgeon's commands.
5. **Real-time visualization:** Advanced imaging technologies, integrated with AI, provide real-time visualization during the procedure. Computer vision algorithms enhance image resolution, reduce noise, and extract relevant information. This enables the surgeon to have a clearer view of the surgical site, visualize critical structures, and make informed decisions throughout the surgery (Islam et al., 2019).

6. **Intelligent assistance:** AI algorithms provide real-time guidance and assistance during surgery. They analyze the surgical data, including imaging, physiological parameters, and instrument movements, to offer insights and recommendations to the surgeon (Thai et al., 2020). For example, AI algorithms can highlight important anatomical landmarks, alert the surgeon about potential risks, or suggest optimal instrument positioning.

7. **Automation and augmentation:** Robotic-assisted surgery using AI enables the automation of certain tasks. AI algorithms can automate repetitive actions, such as suturing or tissue manipulation, with high precision. This reduces the risk of human error and allows surgeons to focus on critical decision-making and complex procedures. AI can also augment the surgeon's skills by eliminating hand tremors, scaling movements, or providing haptic feedback for improved dexterity (Itzkovich et al., 2019).

8. **Real-time monitoring and feedback:** During the surgery, AI algorithms continuously monitor the patient's vital signs, as well as feedback from the robotic system. They analyze the data in real time, providing the surgical team with valuable insights (Qiu et al., 2019). For instance, AI algorithms can detect subtle changes in tissue properties, monitor blood flow, or predict potential complications, helping the surgeon adapt their approach and optimize patient outcomes.

9. **Postoperative analysis and learning** AI algorithms facilitate postoperative analysis by processing and analyzing surgical data. They can compare the surgical procedure to established databases, evaluate outcomes, and identify patterns or trends. This data-driven learning helps refine surgical techniques, optimize future procedures, and contribute to the development of evidence-based best practices (Kinoshita et al., 2021).

It is important to note that the specific steps and considerations may vary depending on the robotic surgical system being used, the type of surgery being performed, and the available AI algorithms and features. Surgeons should receive proper training and certification in robotic-assisted surgery and familiarize themselves with the specific robotic system and AI software they are utilizing. Additionally, adherence to established surgical guidelines, safety protocols, and ethical considerations is paramount throughout the entire process of performing robotic surgery using AI. Robotic-assisted surgery using AI is a transformative approach that revolutionizes the field of surgery. By leveraging the precision of robotics and the intelligence of AI algorithms, it enhances surgical planning, visualization, automation, real-time assistance, and postoperative analysis. This combination of technology and intelligence enables surgeons to perform complex procedures with greater accuracy, improves patient outcomes, and sets the stage for future advancements in the surgical field.

The promise of AI

The integration of AI and biology holds great promise for advancing scientific research and healthcare. By leveraging AI's capabilities in data analysis and pattern recognition, researchers can gain valuable insights from the vast amount of biological data generated. AI algorithms can uncover hidden correlations and patterns, leading to a deeper understanding of biological processes and disease mechanisms. By harnessing the power of AI, biology stands to benefit from enhanced efficiency, improved diagnostics, and transformative breakthroughs, ultimately advancing human health and deepening our understanding of life itself. In the realm of synthetic biology, AI-guided genetic engineering can optimize the production of valuable substances and improve the traits of organisms (Collins and

Curiel, 2021). These integrations have the potential to accelerate scientific advancements, enhance patient care, and drive innovation in the biological sciences.

Limitations of using AI in biology

Despite these challenges, ongoing efforts are being made to integrate AI and biology. Researchers are actively working on developing new AI algorithms and approaches that can better handle the complexities of biological systems. Furthermore, collaborations between AI experts and biologists are growing, fostering a better understanding of the unique requirements of biology and facilitating the development of AI solutions tailored to biological applications.

The complexity of biological systems: Biological systems are incredibly complex, comprising multiple interconnected components and intricate interactions. AI algorithms often rely on simplified models and assumptions, which may not accurately capture the complexity of biological phenomena. Developing AI models that can effectively handle the complexity and heterogeneity of biological data is a significant challenge (Hamadani et al., 2022b).

Limited and noisy data: While biological datasets have grown in size, they are often limited in their scope and quality. Biological data can be noisy, incomplete, and prone to various sources of error. Training AI models on such data can lead to biased or unreliable results. Obtaining large, diverse, and high-quality datasets that represent the full complexity of biological systems is a persistent challenge.

Interpretability and explainability: AI models, especially DL models, can be black boxes, making it difficult to interpret the reasoning behind their predictions. In biology, where understanding the underlying mechanisms is crucial, this lack of interpretability poses a significant hurdle. Biologists need to trust and understand the AI models' decisions for them to be effectively integrated into research and decision-making processes.

Ethics and privacy concerns: Ethical considerations play a crucial role in the integration of AI with biology. As AI becomes increasingly involved in scientific research and healthcare, it is essential to address concerns related to privacy, bias, and responsible use. The collection and analysis of biological data raise significant privacy concerns, as personal information and genetic data must be handled with utmost care and confidentiality (Guan, 2019). Additionally, bias in AI algorithms can inadvertently perpetuate inequalities in healthcare and research, potentially leading to disparities in diagnosis, treatment, and access to resources (Parikh et al., 2019). Therefore, it is vital to ensure that AI models are thoroughly tested for biases and regularly monitored to mitigate any unfairness. Furthermore, the responsible use of AI in decision-making processes must be emphasized, as it is essential to maintain a balance between AI-driven insights and human judgment. Transparent and accountable decision-making frameworks should be established to ensure that AI is utilized ethically, with human values, safety, and well-being at the forefront (McLennan et al., 2022). Ultimately, ethical considerations should guide the integration of AI and biology to promote fairness, privacy, and the betterment of human health. The use of AI in biological research, such as genetic profiling or personalized medicine, can involve sensitive personal information. Ensuring data privacy, informed consent, and responsible use of AI technologies becomes paramount to maintaining public trust and adhering to ethical guidelines. Ethical considerations also come into play when integrating AI and biology (Guan, 2019). Privacy concerns (McLennan et al., 2022), bias in algorithms (Parikh et al., 2019), and the responsible use of AI in decision-making processes are paramount. It is crucial to establish ethical guidelines and frameworks that govern the use of AI in biology, ensuring the protection of individual privacy, mitigating biases, and ensuring fair and responsible decision-making.

Limited domain knowledge: Developing AI models that can fully capture the complexities of biological systems requires a deep understanding of biology. AI researchers often need to collaborate closely with domain experts, such as biologists and clinicians, to develop effective AI solutions. Bridging the gap between AI and biology requires interdisciplinary collaboration, which can be challenging due to differences in language, methodologies, and expertize. Addressing these challenges requires interdisciplinary collaboration between AI experts, biologists, and ethicists. By fostering collaboration, we can develop transparent, reliable, and ethical AI systems that effectively tackle the complexities of the biological domain. Moreover, ongoing research and innovation should focus on advancing the field of AI interpretability, data standardization, and ethical guidelines, enabling the seamless integration of AI and biology.

Conclusion

The integration of AI in the field of biology has significantly transformed and revolutionized research and development processes. AI has emerged as a powerful tool that enables scientists and researchers to efficiently analyze complex biological data, make accurate predictions, and gain deep insights into various biological phenomena. It has revolutionized surgical procedures by enhancing precision, efficiency, and patient outcomes. AI has enabled targeted drug delivery, accelerated the drug discovery process, and resulted in the development of precision medicine techniques. These advancements hold great promise for improving patient outcomes, personalized medicine, ecology, evolution and advancing scientific knowledge. However, ethical considerations and regulatory frameworks must be addressed to ensure responsible implementation. With continued research and collaboration, AI in biology has the potential to transform healthcare and benefit society as a whole. The integration of AI and biology holds immense potential for advancing scientific understanding, healthcare, and synthetic biology applications. AI's analytical power can help researchers analyze vast amounts of data, improve diagnostics, accelerate drug discovery, and optimize genetic engineering processes. However, challenges such as data availability, interpretability, and ethical considerations must be effectively addressed to fully harness the benefits of this integration. Through continued research and interdisciplinary collaboration, AI and biology can synergistically drive innovation, transform our understanding of the natural world, and improve the well-being of individuals and societies as a whole.

References

Ahmad, A., Tariq, A., Hussain, H.K., Gill, A.Y., 2023. Equity and artificial intelligence in surgical care: a comprehensive review of current challenges and promising solutions. Bullet: Journal of Multidisiplinary Ilmu Impact 2, 443−455.

Aminsharifi, A., Irani, D., Tayebi, S., Jafari Kafash, T., Shabanian, T., Parsaei, H., 2020. Predicting the postoperative outcome of percutaneous nephrolithotomy with machine learning system: software validation and comparative analysis with guy's stone score and the CROES nomogram. Journal of Endourology 34 (6), 692−699. https://doi.org/10.1089/end.2019.0475.

Antonelli, A., Veccia, A., Palumbo, C., Peroni, A., Mirabella, G., Cozzoli, A., Martucci, P., Ferrari, F., Simeone, C., Artibani, W., 2019. Holographic reconstructions for preoperative planning before partial nephrectomy: a head-to-head comparison with standard CT scan. Urologia Internationalis 102 (2), 212−217. https://doi.org/10.1159/000495618.

Attanasio, A., Scaglioni, B., De Momi, E., Fiorini, P., Valdastri, P., 2021. Autonomy in surgical robotics. Annual Review of Control, Robotics, and Autonomous Systems 4 (1), 651−679. https://doi.org/10.1146/annurev-control-062420-090543.

Bhardwaj, A., Kishore, S., Pandey, D.K., 2022. Artificial intelligence in biological sciences. Life 12 (9). https://doi.org/10.3390/life12091430.

Bohr, A., Memarzadeh, K., 2020. The rise of artificial intelligence in healthcare applications. Artificial Intelligence in Healthcare 25−60. https://doi.org/10.1016/B978-0-12-818438-7.00002-2.

Braun, M., Hummel, P., Beck, S., Dabrock, P., 2021. Primer on an ethics of AI-based decision support systems in the clinic. Journal of Medical Ethics 47 (12), 0306−6800, e3−e3. https://doi.org/10.1136/medethics-2019-105860.

Chan, H.C.S., Shan, H., Dahoun, T., Vogel, H., Yuan, S., 2019. Advancing drug discovery via artificial intelligence. Trends in Pharmacological Sciences 40 (8), 592−604. https://doi.org/10.1016/j.tips.2019.06.004.

Collins, L.T., Curiel, D.T., 2021. Synthetic biology approaches for engineering next-generation adenoviral gene therapies. ACS Nano 15 (9), 13970−13979. https://doi.org/10.1021/acsnano.1c04556.

Comito, C., Falcone, D., Forestiero, A., 2020. Current trends and practices in smart health monitoring and clinical decision support. Proceedings − 2020 IEEE International Conference on Bioinformatics and Biomedicine, BIBM 2020 2577−2584. https://doi.org/10.1109/BIBM49941.2020.9313449.

Dawoodbhoy, F.M., Delaney, J., Cecula, P., Yu, J., Peacock, I., Tan, J., Cox, B., 2021. AI in patient flow: applications of artificial intelligence to improve patient flow in NHS acute mental health inpatient units. Heliyon 7 (5). https://doi.org/10.1016/j.heliyon.2021.e06993.

Desaire, H., Go, E.P., Hua, D., 2022. Advances, obstacles, and opportunities for machine learning in proteomics. Cell Reports Physical Science 3 (10). https://doi.org/10.1016/j.xcrp.2022.101069.

Dutta, U., Babu, N.D., Setlur, G.S., 2022. Artificial intelligence in biological sciences: a brief overview. Information Retrieval in Bioinformatics: A Practical Approach 19−35. https://doi.org/10.1007/978-981-19-6506-7_2.

Gao, Z., Luo, Z., Zhang, W., Lv, Z., Xu, Y., 2020. Deep learning application in plant stress imaging: a review. AgriEngineering 2 (3), 430−446. https://doi.org/10.3390/agriengineering2030029.

Gao, J., Karp, J.M., Langer, R., Joshi, N., 2023. The future of drug delivery. Chemistry of Materials 35 (2), 359−363. https://doi.org/10.1021/acs.chemmater.2c03003.

Goldberg, D.E., 2013. Genetic Algorithms. Pearson Education India.

Gorpas, D., Phipps, J., Bec, J., Ma, D., Dochow, S., Yankelevich, D., Sorger, J., Popp, J., Bewley, A., Gandour-Edwards, R., Marcu, L., Farwell, D.G., 2019. Autofluorescence lifetime augmented reality as a means for real-time robotic surgery guidance in human patients. Scientific Reports 9 (1). https://doi.org/10.1038/s41598-018-37237-8.

Guan, J., 2019. Artificial intelligence in healthcare and medicine: promises, ethical challenges and governance. Chinese Medical Sciences Journal 34 (2), 76−83. https://doi.org/10.24920/003611.

Gumbs, A.A., Gayet, B., 2022. Why artificial intelligence surgery (AIS) is better than current robotic-assisted surgery (RAS). Artificial Intelligence Surgery 2 (4), 207−213. https://doi.org/10.20517/ais.2022.41.

Hamadani, A., 2023. Exploration of Machine Learning Algorithms for the Evaluation of Factors Affecting COVID-19 Death Rates.

Hamadani, A., Ganai, N.A., 2022. Development of a multi-use decision support system for scientific management and breeding of sheep. Scientific Reports 12 (1). https://doi.org/10.1038/s41598-022-24091-y.

Hamadani, H., Khan, A.A., 2015. Automation in livestock farming − a technological revolution. International Journal of Advanced Research 3, 1335−1344.

Hamadani, A., Ganai, N., Farooq, S., Rather, M., 2019. 'Breeders Toolkit'—a cloud-based breeding toolkit for estimation of various breeding parameters. International Journal of Livestock Research. https://doi.org/10.5455/ijlr.20190220052427.

Hamadani, A., Ganai, N.A., Mudasir, S., Shanaz, S., Alam, S., Hussain, I., 2022a. Comparison of artificial intelligence algorithms and their ranking for the prediction of genetic merit in sheep. Scientific Reports 12 (1). https://doi.org/10.1038/s41598-022-23499-w.

Hamadani, A., Ganai, N.A., Alam, S., Mudasir, S., Raja, T.A., Hussain, I., Ahmad, H.A., 2022b. Artificial intelligence techniques for the prediction of body weights in sheep. Indian Journal of Animal Research. https://doi.org/10.18805/ijar.b-4831.

Hamadani, A., Ganai, N.A., Khan, N.N., Shanaz, S., Rather, M.A., Ahmad, H.A., Shah, R., 2023a. Comparison of various models for the estimation of heritability and breeding values. Tropical Animal Health and Production 55 (4). https://doi.org/10.1007/s11250-023-03665-6.

Hamadani, A., Ganai, N.A., Bashir, J., 2023b. Artificial neural networks for data mining in animal sciences. Bulletin of the National Research Centre 47 (1). https://doi.org/10.1186/s42269-023-01042-9.

Harrer, S., Shah, P., Antony, B., Hu, J., 2019. Artificial intelligence for clinical trial design. Trends in Pharmacological Sciences 40 (8), 577−591. https://doi.org/10.1016/j.tips.2019.05.005.

Hassanzadeh, P., Atyabi, F., Dinarvand, R., 2019. The significance of artificial intelligence in drug delivery system design. Advanced Drug Delivery Reviews 151−152, 169−190. https://doi.org/10.1016/j.addr.2019.05.001.

Islam, M., Atputharuban, D.A., Ramesh, R., Ren, H., 2019. Real-time instrument segmentation in robotic surgery using auxiliary supervised deep adversarial learning. IEEE Robotics and Automation Letters 4 (2), 2188−2195. https://doi.org/10.1109/LRA.2019.2900854.

Itzkovich, D., Sharon, Y., Jarc, A., Refaely, Y., Nisky, I., 2019. Using augmentation to improve the robustness to rotation of deep learning segmentation in robotic-assisted surgical data. Proceedings − IEEE International Conference on Robotics and Automation 5068−5075. https://doi.org/10.1109/ICRA.2019.8793963.

Jaiswal, A.K., Tiwari, P., Kumar, S., Gupta, D., Khanna, A., Rodrigues, J.J.P.C., 2019. Identifying pneumonia in chest X-rays: a deep learning approach. Measurement 145, 511−518. https://doi.org/10.1016/j.measurement.2019.05.076.

Kang, M., Ko, E., Mersha, T.B., 2022. A roadmap for multi-omics data integration using deep learning. Briefings in Bioinformatics 23 (1). https://doi.org/10.1093/bib/bbab454.

Kelleher, J.D., 2019. Deep Learning. MIT Press.

Khan, O., Badhiwala, J.H., Grasso, G., Fehlings, M.G., 2020. Use of machine learning and artificial intelligence to drive personalized medicine approaches for spine care. World Neurosurgery 140, 512−518. https://doi.org/10.1016/j.wneu.2020.04.022.

Kinoshita, T., Sato, R., Akimoto, E., Tanaka, Y., Okayama, T., Habu, T., 2021. Reduction in postoperative complications by robotic surgery: a case-control study of robotic versus conventional laparoscopic surgery for gastric cancer. Surgical Endoscopy 1−10.

Kislov, D.E., Korznikov, K.A., 2020. Automatic windthrow detection using very-high-resolution satellite imagery and deep learning. Remote Sensing 12 (7). https://doi.org/10.3390/rs12071145.

Klang, E., 2018. Deep learning and medical imaging. Journal of Thoracic Disease 10 (3), 1325−1328. https://doi.org/10.21037/jtd.2018.02.76.

Kumar, V., Ram, M., 2021. Predictive Analytics: Modeling and Optimization. CRC Press, 2021.

Larkins, K.M., Mohan, H.M., Gray, M., Costello, D.M., Costello, A.J., Heriot, A.G., Warrier, S.K., 2023. Transferability of robotic console skills by early robotic surgeons: a multi-platform crossover trial of simulation training. Journal of Robotic Surgery 17 (3), 859−867. https://doi.org/10.1007/s11701-022-01475-w.

Lawrence, J., 1993. Introduction to Neural Networks. California Scientific Software.

Li, H., Tian, S., Li, Y., Fang, Q., Tan, R., Pan, Y., Huang, C., Xu, Y., Gao, X., 2020. Modern deep learning in bioinformatics. Journal of Molecular Cell Biology 12 (11), 823−827. https://doi.org/10.1093/jmcb/mjaa030.

Libbrecht, M.W., Noble, W.S., 2015. Machine learning applications in genetics and genomics. Nature Reviews Genetics 16 (6), 321−332. https://doi.org/10.1038/nrg3920.

Lin, J., Chen, H., Li, S., Liu, Y., Li, X., Yu, B., 2019. Accurate prediction of potential druggable proteins based on genetic algorithm and Bagging-SVM ensemble classifier. Artificial Intelligence in Medicine 98, 35−47. https://doi.org/10.1016/j.artmed.2019.07.005.

Lundervold, A.S., Lundervold, A., 2019. An overview of deep learning in medical imaging focusing on MRI. Zeitschrift fur Medizinische Physik 29 (2), 102−127. https://doi.org/10.1016/j.zemedi.2018.11.002.

Mansoor, S., Hameed, A., Anjum, R., Maqbool, I., Masoodi, M., Maqbool, K., Dar, Z.A., Hamadani, A., Mahmoud, A.E.D., 2021. Obesity: causes, consequences, and disease risks for service personnel. Phytochemistry, the Military and Health: Phytotoxins and Natural Defenses 407−425. https://doi.org/10.1016/B978-0-12-821556-2.00004-9.

McClure, E.C., Sievers, M., Brown, C.J., Buelow, C.A., Ditria, E.M., Hayes, M.A., Pearson, R.M., Tulloch, V.J.D., Unsworth, R.K.F., Connolly, R.M., 2020. Artificial intelligence meets citizen science to supercharge ecological monitoring. Patterns 1 (7). https://doi.org/10.1016/j.patter.2020.100109.

McLennan, S., Fiske, A., Tigard, D., Müller, R., Haddadin, S., Buyx, A., 2022. Embedded ethics: a proposal for integrating ethics into the development of medical AI. BMC Medical Ethics 23 (1). https://doi.org/10.1186/s12910-022-00746-3.

Muzio, G., O'Bray, L., Borgwardt, K., 2021. Biological network analysis with deep learning. Briefings in Bioinformatics 22 (2), 1515−1530. https://doi.org/10.1093/bib/bbaa257.

Neftci, E.O., Averbeck, B.B., 2019. Reinforcement learning in artificial and biological systems. Nature Machine Intelligence 1 (3), 133−143. https://doi.org/10.1038/s42256-019-0025-4.

Noé, F., De Fabritiis, G., Clementi, C., 2020. Machine learning for protein folding and dynamics. Current Opinion in Structural Biology 60, 77−84. https://doi.org/10.1016/j.sbi.2019.12.005.

Omisore, O.M., Han, S., Xiong, J., Li, H., Li, Z., Wang, L., 2022. A review on flexible robotic systems for minimally invasive surgery. IEEE Transactions on Systems, Man, and Cybernetics: Systems 52 (1), 631−644. https://doi.org/10.1109/TSMC.2020.3026174.

Ozer, M.E., Sarica, P.O., Arga, K.Y., 2020. New machine learning applications to accelerate personalized medicine in breast cancer: rise of the support vector machines. OMICS: A Journal of Integrative Biology 24 (5), 241−246. https://doi.org/10.1089/omi.2020.0001.

Parikh, R.B., Teeple, S., Navathe, A.S., 2019. Addressing bias in artificial intelligence in health care. JAMA, the Journal of the American Medical Association 322 (24), 2377−2378. https://doi.org/10.1001/jama.2019.18058.

Pisner, D.A., Schnyer, D.M., 2019. Support vector machine. Machine Learning: Methods and Applications to Brain Disorders 101−121. https://doi.org/10.1016/B978-0-12-815739-8.00006-7.

Qiu, L., Li, C., Ren, H., 2019. Real-time surgical instrument tracking in robot-assisted surgery using multi-domain convolutional neural network. Healthcare Technology Letters 6 (6), 159−164. https://doi.org/10.1049/htl.2019.0068.

Rasouli, J.J., Shao, J., Neifert, S., Gibbs, W.N., Habboub, G., Steinmetz, M.P., Benzel, E., Mroz, T.E., 2021. Artificial intelligence and robotics in spine surgery. Global Spine Journal 11 (4), 556−564. https://doi.org/10.1177/2192568220915718.

Richards, B., Tsao, D., Zador, A., 2022. The application of artificial intelligence to biology and neuroscience. Cell 185 (15), 2640−2643. https://doi.org/10.1016/j.cell.2022.06.047.

Rigatti, S.J., 2017. Random forest. Journal of Insurance Medicine 47 (1), 31−39. https://doi.org/10.17849/insm-47-01-31-39.1.

Roberti, A., Piccinelli, N., Meli, D., Muradore, R., Fiorini, P., 2020. Improving rigid 3-D calibration for robotic surgery. IEEE Transactions on Medical Robotics and Bionics 2 (4), 569−573. https://doi.org/10.1109/TMRB.2020.3033670.

Ryo, M., Angelov, B., Mammola, S., Kass, J.M., Benito, B.M., Hartig, F., 2021. Explainable artificial intelligence enhances the ecological interpretability of black-box species distribution models. Ecography 44 (2), 199−205. https://doi.org/10.1111/ecog.05360.

Saraswat, D., Bhattacharya, P., Verma, A., Prasad, V.K., Tanwar, S., Sharma, G., Bokoro, P.N., Sharma, R., 2022. Explainable AI for healthcare 5.0: opportunities and challenges. IEEE Access 10, 84486−84517. https://doi.org/10.1109/ACCESS.2022.3197671.

Schuwirth, N., Borgwardt, F., Domisch, S., Friedrichs, M., Kattwinkel, M., Kneis, D., Kuemmerlen, M., Langhans, S.D., Martínez-López, J., Vermeiren, P., 2019. How to make ecological models useful for environmental management. Ecological Modelling 411. https://doi.org/10.1016/j.ecolmodel.2019.108784.

Serrano, A., Imbernón, B., Pérez-Sánchez, H., Cecilia, J.M., Bueno-Crespo, A., Abellán, J.L., 2018. Accelerating drugs discovery with deep reinforcement learning: an early approach. ACM International Conference Proceeding Series. https://doi.org/10.1145/3229710.3229731.

Sitapati, A., Kim, H., Berkovich, B., Marmor, R., Singh, S., El-Kareh, R., Clay, B., Ohno-Machado, L., 2017. Integrated precision medicine: the role of electronic health records in delivering personalized treatment. Wiley Interdisciplinary Reviews: Systems Biology and Medicine 9 (3). https://doi.org/10.1002/wsbm.1378.

Song, Y.Y., Lu, Y., 2015. Decision tree methods: applications for classification and prediction. Shanghai Archives of Psychiatry 27 (2), 130−135. https://doi.org/10.11919/j.issn.1002-0829.215044.

Spasic, I., Nenadic, G., 2020. Clinical text data in machine learning: systematic review. JMIR Medical Informatics 8 (3). https://doi.org/10.2196/17984.

Spiegel, J.O., Durrant, J.D., 2020. AutoGrow4: an open-source genetic algorithm for de novo drug design and lead optimization. Journal of Cheminformatics 12 (1). https://doi.org/10.1186/s13321-020-00429-4.

Stahlschmidt, S.R., Ulfenborg, B., Synnergren, J., 2022. Multimodal deep learning for biomedical data fusion: a review. Briefings in Bioinformatics 23 (2). https://doi.org/10.1093/bib/bbab569.

Sutton, R.S., Barto, A.G., 2018. Reinforcement Learning: An Introduction. MIT Press.

Suzuki, K., 2017. Overview of deep learning in medical imaging. Radiological Physics and Technology 10 (3), 257−273. https://doi.org/10.1007/s12194-017-0406-5.

Tan, M., Xu, Y., Gao, Z., Yuan, T., Liu, Q., Yang, R., Zhang, B., Peng, L., 2022. Recent advances in intelligent wearable medical devices integrating biosensing and drug delivery. Advanced Materials 34 (27). https://doi.org/10.1002/adma.202108491.

Thai, M.T., Phan, P.T., Hoang, T.T., Wong, S., Lovell, N.H., Do, T.N., 2020. Advanced intelligent systems for surgical robotics. Advanced Intelligent Systems 2 (8). https://doi.org/10.1002/aisy.201900138.

Vamathevan, J., Clark, D., Czodrowski, P., Dunham, I., Ferran, E., Lee, G., Li, B., Madabhushi, A., Shah, P., Spitzer, M., Zhao, S., 2019. Applications of machine learning in drug discovery and development. Nature Reviews Drug Discovery 18 (6), 463−477. https://doi.org/10.1038/s41573-019-0024-5.

Wen, A., Fu, S., Moon, S., El Wazir, M., Rosenbaum, A., Kaggal, V.C., Liu, S., Sohn, S., Liu, H., Fan, J., 2019. Desiderata for delivering NLP to accelerate healthcare AI advancement and a Mayo Clinic NLP-as-a-service implementation. NPJ Digital Medicine 2 (1). https://doi.org/10.1038/s41746-019-0208-8.

Whalen, S., Schreiber, J., Noble, W.S., Pollard, K.S., 2022. Navigating the pitfalls of applying machine learning in genomics. Nature Reviews Genetics 23 (3), 169−181. https://doi.org/10.1038/s41576-021-00434-9.

Yang, D., Martinez, C., Visuña, L., Khandhar, H., Bhatt, C., Carretero, J., 2021. Detection and analysis of COVID-19 in medical images using deep learning techniques. Scientific Reports 11 (1). https://doi.org/10.1038/s41598-021-99015-3.

Yu, K.H., Beam, A.L., Kohane, I.S., 2018. Artificial intelligence in healthcare. Nature Biomedical Engineering 2 (10), 719−731. https://doi.org/10.1038/s41551-018-0305-z.

Zaghloul, M.S., Achari, G., 2022. Application of machine learning techniques to model a full-scale wastewater treatment plant with biological nutrient removal. Journal of Environmental Chemical Engineering 10 (3). https://doi.org/10.1016/j.jece.2022.107430.

Zhang, W., Li, H., Cui, L., Li, H., Zhang, X., Fang, S., Zhang, Q., 2021. Research progress and development trend of surgical robot and surgical instrument arm. International Journal of Medical Robotics and Computer Assisted Surgery 17 (5). https://doi.org/10.1002/rcs.2309.

Understanding life and evolution using AI

3

Tabinda Wani and Nelofar Banday

Division of Floriculture and Landscape Architecture, SKUAST Kashmir, Srinagar, Jammu and Kashmir, India

Introduction

Artificial intelligence (AI) can be described as a branch of computer science that aims to create intelligent computer systems that are capable of carrying out tasks that traditionally demand human intelligence. This field of technology enables machines to understand reason, learn, and make decisions based on data and experiences. There are several ways to exemplify AI. For many people, it is the technology that enabled computers and other devices to work in an intelligent and efficient way. Some believe it to be a machine that would substitute human labor to provide a faster means of task completion for mankind. And the remaining strata it is "a system" that has the ability to analyze external data properly in an accurate way, learning from such information and applying it to specific goals and activities through flexible learning (Kaplan and Haenlein, 2019).

Despite the diversity of definitions, the consensus on AI is that it is used in conjunction with computers and machines to assist humans in problem-solving and simplify operational procedures. AI can be defined as an intellect created by man and displayed by machines. In other words, "artificial intelligence" refers to the characteristics of human-made technologies that imitate the "cognitive" abilities of the inherent intelligence of human minds (Tai, 2020).

Simply described, intelligence is the computational aspect of a person's ability to achieve goals in the real world. The ability to reason, picture, remember, and interpret information, as well as recognize patterns, make judgments, and adjust to change, are all considered to be signs of intelligence. AI is concerned with making computers behave more like humans while taking a small fraction of the time. By applying general knowledge to particular situations, AI seeks to push the boundaries of practical computer science and create systems that are flexible, adaptable, and capable of creating their own analyses and problem-solving methods (Singh and Haju, 2022).

AI has almost completely permeated our daily lives as a result of the prompt developments in cybernetic technology in the past few years. Because it is so commonplace in our daily lives and we are so accustomed to it, some of this technology, such as optical character recognition and the computerized "speech interpretation and recognition interface" (SIRI) information-searching tool, may no longer be regarded as AI (Schank, 1990).

A Biologist's Guide to Artificial Intelligence. https://doi.org/10.1016/B978-0-443-24001-0.00003-8

AI algorithms and techniques
Machine learning

The practice that trains (Schank, 1990) algorithms to see patterns in data and make predictions or judgments without being explicitly programmed is known as machine learning (ML). ML methods are frequently used to classify and predict biological occurrences, construct models, and analyze massive datasets. Generally, it is applied to analyze genomics data, predict protein structure, find new drugs, etc. And methods like supervised learning, unsupervised learning, and reinforcement learning are utilized for this purpose. ML methods have been readily utilized to identify and categorize fossils according to different criteria, such as taxonomy, age, or anatomical features. Researchers may create automated methods for species identification and enhance our understanding of evolutionary links by training ML models on vast databases of fossil records. The construction and analysis of phylogenetic trees—which show the evolutionary links between organisms—use AI methods. ML techniques support the study of evolutionary processes and the reconstruction of ancient ecosystems by inferring phylogenetic connections based on morphological or molecular data (Jordan and Mitchell, 2015).

Some algorithms that are very popular in evolutionary studies are discussed in detail in his chapter. Major AI techniques to study life and evolution are also enlisted in Fig. 3.1.

Deep learning

A sub-branch of ML called deep learning (DL) uses multiple-layer neural networks to extract complicated properties from data. DL has been successfully used in genomics research, protein folding prediction, medical imaging, and image recognition. For understanding life and life processes, DL architectures like convolutional neural networks (CNNs) and recurrent neural networks (RNNs) are frequently employed. CNNs, in particular, are employed in AI algorithms to analyze fossil photographs and automate the identification and categorization of fossils (Rusk, 2016).

With several major benefits over conventional ML techniques like principal component analysis (PCA), Bayesian methods (BM), support vector machines (SVMs), random forests (RF) and decision

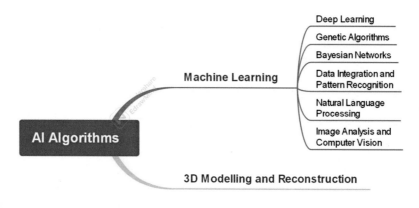

FIGURE 3.1

Different artificial intelligence (AI) algorithms and techniques.

trees (DT), DL methods have recently paved the way for intriguing and stimulating perceptions in fundamental research areas (e.g., image analysis, language analysis, and also omics sciences) (Lecun et al., 2015). The capability of getting classification or prediction results straight from the raw data, is the major benefit of DL over ML techniques, which is precisely referred to as end-to-end learning. Eliminating the possible bias caused by physical involvement in the various data processing processes promotes end-to-end learning. Even this does not save the development from potential sources of bias (such as choosing the input data for the network training phase). Additionally, DL techniques allow the integration of many input data formats, including text, numbers, photos, and audio files. In comparison to conventional ML methodologies, DL architectures offer a significantly better capability of abstraction (Schmidhuber, 2015). A number of frameworks are available today, which make DL very convenient. They are given in Fig. 3.2.

In terms of effectiveness and performance, recent DL architectures like deep neural networks (DNNs), deep belief networks (DBNs), RNNs, deep Boltzmann machines (DBMs), CNNs, auto-encoders (AEs), and generative adversarial networks (GANs) have advanced significantly from Rosenblatt's Perceptron (first neural network system introduced in 1958).

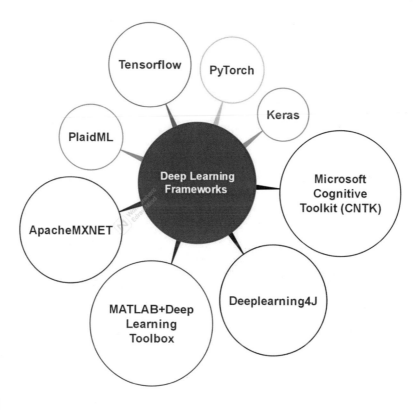

FIGURE 3.2

Popular deep learning frameworks.

Natural language processing

Numerous volumes of scientific literature are mined for information using NLP algorithms. This aids automated literature mining, pertinent information extraction, and knowledge graph construction. Text mining, literature-based discovery, and drug repurposing are made easier with the aid of NLP. Information from databases and publications on palaeontology is extracted and analyzed using natural language processing (NLP) methods. Text mining makes it possible to find new fossil locations, identify new species, and combine massive databases from paleontology for future studies (Chowdhary, 2020).

Image analysis and computer vision

Artificial intelligence techniques are used to analyze biological pictures and extract valuable information. Cell counting, illness diagnosis, and medication screening are made easier by computer vision algorithms' automatic identification and tracking of cells, tissues, or organisms in microscope pictures. Computer vision tools can identify species and conduct morphological research by detecting fossil traits like skeleton or tooth forms (Kodagali and Balaji, 2012).

Genetic algorithms

To improve solutions, genetic algorithms (GAs) imitate natural development. They have been used in biology for activities including DNA sequence design, protein folding, and sequence alignment. By continuously developing and choosing the fittest individuals based on fitness criteria, GAs assist in identifying optimum solutions (Kramer, 2017) for many problems in ecology.

Bayesian networks

Bayesian networks are probabilistic models that show how different variables relate to one another. They are useful in modeling regulatory networks, studying gene expression, and genetics. Using Bayesian networks, it is possible to forecast the course of diseases, uncover genetic variations linked to diseases, and infer gene regulatory networks (Darwiche, 2010). The techniques of ML like linear regression, clustering, and Bayesian networks stand out within AI due to their independence from the requirement to be explicitly programmed to carry out a specific task.

Data integration and pattern recognition

AI makes it possible to combine various information such as geological, climatic, and fossil records. These combined datasets may be analyzed by AI systems to find patterns and correlations that advance our knowledge of paleo-biology, paleoecology, and evolutionary dynamics (Gao et al., 2019).

3D modeling and reconstruction

Artificial intelligence is used to recreate and visualize three-dimensional fossil fossils. The analysis of fossil scans or pictures by computer algorithms can provide precise 3D models that help in the study of fossil anatomy, biomechanics, and paleoecology. This makes it easier to create digital repositories of fossil specimens, increasing their accessibility for study and instruction (Okura, 2022).

Significance of AI in biology

The importance of AI in biology rests in its capacity to manage and interpret the enormous amounts of biological data collected from many sources, such as genomics, proteomics, medical records, and ecological observations. These big datasets are ideal for the analysis, pattern recognition, and insight extraction capabilities of AI techniques like ML, DL, and NLP. These capabilities can increase our understanding of biological processes. With the aid of AI approaches, scientists can decipher enormous volumes of biological data, identify intricate patterns and relationships, and produce insightful findings that advance our understanding of life at all scales.

AI employs a number of strategies that let robots carry out jobs that need intelligence akin to that of a human. Through the development of sophisticated tools for research and interpretation, AI has transformed the study of life and human evolution. The study of enormous volumes of genetic and environmental data is made possible by AI, revealing complex patterns and systems. Our knowledge of evolutionary processes, speciation, and the elements that influence human evolution is improved through computer modeling and simulation powered by AI. For AI to reach its full potential, biologists, computer scientists, and data analysts must work together. Scientific research is accelerated and delivers new insights into the complexity and beginnings of life and human development when AI is combined with human skills.

AI in genomics research

Omics generate data that might quickly become too large and complicated to be examined using statistical correlations or visual analysis. This has promoted the usage of AI, also known as "machine intelligence" (McCarthy et al., 2006), which is capable of managing data volumes that are insurmountable for human brains as well as extracting knowledge that is beyond our current understanding of the system being studied ML employ algorithms to automatically learn from sample data (training data) in order to produce predictions or choices. On genetic data, supervised and unsupervised learning techniques are often used. They may learn to make accurate predictions on the labels of fresh testing data using supervised learning, which supplies them with labeled training data. Contrarily, unsupervised learning employs a variety of algorithms and unlabeled data to uncover patterns in the data, such as clusters or outliers (Lin and Ngiam, 2023). Early in the 1980s, ML techniques in biology saw their initial applications (Stormo et al., 1982). ML programs have lately been used in all functional genomics-related research fields, including genomics, transcriptomics, proteomics, and metabolomics (Zampieri et al., 2019). DL approaches, a group of ML techniques, appear to be the most promising for tackling the complexity of omics data. These techniques function by conducting mathematical operations on a network of neurons, which are coupled to one another and structured in numerous layers to mimic the brain's "computational elements," to process information (Caudai et al., 2021).

The last several decades have seen a significant increase in the application of ML in many 'omics disciplines (genomics, transcriptomics, proteomics), particularly those that produce enormous quantities of data and/or have complicated systems regulated by the synergistic interaction of various elements. Important applications include the detection of transcriptional enhancers, the detection of associations between phenotypes and genotypes, identifying DNA regulatory areas in advance, the discovery of cell morphology and 3-D arrangement of cell organelles, the classification involving

DNA methylation and histone modifications, the diagnosis of cancer, and the analysis of evolutionary mechanisms (Alrefaei et al., 2022; Miotto, 2016).

The initial attempt to use a supervised training method in "omics" sciences dates back to the 1980s. In order to extract the *Escherichia coli* translational start sites from all other sites using the Perceptron approach, (Stormo et al., 1982) used a collection of mRNA sequences consisting of 78,000 nucleotides to identify connections between phenotypes and genotypes. To anticipate the protein secondary structure, a neural network was constructed by Rost and Sander in 1993 (Rost and Sander, 1993). It was in the second decade of the 2000s, when the performance of PCs improved and consequently the cost of sequencing for genome fell and then these DL approaches start to be widely employed in functional genomics (Asgari et al., 2015).

A noteworthy deep architecture were put into use in functional genomics is a completely automated stand-alone program called DeepBind, which is used to estimate the sequence specificities of proteins that bind to DNA and RNA. Getting acquainted with regulatory sequences from massive amounts of chromatin-profiling data, and this was further aided by DeepSEA or DL-based sequence analyzer predicts chromatin effects of sequence modifications with one-nucleotide resolution (Zhou and Troyanskaya, 2015). The treating of multimillions of sequences, the generalization between data gathered through various technologies, the tolerance of missing data, and the end-to-end and fully automatic learning, without the need for physical manipulation and refinement, are just a few of the difficulties that both methods, based on deep architectures, have helped to overcome. These techniques performed better than other cutting-edge ones, which inspired many scientists to pursue related intriguing directions.

AI and the Human Genome Project

The Human Genome Project (HGP) was a significant turning point in our knowledge of human genetics. But the abundance of genetic data this effort generated brought further difficulties in data analysis and interpretation (Chew, 2000). The development of AI in recent years has helped us comprehend the human genome more thoroughly. By facilitating effective analysis and interpretation of the enormous amount of genetic data, AI plays a significant role in the HGP. In order to analyze, understand, and get useful insights from the enormous quantity of genomic data produced by the HGP, AI approaches have been crucial. AI has improved human genetics research, expedited drug development, enabled personalized therapy, and expanded the area of genomics study. Our understanding of the human genome, clinical practice, and medical research are all being advanced by the combination of AI and genomics. ML algorithms are excellent at finding patterns, finding relationships, and drawing out important insights from large, complicated information. AI algorithms have fundamentally changed how we understand the human genome and how it affects human health in the setting of genomics (Cavalli-Sforza, 2005).

The significance of AI in the HGP extends beyond scientific advancements. Genomic data may now be integrated and shared globally, thanks to AI technology, fostering research collaborations and information sharing. The ingenuity, multidisciplinary research, and discoveries sparked by this group effort have the potential to advance human health (Alrefaei et al., 2022).

AI in ecology

In order to comprehend complicated methods or to make predictions in a world that is slowly evolving, ecologists frequently need robust and precise predictive models. DL techniques are no different from other ML techniques in that sense. Recent research shows that DNNs can effectively forecast the distribution of species based on their ecological interactions with other species. If there is adequate information, these techniques may also be used to explore ecological interactions (Christin et al., 2019).

Although deep networks have not yet been used in this manner, they offer the potential to mimic how environmental factors affect biological creatures. According to studies in the medical profession, environmental contaminants can predict gastrointestinal morbidity in people (Song et al., 2017). This technique might be readily applied to wild animals. (Jeong et al., 2001) stated that recurrent networks successfully help in predicting abundance and community dynamics for both phytoplankton and benthic communities dependent on environmental conditions. Overall, DL might join the toolkit for ecological niche models because of its ability for forecasting species distribution from environmental factors.

Ecologists have encountered considerable difficulties in observing and understanding ecosystems and their changes for management and conservation reasons since human activity has an influence on all ecosystems (Ellis, 2015). Here, we make the case that DL technologies are the best methodologies for achieving these goals. (Villon et al., 2018) exemplified that it is possible to assess the biodiversity of a particular location by identifying the species collected during automated recordings. (Norouzzadeh et al., 2018) stated that the timing of a species' presence in a specific location may be estimated by using time labels created for species' life cycles. The functionality and the stability of ecosystems may be evaluated by including all of these species' interactions and data in food web models, as well as focusing on indicator species like bats, which are especially sensitive to climate, particularly habitat changes (Mac Aodha et al., 2018). Additionally, decision-makers can evaluate the significance of ecosystem services to assist them in making policies or decisions regarding management (Lee et al., 2019).

Landscape analysis for extensive surveillance is another task for which DL is ideal. CNNs have been trained to monitor coral reefs and can determine the ratio for significant benthic substrates from high-resolution pictures (Beijbom, 2015). Using convolutional networks and aerial photos, it is possible to identify events that alter the environment, such as cotton blooms (Xu et al., 2018). Additionally, the above-ground carbon density was measured in order to identify regions of high conservation importance in Borneo's forests using a combination of satellite imagery, data from LIDAR, and multilayered neural networks (Asner et al., 2018).

Deep learning offers a varied and broad range of possible applications for tracking the effects of human actions, in addition to mapping species and regions highly significant for ecosystems and their conservation. DNNs have used data obtained by monitoring commercial fishing vessels to map the footprint of fisheries (Kroodsma et al., 2018). Additionally, it has been recommended that algorithms involving DL be used for tracking such activities on social media in order to spontaneously recognize images of illicit wildlife items to decrease illegal trafficking. (Di Minin et al., 2018; Hart et al., 2018) stated that social media mining has been shown effective for ecological research, such as phenological studies and the utilization of DL for data mining might be readily expanded to other fields.

To take it a step further, DL has already been conceptualized as the foundation of a completely automated system for managing ecosystems that makes use of automated sensors, drones, and robotics. This plural system would enable ongoing ecosystem management without a lot of manual involvement (Cantrell et al., 2017).

AI in evolutionary biology

How and where did life first appear on Earth has long been a topic of intense interest, and this trend has only been stronger in recent years. Origins of life research is a fascinating and extremely interdisciplinary area of study that includes contributions from a variety of academic disciplines, including geology, physics, biology, arithmetic, chemistry, and computer science. Thus, computational modeling and simulation are powerful tools used in evolutionary biology to study the evolution of organisms and their interactions with the environment. These tools allow researchers to simulate the evolution of populations over time and to test hypotheses about how different factors affect the evolution of traits. Numerical approaches or simulations are frequently used in theoretical models of evolutionary games in limited populations. This is true even in cases when analytical results are present since they are frequently imprecise or limited. Therefore, in the field of evolutionary biology, simulations and numerical approximations are widespread. Unsupervised and unsupervised techniques can significantly reduce (but not completely eliminate) the amount of subjectivity in discrete fossiliferous level identification, and AIAs have been demonstrated to be a highly effective and objective method of detecting spatial patterns in paleontological sites (Martín-Perea et al., 2020).

Evolutionary computation (EC) has demonstrated potential in leveraging the concepts of biological evolution for more than 50 years. With the help of modern evolutionary theory and a million-fold increase in computing capacity, it is now possible to accurately model evolutionary processes at unprecedented scales. Neutrality and genotype-to-phenotype mappings are examples of areas where using current biological knowledge might result in significant advancements. Major transitions, however, might not appear in EC until they are well understood in biology and computational methods are modified as a result. With such advancements, EC may take the lead in fostering machine innovation and supporting paradigm shifts in approaches. With such advancement, EC can play a key role in machine creativity, promote technical discoveries, and give insight into evolutionary theory (Miikkulainen and Forrest, 2021).

Conclusion

In conclusion, AI has a significant impact on our knowledge of life and evolution. AI enables researchers and scientists to examine intricate biological systems, sift through massive volumes of data, and derive insightful patterns that further our knowledge of the complexities of life. By using AI methods like ML, DL, and NLP, we may forecast evolutionary patterns, speed up scientific research, and find undiscovered connections in genetic, ecological, and medical data. AI improves human capacities, changes industries, and promotes developments in biotechnology, healthcare, and other areas. As we embrace the promise of AI, it is essential to address ethical issues and guarantee that ethical AI practices are followed that support justice and human values. Incorporating AI into the study of life and evolution opens new vistas of knowledge, empowering us to tackle critical issues, save biodiversity, enhance human health, and negotiate the intricate dynamics of our always-changing planet. We set out on a voyage of exploration and discovery with AI by our side that advances our knowledge of life and evolution to new heights.

References

Alrefaei, A.F., Hawsawi, Y.M., Almaleki, D., Alafif, T., Alzahrani, F.A., Bakhrebah, M.A., 2022. Genetic data sharing and artificial intelligence in the era of personalized medicine based on a cross-sectional analysis of the Saudi human genome program. Scientific Reports 12 (1). https://doi.org/10.1038/s41598-022-05296-7.

Asgari, E., Mofrad, M.R.K., Kobeissy, F.H., 2015. Continuous distributed representation of biological sequences for deep proteomics and genomics. PLoS One 10 (11). https://doi.org/10.1371/journal.pone.0141287.

Asner, G.P., Brodrick, P.G., Philipson, C., Vaughn, N.R., Martin, R.E., Knapp, D.E., Heckler, J., Evans, L.J., Jucker, T., Goossens, B., Stark, D.J., Reynolds, G., Ong, R., Renneboog, N., Kugan, F., Coomes, D.A., 2018. Mapped aboveground carbon stocks to advance forest conservation and recovery in Malaysian Borneo. Biological Conservation 217, 289–310. https://doi.org/10.1016/j.biocon.2017.10.020.

Beijbom, O., 2015. Quantification in The-Wild: Data-Sets and Baselines.

Cantrell, B., Martin, L.J., Ellis, E.C., 2017. Designing autonomy: opportunities for new wildness in the anthropocene. Trends in Ecology & Evolution 32 (3), 156–166. https://doi.org/10.1016/j.tree.2016.12.004.

Caudai, C., Galizia, A., Geraci, F., Le Pera, L., Morea, V., Salerno, E., Via, A., Colombo, T., 2021. AI applications in functional genomics. Computational and Structural Biotechnology Journal 19, 5762–5790. https://doi.org/10.1016/j.csbj.2021.10.009.

Cavalli-Sforza, L.L., 2005. The human genome diversity project: past, present and future. Nature Reviews Genetics 6 (4), 333–340. https://doi.org/10.1038/nrg1596.

Chew, M., 2000. Cracking the code: how will the Human Genome Project affect life as we know it? Medical Journal of Australia 173 (11–12), 590. https://doi.org/10.5694/j.1326-5377.2000.tb139351.x.

Chowdhary, K.R., 2020. Natural Language Processing. Springer Science and Business Media LLC, ISBN 978-81-322-3970-3, pp. 603–649. https://doi.org/10.1007/978-81-322-3972-7_19.

Christin, S., Hervet, É., Lecomte, N., Ye, H., 2019. Applications for deep learning in ecology. Methods in Ecology and Evolution 10 (10), 1632–1644. https://doi.org/10.1111/2041-210x.13256.

Darwiche, A., 2010. Bayesian networks. Communications of the ACM 53 (12), 80–90. https://doi.org/10.1145/1859204.1859227.

Di Minin, E., Fink, C., Tenkanen, H., Hiippala, T., 2018. Machine learning for tracking illegal wildlife trade on social media. Nature Ecology and Evolution 2 (3), 406–407. https://doi.org/10.1038/s41559-018-0466-x.

Ellis, E.C., 2015. Ecology in an anthropogenic biosphere. Ecological Monographs 85 (3), 287–331. https://doi.org/10.1890/14-2274.1.

Gao, X., Shen, S., Hu, Z., Wang, Z., 2019. Ground and aerial meta-data integration for localization and reconstruction: a review. Pattern Recognition Letters 127, 202–214. https://doi.org/10.1016/j.patrec.2018.07.036.

Hart, A.G., Carpenter, W.S., Hlustik-Smith, E., Reed, M., Goodenough, A.E., 2018. Testing the potential of Twitter mining methods for data acquisition: evaluating novel opportunities for ecological research in multiple taxa. Methods in Ecology and Evolution 9 (11), 2194–2205. https://doi.org/10.1111/2041-210X.13063.

Jeong, K.-S., Joo, G.-J., Kim, H.-W., Ha, K., Recknagel, F., 2001. Prediction and elucidation of phytoplankton dynamics in the Nakdong River (Korea) by means of a recurrent artificial neural network. Ecological Modelling 146 (1–3), 115–129. https://doi.org/10.1016/s0304-3800(01)00300-3.

Jordan, M.I., Mitchell, T.M., 2015. Machine learning: trends, perspectives, and prospects. Science 349 (6245), 255–260. https://doi.org/10.1126/science.aaa8415.

Kaplan, A., Haenlein, M., 2019. Siri, Siri, in my hand: who's the fairest in the land? On the interpretations, illustrations, and implications of artificial intelligence. Business Horizons 62 (1), 15–25. https://doi.org/10.1016/j.bushor.2018.08.004.

Kodagali, J., Balaji, S., 2012. Computer vision and image analysis based techniques for automatic characterization of fruits A review. International Journal of Computer Applications 50 (6), 6–12. https://doi.org/10.5120/7773-0856.

Kramer, O., 2017. Genetic Algorithms, vol. 679. Springer Nature, pp. 11–19. https://doi.org/10.1007/978-3-319-52156-5_2.

Kroodsma, D.A., Mayorga, J., Hochberg, T., Miller, N.A., Boerder, K., Ferretti, F., Wilson, A., Bergman, B., White, T.D., Block, B.A., Woods, P., Sullivan, B., Costello, C., Worm, B., 2018. Tracking the global footprint of fisheries. Science 359 (6378), 904–908. https://doi.org/10.1126/science.aao5646.

Lecun, Y., Bengio, Y., Hinton, G., 2015. Deep learning. Nature 521 (7553), 436–444. https://doi.org/10.1038/nature14539.

Lee, H., Seo, B., Koellner, T., Lautenbach, S., 2019. Mapping cultural ecosystem services 2.0 – potential and shortcomings from unlabeled crowd sourced images. Ecological Indicators 96, 505–515. https://doi.org/10.1016/j.ecolind.2018.08.035.

Lin, J., Ngiam, K.Y., 2023. How data science and AI-based technologies impact genomics. Singapore Medical Journal 64 (1), 59–66. https://doi.org/10.4103/singaporemedj.SMJ-2021-438.

Mac Aodha, O., Gibb, R., Barlow, K.E., Browning, E., Firman, M., Freeman, R., Harder, B., Kinsey, L., Mead, G.R., Newson, S.E., Pandourski, I., Parsons, S., Russ, J., Szodoray-Paradi, A., Szodoray-Paradi, F., Tilova, E., Girolami, M., Brostow, G., Jones, K.E., Fenton, B., 2018. Bat detective—deep learning tools for bat acoustic signal detection. PLoS Computational Biology 14 (3). https://doi.org/10.1371/journal.pcbi.1005995.

Martín-Perea, D.M., Courtenay, L.A., Domingo, M.S., Morales, J., 2020. Application of artificially intelligent systems for the identification of discrete fossiliferous levels. PeerJ 8 (3), e8767. https://doi.org/10.7717/peerj.8767.

McCarthy, J., Minsky, M.L., Rochester, N., Shannon, C.E., 2006. A proposal for the Dartmouth summer research project on artificial intelligence. AI Magazine 27 (4), 12–14.

Miikkulainen, R., Forrest, S., 2021. A biological perspective on evolutionary computation. Nature Machine Intelligence 3 (1), 9–15. https://doi.org/10.1038/s42256-020-00278-8.

Miotto, R., 2016. Deep patient: an unsupervised representation to predict the future of patients from the electronic health records. Scientific Reports 6 (1), 1–10.

Norouzzadeh, M.S., Nguyen, A., Kosmala, M., Swanson, A., Palmer, M.S., Packer, C., Clune, J., 2018. Automatically identifying, counting, and describing wild animals in camera-trap images with deep learning. Proceedings of the National Academy of Sciences of the United States of America 115 (25), E5716–E5725. https://doi.org/10.1073/pnas.1719367115.

Okura, F., 2022. 3D modeling and reconstruction of plants and trees: a cross-cutting review across computer graphics, vision, and plant phenotyping. Breeding Science 72 (1), 31–47. https://doi.org/10.1270/jsbbs.21074.

Rost, B., Sander, C., 1993. Secondary structure prediction of all-helical proteins in two states. Protein Engineering Design and Selection 6 (8), 831–836. https://doi.org/10.1093/protein/6.8.831.

Rusk, N., 2016. Deep learning. Nature Methods 13 (1), 35. https://doi.org/10.1038/nmeth.3707.

Schank, R.C., 1990. Tell Me a Story: A New Look at Real and Artificial Memory.

Schmidhuber, J., 2015. Deep learning in neural networks: an overview. Neural Networks 61, 85–117. https://doi.org/10.1016/j.neunet.2014.09.003.

Singh, A.M., Haju, W.B., 2022. Artificial intelligence. International Journal for Research in Applied Science and Engineering Technology 10 (7), 1210–1220. https://doi.org/10.22214/ijraset.2022.44306.

Song, Q., Zheng, Y.-J., Xue, Y., Sheng, W.-G., Zhao, M.-R., 2017. An evolutionary deep neural network for predicting morbidity of gastrointestinal infections by food contamination. Neurocomputing 226, 16–22. https://doi.org/10.1016/j.neucom.2016.11.018.

Stormo, G.D., Schneider, T.D., Gold, L., Ehrenfeucht, A., 1982. Use of the 'perceptron' algorithm to distinguish translational initiation sites in *E. coli*. Nucleic Acids Research 10 (9), 2997–3011. https://doi.org/10.1093/nar/10.9.2997.

Tai, M.C.T., 2020. The impact of artificial intelligence on human society and bioethics. Tzu Chi Medical Journal 32 (4), 339–343. https://doi.org/10.4103/tcmj.tcmj_71_20.

Villon, S., Mouillot, D., Chaumont, M., Darling, E.S., Subsol, G., Claverie, T., Villéger, S., 2018. A deep learning method for accurate and fast identification of coral reef fishes in underwater images. Ecological Informatics 48, 238–244. https://doi.org/10.1016/j.ecoinf.2018.09.007.

Xu, R., Li, C., Paterson, A.H., Jiang, Y., Sun, S., Robertson, J.S., 2018. Aerial images and convolutional neural network for cotton bloom detection. Frontiers in Plant Science 8. https://doi.org/10.3389/fpls.2017.02235.

Zampieri, G., Vijayakumar, S., Yaneske, E., Angione, C., Nielsen, J., 2019. Machine and deep learning meet genome-scale metabolic modeling. PLoS Computational Biology 15 (7). https://doi.org/10.1371/journal.pcbi.1007084.

Zhou, J., Troyanskaya, O.G., 2015. Predicting effects of noncoding variants with deep learning-based sequence model. Nature Methods 12 (10), 931–934. https://doi.org/10.1038/nmeth.3547.

Decoding life: Genetics, bioinformatics, and artificial intelligence

4

Parvaze A. Sofi[1], Sajad Majeed Zargar[1], Ambreen Hamadani[2], Sadiah Shafi[1], Aaqif Zaffar[1], Ishrat Riyaz[1], Deepak Bijarniya[1] and P.V. Vara Prasad[3]

[1]SKUAST Kashmir, Srinagar, India; [2]National Institute of Technology, Srinagar, Jammu and Kashmir, India; [3]Kansas State University, Manhattan, KS, United States

Introduction

Biology, the science of life, is probably one of the most fascinating areas of scientific endeavor. In particular, the science of genetics, which deals with the heredity and variations and inheritance of traits in biological populations, driven by action and interaction of a complex network of genes is one of the most advanced areas of scientific advances resulting in path-breaking discoveries Interestingly, a significant portion of such a big world is investigated at much smaller levels including the DNA strands (genome), RNA (transcriptome), proteins (proteome), metabolites (metabolome), ions (ionome), as well as their interactions (interactome), creating massive amounts of data. In fact, research in biological sciences including genetics has become more data-intensive, with a higher throughput of studies, cases and assays, in order to derive meaningful insights from studying small components of a larger system. This creates an enabling ecosystem for a field like data science to flourish, resulting in advanced toolkits for the analyses, interpretation, and integration of a vast amount of data. This change of paradigm in genetics research is accompanied by a transformation, in which disruptive technologies have emerged to accommodate big data and artificial intelligence (AI) techniques. The science of genetics also relies on its advances in disciplines like biochemistry, molecular biology, advanced applied mathematics, and computer science. Bioinformatics has emerged as a potential scientific tool having a significant role in providing meaningful insights into various questions in genetics especially population and evolutionary genetics, especially in plants and animals having huge diversity and complexity such as humans. This is possible on account of its predictive power, especially, if tools like AI and machine learning (ML) are used for helping biologists answer certain tricky questions in evolutionary and population genetics. Similarly, AI seeks to mimic some aspects of human intelligence through technological means (Yang and Siau, 2018). AI and ML toolkits help overcome a major challenge in bioinformatics. The interpretation of complicated cases based on huge data sets in bioinformatics is challenging, mainly due to the nature of biological data which is noisy. There are also high-resolution/high-dimensionality issues including multicollinearity, sparsity, and overfitting, that further make analyses a challenging task. Although new and improved ML tools to handle big data have increased our understanding of biological phenomena, most of the models still suffer from a lack of knowledge of hidden biological factors, which results in the overfitting of models and false-positive discoveries.

Genetics: Bioinformatics and artificial intelligence interface

Genetics as a science has progressed beyond imagination, from the Mendelian ratios to more of a predictive analysis that has touched every aspect of human existence such as the origin and evolution of life (plants, animals, microbes), evolutionary trends, plant and human health, understanding the genomic basis of biological diversity and synteny relationships (structural, functional, and comparative genomics) to precise manipulation of genes (gene editing). Over the period of phenomenal growth and outstanding discoveries, genetics has been transformed by certain key technological developments such as genome sequencing, gene manipulation, and reverse genetic approaches that have created massive data and information that needed toolkits for interpretation using AI/ML algorithms, statistical techniques, and data mining to mine historical data and to forecast events.

The genetic data and the related gene expression data are the main raw material of bioinformatics. The approaches to understanding and interpreting genetic data have come a long way from Mendel's era, where these properties were studied mostly indirectly through simple genetic ratios, karyotyping and linkage analysis, etc. to the advanced genome sequencing platforms to know and interpret the arrangement and variation of A, T, C, G, and U in DNA. Genes and genetic elements are specific to the species. Conventional genetic analysis and modern sequencing technology can be suitably used in an integrated analysis using bioinformatics as well as AI/ML tools to improve the quality of life. Its application ranges from evolutionary history, healthcare and drug discovery, and population studies, as well as enormous industrial applications.

DNA sequencing has advanced from Sanger sequencing, which made use of radio or fluorescently labeled dideoxy nucleotides (ddNTPs) in a chain termination method, to Maxam–Gilbert's chemical sequencing method using radioactive P^{32} for incorporation in $5'$ phosphate moiety, to modern day faster and cheaper sequencing platforms. Currently, NGS technologies, also called massive parallel sequencing (MPS) perform sequencing. In addition, they can run multiple reactions simultaneously to gather data from them. NGS essentially follows the principle of Sanger sequencing; however, in NGS, array-based technologies are used (Ilyas, 2018).

Bioinformatics, within the realm of genetics, utilizes computer technology to gather, store, scrutinize, and distribute biological data and information. This includes DNA and amino acid sequences, as well as annotations pertaining to these sequences that encompass vital biological details. The primary objective is to enhance our comprehension of life's evolution and biological diversity. Big data handled through AI gives better interpretation and helps overcome inherent noise that falsifies predictions (Mesko, 2017). The great heterogeneity and complexity of biological processes need more precise elucidations.

Recent advances in whole genome sequencing of an individual generate roughly 100 gigabytes of raw data. The use of various complex algorithms, as well as applications such as deep learning (DL) and natural language processing (NLP), nearly doubles it. The amount of data is going to be exponential with a progressive decrease in the cost of genome sequencing ($\sim$ 200–500 US$).

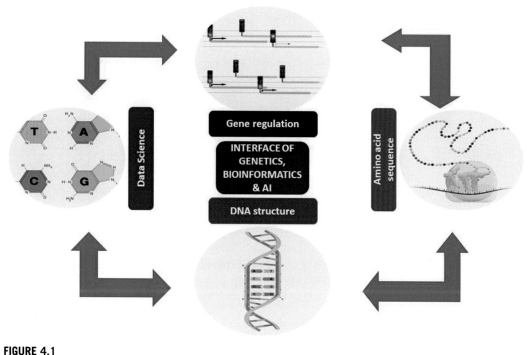

FIGURE 4.1

The interface of genetics, bioinformatics, and artificial intelligence.

In fact, many genome analysis pipelines are struggling to handle the enormous levels of raw data being generated, as such there is a need for an effective interface between genetics, genomics, proteomics, data science, and AI (Fig. 4.1).

Bioinformatics: a boon for new-age genetics research

The field of modern genetics has generated an extensive amount of biological data, particularly through multi-omics research in crops and animals. However, the value of these data is limited without appropriate analysis and interpretation tools. Bioinformatics has emerged as a highly sought-after discipline that addresses the challenges posed by the vast amount of data being generated. Bioinformatics involves the application of computational tools to analyze, capture, and interpret biological data. It is an interdisciplinary science that combines computer science, mathematics, physics, and biology (Fig. 4.2). It is important to note that bioinformatics and computational biology, although related, have distinct focuses. Bioinformatics aims to solve biological issues by analyzing and interpreting biological data using codes, algorithms, and models, while computational biology focuses on finding solutions to issues that arise from bioinformatics studies. In terms of applications, bioinformatics plays a crucial role in fields such as molecular medicine, personalized medicine, microbial genome applications, preventive medicine, drug development, and climate change studies. On the other hand, computational biology finds major applications in areas such as stochastic modeling. The potential applications of bioinformatics in biological research are vast and varied.

FIGURE 4.2

Bioinformatics as an interface of physics, mathematics, computer, and biology.

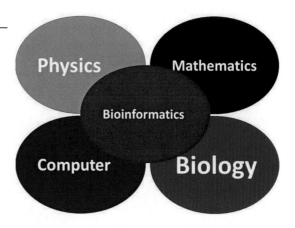

Genome analysis

The genome analysis is the process analysing all the genes of an individual. Technologies such as high-throughput sequencing (NGS) and CRISPR-Cas9 have helped us identify genome synteny and variations in coding and noncoding regions (Bayat, 2002). Genome analysis is comprised of the following aspects.

Nucleotide sequencing and analysis

The advances in molecular biology in recent times are mostly attributable to advances in NGS like second-generation platforms (Roche/454, Illumina/Solexa, AB/SOLiD, and LIFE/Ion Torrent) or third-generation platform Oxford Nanopore, Genia, NABsys, and GnuBio, that has allowed us to sequence genomes in an efficient and precise way with significant cost reduction, creating a huge amount of raw data. In order to understand the output, multiple types of bioinformatics tools and programs are used. The pipeline of computational analysis of next-generation sequencing data involves.

1. **Preprocessing**: Preprocessing of sequences using FASTQC to ensure the quality and precision of analysis.
2. **Alignment**: This refers to mapping sequence reads to a reference genome, by aligning short reads with millions of possible positions in the reference genome. This step uses complex computations due to the enormity of the short sequence, variations in base quality, and also unique versus nonunique mapping. This is a very critical step as any errors at this step will be reflected in the final analysis. Read alignments file formats are sequence alignment map (SAM) and binary alignment map (BAM). Various software used for alignment is Bowtie, Bowtie-2, SEAL, SOAP, SOAP3, BWA, mrFAST, Novoalign, SHRIMP, MAQ, Stampy, Eland, LAST Aligner, SARUMAN, HISAT-2, TOPHAT-2, and StringTie. The alignment algorithms are based on an indexing method.
3. **Variant calling:** This refers to the identification of variants in the comparison of the sample genome with the reference genome. These variants are responsible for traits and help us to identify the molecular basis of a trait. Variant calling is done through standard variant call format

(VCF) identifying sequence variations such as SNPs, indels, annotations, and large structural variants. The most possible causes of errors in variant calling are the presence of indels and errors in library preparation. The most commonly used variant calling software are SOAPsnp, VarScan/VarScan2 and ATLAS 2.

4. **Filtering and annotation:** This step involves identifying differences present between the sample genome and the reference genome to identify the variants that are involved in the trait. Some of the common tools used in filtering and annotation are sorting intolerant from tolerant (SIFT), PolyPhen/PolyPhen2, VariBench, snpEFF, SeattleSeq, ANNOVAR, variant annotation, analysis and search tool (VAAST), and VARIANT.

Gene prediction

This step helps in locating the protein-coding regions as well as the prediction of the functional site of genes. The programs used for gene prediction are categorized into four generations: (i) The first generation of gene prediction programs such as GRAIL and TestCode located the coding regions present in genomic DNA approximately, (ii) the second generation of gene prediction programs such as Xpound and SORFIND predicted potential exons by combining splice signals and coding region identification but failed to combine the exons into a complete gene, (iii) the third generation of gene prediction programs such as GeneID, GenLang, GeneParser, and FGENEH compiled the complete gene structure from exons, but their precision and efficiency were not high, and (iv) the currently used fourth-generation programs such as GENSCAN and AUGUSTUS that are highly precise and accurate and make use of multiple algorithms. There are two main methods used for gene prediction.

1. **Similarity-based searches:** This method uses a simple similarity search and finds gene sequence similarity between expressed sequence tags (ESTs), proteins, or the other genomes with input sequence. This method is based on the premise that exons are functional genome regions and are invariably more conserved than introns. Similarity-based search is accomplished either by local alignment tools viz., BLAST or global alignment software like PROCRUSTES and GeneWise. Also, a heuristic method can be used for pairwise comparison using CSTfinder software.

2. Ab initio **prediction:** In this approach, gene prediction is based on computational tools using gene structure as a template. It depends upon two types of sequence information viz., signal sensors and content sensors. The former is a short sequence motif like branch points, splice sites, start codons, and stop codons, while the latter refers to a codon usage pattern that is exclusive or unique to a species, and thus allows the discrimination between coding sequencing and surrounding noncoding sequences.

Their gene structure is modeled using various algorithm,s such as dynamic programming (DP), linear discriminant analysis (LDA), linguist method (LM), neural network (NN), and Hidden Markov model (HMM).

Genome annotation

Sequencing of the genome is followed by genome annotation in which biological information, especially the functional elements of the genome, are identified (Fig. 4.3). It is done by various specialized databases for the determination of function, such as Pfam, SMART, CDD, etc.

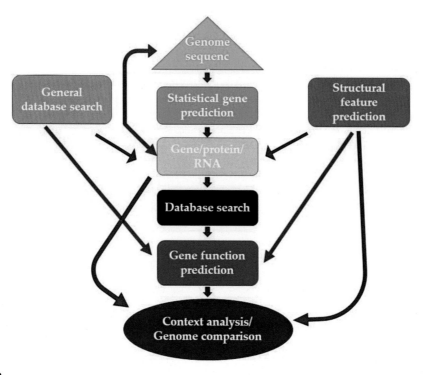

FIGURE 4.3

A generalized genome annotation pipeline.

Transcriptome analysis

Transcriptome is the expressed part of the genome, viz., complete sets of RNA transcripts that are made by the genomes under specific situations, in a specific cell. Transcriptomics refers to the analysis of transcriptomes by employing high-throughput methods. The profiling of transcriptome is imperative for elucidating gene expression, which leads to the discovery and validation of genes as well as the study of loss or gain of function mutations and their association with a particular phenotype. Transcriptomics has come a long way from microarray-based approaches to high-throughput sequencing technology, that can detect all types of transcripts such as miRNA, regulatory siRNA, and lncRNA in a given sample. RNA sequence essentially includes RNA extraction from the sample, followed by reverse transcription through RT-PCR to develop ds-cDNA, high-throughput sequencing, alignment of sequences to the reference genome, and identification of transcribed genes. It generates information in terms of quantitative expression of the transcribed genes. The transcriptome data are generated by RNA-seq and analyzed using tools such as FastQC and HTQC. Similarly, alignment is done using HISAT2, TopHat, MapSplice, STAR, and GSNAP. Similarly, the quantification of expression is done using tools such as ALEXA-seq, ERANGE, and NEUMA. In order to elucidate and quantify differential expressions, tools such as DESeq2, SAMseq, edgeR, Cuffdiff, NOIseq, and EBSeq, are employed.

Protein analysis

Bioinformatics also plays a critical role in the structural and functional analysis of proteins. This includes the analysis of protein sequence to predict protein structure and its function. Many protein databases are employed for protein sequence analysis, such as Prosite databases, which contain protein sequence patterns and profiles from diverse families. This is followed by the prediction of 3D protein structure through various complex computational techniques including CASP, that use comparative modeling or homology modeling, fold recognition, and protein–protein interactions.

Artificial intelligence in biological research

Artificial intelligence seeks to mimic some aspects of human intelligence through technological means (Yang and Siau, 2018). The purpose of this field of study and practice is to create technological systems that can solve problems and carry out jobs or duties that would typically be handled by the human intellect (Amit, 2018). This field focuses on computer science and the mathematical components of statistical modeling. The rising interest in AI in the biological sciences is a result of the technological maturity gained or the capacity to quickly and efficiently analyze massive amounts of data to uncover unexpected relationships (Khan et al., 2022).

ML has three main branches (Fig. 4.4).

AI and ML have found numerous applications in biological systems, including:

Genome sequencing: ML techniques have revolutionized DNA sequencing methods like NGS, enabling cost-effective and high-throughput sequencing of plant and animal genomes.

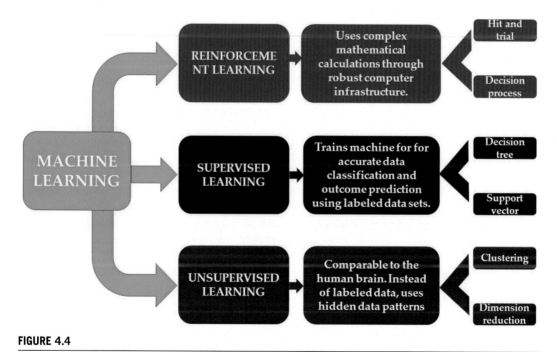

FIGURE 4.4

Different branches of machine learning with functional structure.

Gene editing: ML algorithms aid in comparing gene expression levels in different tissues, detecting mutations, and predicting DNA repair outcomes in gene editing processes such as CRISPR-Cas. AI-enabled CRISPR/Cas9 systems have facilitated gene mutation modification, molecular simulation, antigen prediction, and vaccine design.

Clinical workflow: ML has greatly impacted healthcare by facilitating access to patient data from electronic records, paper charts, and other sources. ML-based technologies like Intel's Analytics Toolkit improve data management and accelerate end-to-end data science and analytics pipelines.

Proteomics: ML helps analyze and interpret large proteomic datasets generated through techniques like mass spectrometry. Tools like Prosit enable quick and accurate protein pattern recognition.

Microarrays: AI and ML assist in studying genome organization, gene expression, and chromatin structures using microarray technology.

Systems biology: AI and ML techniques, such as probabilistic graphical modeling and genetic algorithms, are employed in systems biology to understand the relationships between variables and model genetic networks. Techniques like Markov chain optimization aid in identifying transcription binding sites.

Healthcare: AI and ML play a crucial role in improving patient care and quality of life. They utilize real-time data from multiple healthcare systems for efficient treatment. Key applications include drug discovery and design, medical imaging and diagnosis, and personalized medicine.

By leveraging AI and ML, these areas have witnessed significant advancements in biological research, diagnosis, and treatment, enabling breakthroughs and enhancing healthcare outcomes.

AI and ML in plant breeding

Applying next-gen AI to plant breeding involves efficient and intelligent mining of breeding datasets using relevant models and algorithms (Harfouche et al., 2019). The aim is to accelerate the crop development process by enabling high-resolution image identification and analysis of complex datasets (Godwin et al., 2019). Researchers are continuously innovating and improving AI techniques to enhance the accuracy and efficiency of multiomics data analysis. NNs and DL technologies, with their nonlinear hierarchical building techniques, replicate brain neurons and facilitate the classification of datasets.

Plant breeders are actively working on developing a next-gen AI system that can evaluate breeding values and provide a comprehensive analysis of complex traits under changing environmental conditions (Parmley et al., 2019). Through iterative training and enhancement, AI will improve data mining efficiency and accuracy, enabling more precise identification of underlying factors related to disease resistance and agronomic features. The extensive hybridization and rigorous selection criteria used in crop plant breeding have significantly influenced the phenotypic plasticity of economically important traits. Genotypic heterogeneity across different genotypes further limits the phenotypic flexibility resulting from interactions with the environment.

To bridge the gap between genotype and phenotype caused by changes in phenotypic plasticity, current breeding initiatives focus on improving abiotic stress tolerance in crop plants. Researchers now integrate genotypic, environmental, and observable phenotypic data to enhance the agronomic abiotic stress breeding program and identify the best genotypes with crucial agronomic traits (Niazian and Niedbała, 2020). The broader framework of the application of AI/ML in plant breeding encompasses these efforts (van Dijk et al., 2021).

Researchers have addressed the challenge of detecting small changes in plants' relationship to soil and atmosphere by developing the soil-plant-atmosphere continuum (SPAC) an AI-based physiological gravimetric system. This system enables plant scientists to easily phenotype complex features at various stages of plant growth and development, even identifying the smallest differences (Godwin et al., 2019; Harfouche et al., 2019). Ongoing and systematic monitoring of these phenotypic data, coupled with the application of next-gen AI techniques, can facilitate the identification of stress-responsive quantitative trait loci (QTLs) and QTLs associated with important agronomic traits (Niazian and Niedbała, 2020).

To expedite breeding programs, a field phenomics suite has been created, allowing for the capture of high-resolution photos that aid in distinguishing genotypes with superior performance in large populations. Ground-based tools and unmanned aerial vehicles (UAVs) equipped with high-resolution cameras and sensors collect high-throughput phenotypic data relevant to breeding operations. By analyzing the data provided by AI or specialized tools, breeders can select superior genotypes that exhibit desired agronomic and disease-resistant features. This advanced phenotypic technique also enables the discovery of novel genes and QTLs through molecular-level analysis of plants.

The use of field phenomics has made significant advancements in studying stress-responsive characteristics, as evidenced by recent applications in glycine max. However, linking AI-generated phenomics data to the genotype to identify genotypes with higher genetic gain remains a challenge. Additionally, understanding the relationship between complex traits and environmental factors is crucial for overcoming this challenge. To bridge the gap between phenotype and genotype and support crop improvement initiatives, future research should focus on developing next-generation AI approaches.

Using AI to study biochemical phenotype

Technological advancements have significantly enhanced the precision and sophistication of recording genotypic and phenotypic variations in plants, enabling the extraction of crucial information from large datasets (van Dijk et al., 2021). In response to abiotic stress conditions, which induce real-time changes at the molecular level, researchers are on the brink of utilizing AI to analyze complex biochemical pathway datasets, aiding in the understanding of these changes. Distinct changes at the genomics (gene expression), proteomics (protein distribution), metabolomics (metabolite expression), and epigenomics (DNA/histone modification) levels govern various biochemical and metabolic alterations. With the advancement of technology, sophisticated tools and technologies have been developed, facilitating the quantification of important metabolic features at the OMICS level (Parmley et al., 2019).

Analyzing the data generated by instruments such as Next Generation Sequencing (NGS), Chromatin Immunoprecipitation (ChIP), and Matrix-Assisted Laser Desorption and Ionization-Time of Flight-Mass Spectrometry (MALDI-TOF-MS) would be time-consuming and laborious. To overcome the challenge of accurately interpreting complex biochemical traits from these large and intricate datasets, researchers have started employing AI (van Dijk et al., 2021). Numerous studies have demonstrated the potential of AI in analyzing biochemical data and enhancing our understanding of plant stress biology. For example, AI has aided in the identification of genomic crossovers between maternal and parental maize plants, as well as the identification of genomic regions with high mutation rates (Demirci et al., 2021).

Similarly, AI has been employed to classify and define genomic regions by examining the DNA methylation patterns of stressed maize plants and identifying functional genes and pseudogenes (Sartor et al., 2019). Researchers have also utilized AI to identify gene promoters and cis-regulatory elements in maize and Arabidopsis plants by analyzing the expression patterns of essential genes, as demonstrated by Uygun et al., in 2019. Several studies have confirmed the value of AI in elucidating plant metabolic regulatory networks by analyzing tissue-specific changes in biosynthetic genes, such as nitrogen use efficiency, starch biosynthesis, and other secondary metabolites in Arabidopsis and rice.

Furthermore, AI has been utilized to enhance biofuel production by increasing biomass output through the use of plant species and algal bloom, underscoring its crucial role in regulating bioenergy production. The versatile application of AI in studying single-cell RNA sequencing, DNA methylation, and post-translational modifications has been extensively documented. These studies provide actionable insights into specific genomic regions or candidate genes that govern secondary metabolite production under stress conditions (Varala et al., 2018). Additionally, scientists are experimenting with AI to predict complex genetic features, including photosynthesis, hormonal changes, and yield.

In breeding programs, AI can facilitate the identification of QTLs by enabling marker-assisted selection (MAS) and genome-wide association studies (GWAS), particularly in complex genomic regions associated with specific traits (Gupta et al., 2010). AI in breeding also has the potential to provide a comprehensive understanding of the genetic architecture by revealing the location of critical genes that influence economically important traits. Integrating genomic, transcriptomic, proteomic, and metabolomic data with AI can further be employed to investigate the macroscopic biochemical factors that regulate plant growth and development in response to environmental cues.

How does AI aid crop improvement efforts by changing the breeding paradigm?

The entire value chain of plant breeding is seeing new opportunities because of technology. One of the technologies transforming agriculture for the better is AI. Computers may learn from data with the aid of a wide range of tools known as AI and replicating human intelligence is one of AI's primary goals in daily life (Vinuesa et al., 2020).

As a result, we may use more intricate and comprehensive data sets to make informed decisions. A kind of AI called ML seeks out patterns in data. With exposure to additional incoming data, ML algorithms' performance can be enhanced over time. DL, a more specialized branch of ML, has demonstrated efficiency comparable to or superior to that of humans in tasks like image recognition. Based on extensive environmental data, consider making better choices about the plant kinds to use, the detection and forecasting of insect outbreaks, and the selection and timing of pest management agents. Plant breeder's decision-making can safeguard production and biodiversity simultaneously by incorporating more data. Crop breeding's main objective is to enhance traits with commercial value (Breseghello and Coelho, 2013). How many AI aid crop development efforts by changing the breeding paradigm? Breeding is a long-term investment. Therefore, breeders and geneticists must quickly determine which breeding materials are the greatest parents to achieve a specific breeding aim to meet the expectations of the market promptly. Then, they will cross them to produce offspring that will

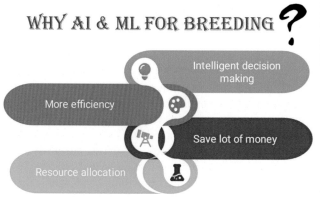

WHY AI & ML FOR BREEDING ?

Intelligent decision making

More efficiency

Save lot of money

Resource allocation

FIGURE 4.5

The reason behind the integration of artificial intelligence (AI) and machine learning (ML) in plant breeding.

perform better than the parents. Reduce labor and time costs associated with assessing breeding materials and increase the scale of breeding experiments by using AI to extract data from photos of field-grown plants. Utilize AI to assist breeders in managing more effective and focused breeding programs that can concurrently improve crops in numerous ways while keeping the flexibility to address future problems (Fig. 4.5).

Machine learning for biochemical phenotypes

The term "-omics" refers to molecular data sources at the biochemical level, including genomics (sequencing genomic DNA), epigenomics (genome alterations such as methylation affecting genomic activity), protein binding to DNA (using techniques like Y1H, ChIP-seq, or DAP-seq), transcriptomics (identifying and analyzing transcript expression levels), proteomics (identifying and analyzing protein expression levels and modifications), and metabolomics (measuring levels of small molecules). These data sources, combined with different microscopy techniques, allow the exploration of biomolecule quantity and location at the cellular and subcellular levels. Recent technological advancements, mainly based on high-throughput DNA sequencing, have significantly increased the scale at which molecules and interactions can be quantified. However, the large datasets, lack of mechanistic understanding, and complexity of the data require the use of ML to assess and interpret the results (Mahood et al., 2020).

This review focuses on recent advancements where ML is used to investigate the influence of molecular factors on plant phenotypes. ML is increasingly used to analyze biochemical data and advance our understanding of plant biology. One application of ML is differentiating between various genomic regions. For example, ML has successfully divided DNA sequence regions in maize into active genes and inactive pseudogenes based on characteristics like DNA methylation. ML has also been used to predict the locations where genome crossovers occur, where genetic material from paternal and maternal genomes is exchanged. While ML in population genetics is primarily applied to human data, it can identify areas in the plant genome where natural selection has almost completely fixed specific mutations. These examples demonstrate how ML complements comparative genomics by focusing on uncovering predicted patterns and exploring genome function.

ML algorithms are created through supervised and unsupervised learning. Supervised learning predicts an output (either a discrete label or a numerical value) for a given object using input features. The prediction model's parameters are refined by adjusting them to perform well on labeled training data, allowing the model to predict new, unseen test data. Overfitting, where a model performs well on training data but struggles to generalize to new data, should be avoided. Unsupervised learning, on the other hand, looks for patterns in unlabeled data, making it more challenging to evaluate model performance. Unsupervised algorithms can enhance our understanding of large datasets by identifying relevant patterns that can aid in more effective supervised learning. There are various supervised and unsupervised ML algorithms, each with its advantages and disadvantages. DL, using neural networks composed of interconnected "neurons," has achieved remarkable success in various domains. Convolutional neural networks (CNNs) excel at analyzing images by considering spatial relationships, while recurrent neural networks are valuable for sequential data like text or time-dependent signals. Other effective algorithms include ensemble methods like random forests or boosting, support vector machines (SVMs), and DL. Data preprocessing depends on the data type and chosen algorithm, and careful selection of measurements or calculations is crucial to best represent the objects. The availability of well-measured and accurately labeled training data is a significant determinant of ML application success. Model selection is also important, as simpler models are less prone to overfitting but may struggle to capture complex relationships. It is advisable to start with a simple model and increase complexity if necessary based on the available data.

Machine learning for genomic prediction

In plant science and breeding, explaining complex features such as yield based on genetic, phenotypic, and environmental data is a crucial objective. ML and other methods are employed to achieve this goal. These methods aim to:

Identify QTLs or genetic regions associated with specific traits. This is done through QTL mapping or genome-wide association mapping for experimental populations and diversity panels, respectively.

Evaluate the genetic architecture by determining the contributions of each locus and the percentage of trait variance explained by the sum of all loci. Genetic correlations between traits provide insights into the degree of overlap in genetic signals.

Predict trait values for new genotypes when only marker data are available. Marker-assisted selection, genomic selection, and marker-assisted breeding rely on such predictions to acquire advantageous alleles at specific loci, necessitating QTL mapping.

Understanding the genetic architecture and the genes underlying QTLs is of particular interest to biologists. Objectives one to three mentioned above involve variable selection, estimation, and prediction from a data-analytic perspective. ML methods, such as SVMs, gradient tree boosting, and random forests, have been widely used for prediction. Recent advancements have also explored the combination of genetic programming (GP) with DL to develop various network topologies.

While some researchers have investigated ML approaches for QTL mapping (primarily for prescreening purposes), their application is somewhat limited since practitioners often require P-values or other forms of statistical assurance. In this context, GP is particularly relevant. Two notable recent advancements in genotype-by-environment (G3E) interactions in plants are:

Prediction for unseen environments: GP can be extrapolated to target environments of interest by describing different training contexts.

Use of secondary traits or phenotypes: Incorporating simple-to-measure phenotypes, which may not be directly relevant but are indicative of the target feature (e.g., yield or stress tolerance), can improve prediction accuracy. These approaches, as described by Montesinos et al., enhance the understanding and application of GP in addressing G3E interactions.

Potential applications of AI and ML in classical and modern plant breeding

Even though researchers still frequently utilize more traditional statistical genetics techniques, in recent years, there has been a paradigm shift toward the more widespread use of ML algorithms in biological studies for both prediction and discovery. The use of ML techniques, particularly DL strategies, has grown in importance with the availability of more and more diverse types of omics large data (Gakhar, 2021). The application of ML in several plant breeding domains is depicted in Fig. 4.5.

Assessments of biotic and abiotic stress

One of the crucial factors affecting plant fitness and food production is plant stress. Plant stress can also have positive effects due to its hormetic behavior, where low doses of stressors stimulate the production of specialized metabolites and enhance stress tolerance in crops. Managing hormesis, which involves applying controlled doses of stressors to crops, holds promise for increasing crop productivity and quality. However, understanding and controlling hormesis are challenging due to the complex physiological responses of plants to stress.

Recent technological advancements have enabled plant stress research to overcome these challenges by generating large datasets that capture various aspects of the plant defense response. AI methods, particularly ML and DL, have become indispensable for effectively analyzing and interpreting data related to plant stress (Rico-Chávez et al., 2022). These AI technologies are valuable for modeling plant distribution, species identification, disease and stress detection, assessment of nutritional deficiencies, and optimizing the use of agrochemicals in precision agriculture.

The use of ML techniques in plant stress research has significantly advanced our understanding of complex biological processes. ML can predict the outcomes of various biological processes, such as gene function, gene networks, protein interactions, and optimal growth conditions. To categorize the applications of plant stress research, the four ICQP paradigm categories are employed, encompassing image-based phenotyping and remote sensing, genomics and transcriptomics, metabolomics and proteomics, and crop modeling and decision support systems.

These advancements in AI technologies, combined with large-scale data analysis, have provided valuable insights into plant responses to stress. They have the potential to improve crop productivity and enhance the quality of agricultural production Fig. 4.6.

Artificial intelligence in crop genomics

Long before the principles of genetics were understood, the art of plant breeding was being practiced. Breeding has developed into a science-based technology that has significantly improved crop plants since the introduction of genetics principles at the turn of the previous century. Grain yields of various cereal crops were significantly boosted, largely due to the effective use of hybrid vigor. When combined

Machine learning Applications

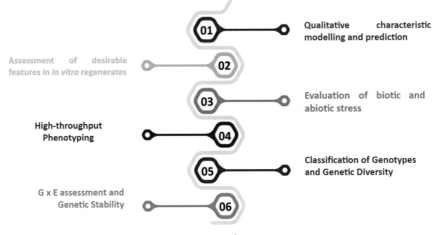

01	Qualitative characteristic modelling and prediction
Assessment of desirable features in *in vitro* regenerates	02
03	Evaluation of biotic and abiotic stress
High-throughput Phenotyping	04
05	Classification of Genotypes and Genetic Diversity
G x E assessment and Genetic Stability	06

www.presentationgo.com

FIGURE 4.6

Potential applicability of machine learning techniques in conventional and modern plant breeding.

with the right cytogenetic alterations, intervarietal and interspecific hybridizations have been successful in transferring disease and insect pest resistance genes from suitable foreign donors into crop cultivars. The ability to create new features in plants that are exceedingly challenging to introduce through traditional breeding has further expedited plant progress (Jauhar, 2006). As of 2021, 20 years have passed since the landmark completion of the draft human genome sequence. Estimates predict that genomics research will generate between 2 and 40 exabytes of data within the next decade. Crop types that provide the desired features specifically for the areas where they are grown are the first step in developing high-quality and sustainable crops for the future. Using data-driven, molecular breeding techniques, the plant breeding cluster will concentrate on creating AItools to assist plant breeders in creating the next generation of high-yielding, high-quality, nutrient-rich, and flavorful varieties of different crops.

The field of genomics, which includes the sequencing, assembly, and functional study of a plant's genome, has advanced more quickly than any other omics in plant research (Mir et al., 2019). The sequences of more than 100 plant genomes have just recently been revealed, and subsequent genomics' technical advancements have deepened our understanding of plant biology, resulting in significant agricultural advancements (Purugganan and Jackson, 2021). The reconstruction of gene regulatory networks and the crosstalk by which they interact during particular physiological processes precede the computational analysis of transcriptomic datasets (García-Gómez et al., 2020). In particular, machine learning techniques are quite accurate at predicting gene interactions (Haque et al., 2019). Biological processes other than transcription, such as epigenome dynamics, which are dependent on chromatin structural alterations, such as DNA methylation and histone modifications, are nevertheless involved in the control of genes during hormetic responses. To improve model resolution and accurately reflect the mechanisms of hormesis, it is crucial to integrate more types of datasets and to incorporate geographical and temporal information (Qian and Huang, 2020).

Prediction of functional genomic regions

A genome may now be stitched together and genic characteristics can be annotated with remarkable ease and regularity (Mahood et al., 2020). The initial stage in genome annotation is the discovery of the genomic sequences that make up functional areas. To identify various features, such as microRNAs, protein-coding genes, long noncoding RNAs, polyadenylation sites (Gao et al., 2018), DNAse I hypersensitive sites (Lyu et al., 2018), cis-regulatory elements, and chromatin states, ML approaches have been developed (Ernst and Kellis, 2017). ML may be used to identify these genic elements in a classification context, where they may be categorized as "protein-coding genes," "miRNA," or "CRE" or not. When applied in certain circumstances, ML's capacity to integrate significant amounts of diverse data may result in more accuracy than non-ML techniques. SVMs have more tuning parameters but have been demonstrated to be effective in binary classification issues and multiclass problems. Additionally, most jobs may perform better with RFs (Caruana and Niculescu-Mizil, 2006).

Application of AI in phenomics

Phenotyping is characterized as the evaluation of traits like growth, development, tolerance, resistance, architecture, physiology, ecology, and yield, as well as the fundamental quantification of individual quantitative parameters individually that serve as the foundation for complex trait assessment (Costa et al., 1933). The utilization of phenomic-level data to support the association between genetics and the variance in crop yields and plant health is of growing interest to scientists (Arvidsson et al., 2011). The last few years have seen a significant advancement in plant phenomics. In recent years, the use of AI in numerous scientific fields has expanded tremendously. Notably, the AI components of computer vision, ML, and DL have been successfully incorporated into noninvasive imaging systems. Through the use of machines and DL for reliable picture analysis, this integration is gradually increasing the efficiency of data gathering and analysis. Additionally, the development of software and technologies used in field phenotyping for data management and gathering has been aided by AI (Nabwire et al., 2021). To comprehend the fundamental mechanics of how plants interact with their environment, plant phenotyping measures phenotypic features at the cell, organ, or entire plant level (Pieruschka and Schurr, 2019). Data collection is currently made possible by high-throughput phenotyping methods in both lab and field settings. They work in the areas of data management, data analysis, and data collection. They include growing chambers, data management, and analysis software, imaging sensors, and more (White et al., 2012; Bai et al., 2016). The development of the noninvasive imaging component of phenomics has been aided by the incorporation of AI into these technologies. Data on plant images are rapidly being collected and analyzed using computer vision and ML techniques, both of which are forms of AI. For the goal of object recognition, computer vision systems analyze digital photographs of plants to find certain properties (Casanova et al., 2014).

Another significant issue with current plant breeding is the genotype-to-phenotype gap. While genomics research has access to cutting-edge methods to provide details about the genetic makeup of various plant species, processing and analyzing the generated huge data has always been a source of worry. In contrast to the usual short reads, advanced sequencing technologies allow longer readings of sequences. However, a higher percentage of sequencing errors (5%−15%) is typically existential. Artificial neural networks, for example, are important ML technologies that can be used to get around this restriction (Fig. 4.5). Using aligned long reads and a CNN model, indel variants or SNP, zygosity, and indel length are predicted. ML under supervision has been used to examine recombination rates in a target genome and accurate prediction of the presence of a candidate variant.

Application of ML in image processing

The intricate interactions between a plant's genotype and environment lead to its phenotypic dependence on accurate assessment of factors including soil composition, weather, and water availability. It is important to comprehend the connection between gene function, plant performance, and environmental response. The potential for image-based techniques to significantly expand the scope and output of plant phenotyping efforts is enormous. Innovative applications of sensing technology in agriculture provide a stronger framework for functioning as an exemplary instrument to correlate and anticipate the conditions of indoor and outdoor cultivation to the important physiological changes brought on in plants by external stimuli (Lee et al., 2019). Simple image processing algorithms might occasionally fall short when dealing with difficult nonlinear, nongeometric phenotyping jobs. The use of increasingly advanced image processing techniques is required due to the increased level of obstruction in activities including vigor assessments, disease detection, pod/fruit/leaf counts, injury ratings, age estimation, and mutant classification (Pape and Klukas, 2015). For instance, deep CNNs combine regression and visual feature extraction in ML methods. The accuracy of image-based phenotyping may be improved through DL, ML, and linked components analysis. It makes it possible to process massive amounts of data from sensors and phenotyping tools, which enhances output and analytical precision.

In many plant breeding initiatives, analyzing big datasets made up of spectral reflectance data involves intense computing and statistical analysis, which is still difficult (Lopes et al., 2012). Researchers are now focused on creating model-based breeding techniques that can increase the effectiveness of breeding operations because of ML algorithms. The multilayer perceptron (MLP), created by Pal and Mitra in 1992, is one of the most popular artificial neural networks (ANNs) that has recently been widely employed for modeling and forecasting complicated qualities, such as yield, in various breeding programs.

MLP is a powerful nonlinear computational technology widely employed for various tasks, such as complex system classification and regression. It leverages the inherent knowledge within datasets to establish connections between input and output variables. MLP consists of interconnected processing neurons that work in parallel to address specific problems. SVMs are renowned for their effectiveness in identifying nonlinear relationship patterns and behaviors. Compared to MLP, SVMs offer certain advantages due to the complexity of their networks. SVMs utilize multiple learning problem formulations, enabling them to solve quadratic optimization problems. Theoretically, SVMs should outperform MLP because they employ structural risk minimization inductive principles instead of empirical risk minimization. Random forest (RF) is another alternative to MLP and SVM. It is a data modeling method known for its computationally efficient training phase and high generalization accuracy. RF has found applications in various fields, including object identification, skin detection, plant phenomics, and genomics. ML algorithms, including MLP, SVM, and RF, are susceptible to overfitting when relying on a single prediction model and limited training data. To address this issue, ensemble approaches are employed, where multiple algorithms are combined to integrate various predictions into a final forecast.

Research challenges

Research in genetics and bioinformatics using AI presents several challenges, some of which are listed below.

- Data integration and preprocessing: Genetics and bioinformatics research involve working with diverse and large-scale datasets from various sources such as genomic sequencing, gene expression, protein–protein interactions, and clinical data. Integrating and preprocessing these heterogeneous data types to extract meaningful insights can be challenging due to differences in data formats, quality, and standards.
- Dimensionality and scalability: Genetic and biological data are high-dimensional, with thousands of features (genes, proteins, etc.) to consider. Analyzing and interpreting such high-dimensional data poses challenges in terms of computational scalability, as traditional algorithms may become inefficient or infeasible. Developing efficient algorithms and techniques to handle high-dimensional data is crucial.
- Algorithm selection and optimization: AI techniques such as ML and DL have shown promise in genetics and bioinformatics research. However, selecting the appropriate algorithms and optimizing them for specific biological problems can be challenging. Different algorithms may perform differently on different types of biological data, and finding the most suitable approach requires careful evaluation and experimentation.
- Interpretability and explainability: Genetic and bioinformatics research often deals with complex models and algorithms, such as deep neural networks, which can be difficult to interpret and explain. Understanding the underlying biological mechanisms and reasoning behind AI predictions is crucial for gaining trust and making informed decisions. Developing interpretable AI models and visualization techniques is an ongoing challenge.
- Handling noisy and incomplete data: Biological data can be noisy and incomplete due to various factors such as experimental errors, measurement limitations, and missing values. Dealing with noisy data and developing robust algorithms that can handle missing information is a challenge. Additionally, integrating multiple data sources while accounting for data quality and reliability adds complexity to the analysis.
- Biological context and domain knowledge: Genetic and bioinformatics research requires domain expertise and knowledge of biological processes. Integrating AI techniques with domain-specific knowledge and incorporating biological context into the analysis is crucial for accurate interpretation and meaningful discoveries. Bridging the gap between AI researchers and biologists to collaborate effectively is a challenge.
- Privacy and security: Genetic and health data are sensitive and require strict privacy and security measures. AI methods need to address privacy concerns and ensure data protection while extracting insights. Developing privacy-preserving AI techniques that can work effectively on encrypted or anonymized data is an ongoing challenge.
- Reproducibility and benchmarking: Reproducibility is a key aspect of scientific research. However, in AI-driven genetics and bioinformatics, reproducing and benchmarking results can be challenging due to the complexity of algorithms, data availability, and differences in experimental setups. Developing standardized benchmarks, open datasets, and reproducible workflows is important for advancing research in this field.

- Ethical considerations: Genetics and bioinformatics research using AI raise ethical concerns related to privacy, consent, and potential misuse of genetic information. Ensuring ethical guidelines and regulations are in place to protect individuals' rights and prevent misuse of data is a challenge that needs to be addressed.
- Addressing these challenges requires interdisciplinary collaboration between AI researchers, biologists, and clinicians. It also involves developing novel algorithms, tools, and frameworks specifically tailored to the unique characteristics of genetic and biological data, while considering the ethical implications and ensuring responsible use of AI in this field.

Conclusion

In conclusion, the intersection of genetics, bioinformatics, and AI holds immense potential for advancing our understanding of complex biological systems and revolutionizing healthcare. While this field presents numerous challenges, including data integration, algorithm selection, interpretability, and ethical considerations, researchers are actively addressing these hurdles through interdisciplinary collaborations and innovative solutions. By leveraging AI techniques, scientists can analyze and interpret vast amounts of genetic and biological data, uncovering crucial insights into disease mechanisms, drug discovery, and personalized medicine. As the field progresses, the integration of AI into genetics and bioinformatics research will continue to reshape our understanding of life sciences, leading to transformative breakthroughs and improving human health on a global scale.

References

Amit, K., 2018. Artificial Intelligence and Soft Computing: Behavioral and Cognitive Modeling of the Human Brain. CRC Press, 2018.

Arvidsson, S., Pérez-Rodríguez, P., Mueller-Roeber, B., 2011. A growth phenotyping pipeline for Arabidopsis thaliana integrating image analysis and rosette area modeling for robust quantification of genotype effects. New Phytologist 191 (3), 895−907. https://doi.org/10.1111/j.1469-8137.2011.03756.x.

Bai, G., Ge, Y., Hussain, W., Baenziger, P.S., Graef, G., 2016. A multi-sensor system for high throughput field phenotyping in soybean and wheat breeding. Computers and Electronics in Agriculture 128, 181−192. https://doi.org/10.1016/j.compag.2016.08.021.

Bayat, A., 2002. Science, medicine, and the future. BMJ British Medical Journal 324 (7344).

Breseghello, F., Coelho, A.S.G., 2013. Traditional and modern plant breeding methods with examples in rice (Oryza sativa L.). Journal of Agricultural and Food Chemistry 61 (35), 8277−8286. https://doi.org/10.1021/jf305531j.

Caruana, R., Niculescu-Mizil, A., 2006. An empirical comparison of supervised learning algorithms. In: ICML 2006 − Proceedings of the 23rd International Conference on Machine Learning, pp. 161−168.

Casanova, J.J., O'Shaughnessy, S.A., Evett, S.R., Rush, C.M., 2014. Development of a wireless computer vision instrument to detect biotic stress in wheat. Sensors 14 (9), 17753−17769. https://doi.org/10.3390/s140917753.

Costa, C., Schurr, U., Loreto, F., Menesatti, P., Carpentier, S., 1933. Plant phenotyping research trends, a science mapping approach. Frontiers in Plant Science 9.

Demirci, M., Gozde, H., Taplamacioglu, M.C., 2021. Comparative dissolved gas analysis with machine learning and traditional methods. In: HORA 2021 − 3rd International Congress on Human-Computer Interaction, Optimization and Robotic Applications, Proceedings. Institute of Electrical and Electronics Engineers Inc., Turkey https://doi.org/10.1109/HORA52670.2021.9461371.

Ernst, J., Kellis, M., 2017. Chromatin-state discovery and genome annotation with ChromHMM. Nature Protocols 12 (12), 2478−2492. https://doi.org/10.1038/nprot.2017.124.

Gakhar, A., 2021. Machine Learning Assisted Plant Breeding−A Key Step to Overcome Agricultural Production Challenges, vol 15.

Gao, X., Zhang, J., Wei, Z., Hakonarson, H., 2018. DeepPolyA: a convolutional neural network approach for polyadenylation site prediction. IEEE Access 6, 24340−24349. https://doi.org/10.1109/ACCESS.2018.2825996.

García-Gómez, M.L., Castillo-Jiménez, A., Martínez-García, J.C., Álvarez-Buylla, E.R., 2020. Multi-level gene regulatory network models to understand complex mechanisms underlying plant development. Current Opinion in Plant Biology 57, 171−179. https://doi.org/10.1016/j.pbi.2020.09.004.

Godwin, I.D., Rutkoski, J., Varshney, R.K., Hickey, L.T., 2019. Technological perspectives for plant breeding. Theoretical and Applied Genetics 132 (3), 555−557. https://doi.org/10.1007/s00122-019-03321-4.

Gupta, P.K., Kumar, J., Mir, R.R., Kumar, A., 2010. Marker-assisted selection as a component of conventional plant breeding. Plant Breeding Reviews 33.

Haque, S., Ahmad, J.S., Clark, N.M., Williams, C.M., Sozzani, R., 2019. Computational prediction of gene regulatory networks in plant growth and development. Current Opinion in Plant Biology 47, 96−105. https://doi.org/10.1016/j.pbi.2018.10.005.

Harfouche, A.L., Jacobson, D.A., Kainer, D., Romero, J.C., Harfouche, A.H., Scarascia Mugnozza, G., Moshelion, M., Tuskan, G.A., Keurentjes, J.J.B., Altman, A., 2019. Accelerating climate resilient plant breeding by applying next-generation artificial intelligence. Trends in Biotechnology 37 (11), 1217−1235. https://doi.org/10.1016/j.tibtech.2019.05.007.

Ilyas, M., 2018. Next-generation sequencing in diagnostic pathology. Pathobiology 84 (6), 292−305. https://doi.org/10.1159/000480089.

Jauhar, P.P., 2006. Modern biotechnology as an integral supplement to conventional plant breeding: the prospects and challenges. Crop Science 46 (5), 1841−1859. https://doi.org/10.2135/cropsci2005.07-0223.

Khan, M.H.U., Wang, S., Wang, J., Ahmar, S., Saeed, S., Khan, S.U., Xu, X., Chen, H., Bhat, J.A., Feng, X., 2022. Applications of artificial intelligence in climate-resilient smart-crop breeding. International Journal of Molecular Sciences 23 (19). https://doi.org/10.3390/ijms231911156.

Lee, S.S., Jeong, Y.N., Son, S.R., Lee, B.K., 2019. A self-predictable crop yield platform (SCYP) based on crop diseases using deep learning. Sustainability 11 (13). https://doi.org/10.3390/su11133637.

Lopes, M.S., Reynolds, M.P., Manes, Y., Singh, R.P., Crossa, J., Braun, H.J., 2012. Genetic yield gains and changes in associated traits of CIMMYT spring bread wheat in a "Historic" set representing 30 years of breeding. Crop Science 52 (3), 1123−1131. https://doi.org/10.2135/cropsci2011.09.0467.

Lyu, C., Wang, L., Zhang, J., 2018. Deep learning for DNase I hypersensitive sites identification. BMC Genomics 19, 155. https://doi.org/10.1186/s12864-018-5283-8.

Mahood, E.H., Kruse, L.H., Moghe, G.D., 2020. Machine learning: a powerful tool for gene function prediction in plants. Applications in Plant Sciences 8 (7). https://doi.org/10.1002/aps3.11376.

Mesko, B., 2017. The role of artificial intelligence in precision medicine. Expert Review of Precision Medicine and Drug Development 2 (5), 239−241. https://doi.org/10.1080/23808993.2017.1380516.

Mir, R.R., Reynolds, M., Pinto, F., Khan, M.A., Bhat, M.A., 2019. High-throughput phenotyping for crop improvement in the genomics era. Plant Science 282, 60−72. https://doi.org/10.1016/j.plantsci.2019.01.007.

Nabwire, S., Suh, H.K., Kim, M.S., Baek, I., Cho, B.K., 2021. Application of artificial intelligence in phenomics. Sensors 21 (13), 4363.

Niazian, M., Niedbała, G., 2020. Machine learning for plant breeding and biotechnology. Agriculture 10 (10), 436. https://doi.org/10.3390/agriculture10100436.

Pape, J.M., Klukas, C., 2015. 3-d histogram-based segmentation and leaf detection for rosette plants. Lecture Notes in Computer Science (including subseries Lecture Notes in Artificial Intelligence and Lecture Notes in Bioinformatics) 8928, 61−74. https://doi.org/10.1007/978-3-319-16220-1_5.

Parmley, K.A., Higgins, R.H., Ganapathysubramanian, B., Sarkar, S., Singh, A.K., 2019. Machine learning approach for prescriptive plant breeding. Scientific Reports 9 (1). https://doi.org/10.1038/s41598-019-53451-4.

Pieruschka, R., Schurr, U., 2019. Plant phenotyping: past, present, and future. Plant Phenomics 1–6. https://doi.org/10.1155/2019/7507131.

Purugganan, M.D., Jackson, S.A., 2021. Advancing crop genomics from lab to field. Nature Genetics 53 (5), 595–601. https://doi.org/10.1038/s41588-021-00866-3.

Qian, Y., Huang, S.S.C., 2020. Improving plant gene regulatory network inference by integrative analysis of multi-omics and high resolution data sets. Current Opinion in Systems Biology 22, 8–15. https://doi.org/10.1016/j.coisb.2020.07.010.

Rico-Chávez, A.K., Franco, J.A., Fernandez-Jaramillo, A.A., Contreras-Medina, L.M., Guevara-González, R.G., Hernandez-Escobedo, Q., 2022. Machine learning for plant stress modeling: a perspective towards hormesis management. Plants 11 (7). https://doi.org/10.3390/plants11070970.

Sartor, R.C., Noshay, J., Springer, N.M., Briggs, S.P., 2019. Identification of the expressome by machine learning on omics data. Proceedings of the National Academy of Sciences of the United States of America 116 (36), 18119–18125. https://doi.org/10.1073/pnas.1813645116.

van Dijk, A.D.J., Kootstra, G., Kruijer, W., de Ridder, D., 2021. Machine learning in plant science and plant breeding. iScience 24 (1). https://doi.org/10.1016/j.isci.2020.101890.

Varala, K., Marshall-Colón, A., Cirrone, J., Brooks, M.D., Pasquino, A.V., Léran, S., Mittal, S., Rock, T.M., Edwards, M.B., Kim, G.J., Ruffel, S., Richard McCombie, W., Shasha, D., Coruzzi, G.M., 2018. Temporal transcriptional logic of dynamic regulatory networks underlying nitrogen signaling and use in plants. Proceedings of the National Academy of Sciences of the United States of America 115 (25), 6494–6499. https://doi.org/10.1073/pnas.1721487115.

Vinuesa, R., Azizpour, H., Leite, I., Balaam, M., Dignum, V., Domisch, S., Felländer, A., Langhans, S.D., Tegmark, M., Fuso Nerini, F., 2020. The role of artificial intelligence in achieving the Sustainable Development Goals. Nature Communications 11 (1), 1. https://doi.org/10.1038/s41467-019-14108-y.

White, J.W., Andrade-Sanchez, P., Gore, M.A., Bronson, K.F., Coffelt, T.A., Conley, M.M., Feldmann, K.A., French, A.N., Heun, J.T., Hunsaker, D.J., Jenks, M.A., Kimball, B.A., Roth, R.L., Strand, R.J., Thorp, K.R., Wall, G.W., Wang, G., 2012. Field-based phenomics for plant genetics research. Field Crops Research 133, 101–112. https://doi.org/10.1016/j.fcr.2012.04.003.

Yang, Y., Siau, K.L., 2018. A Qualitative Research on Marketing and Sales in the Artificial Intelligence Age, 2018.

AI in healthcare: Pioneering innovations for a healthier tomorrow

5

Abida Yousuf[1], Burhan Khursheed[2], Rukia Rahman[3], Henna Hamadani[4] and Ambreen Hamadani[1]

[1]*National Institute of Technology, Srinagar, Jammu and Kashmir, India;* [2]*Department of Electronics and Communication, NIT, Srinagar, India;* [3]*University of Kashmir, Srinagar, Jammu and Kashmir, India;* [4]*Sher-e-Kashmir University of Agricultural Sciences and Technology of Kashmir, Srinagar, Jammu and Kashmir, India*

Introduction

One of the most intriguing and amazing inventions in the history of humans is undoubtedly artificial intelligence (AI). Since its creation, a significant amount of the universe has remained unexplored. In reality, the applications it has seen in everyday life so far are probably just the beginning. The field of AI has experienced remarkable growth in recent years, thus becoming one of the most important sectors. One of the most prominent fields that have been highly benefited by AI is the health sector. AI is praised for its ability to handle significant health issues, such as addressing the care requirements of the elderly, detecting diseases, etc. Practically speaking, AI tools may not be able to entirely substitute human doctors, but they can help them get better outcomes and more correctness in medical care. The availability of healthcare data is a key factor supporting the development of AI solutions in the medical area (Ali et al., 2023). AI and machine learning (ML) have proven effective tools in healthcare, especially for the detection and treatment of cardiovascular diseases. ML and AI algorithms can process enormous quantities of clinical and imaging data to deliver a precise and customized diagnosis, risk classification, surgery, and therapy planning (Bohr and Memarzadeh, 2020). Clinical judgment and illness diagnosis are two more crucial areas of medicine where AI is having a profound effect. In order to diagnose disease and direct clinical decisions, AI technologies can ingest, evaluate, and report massive volumes of data from many modalities. Applications of AI can handle the enormous amount of data generated in the medical field and uncover novel details that would otherwise be concealed in the sea of medical big data. These technologies are additionally capable of discovering new medications for managing healthcare services and treating patients. These innovations enable AI to undertake activities that people occasionally are unable to do with the simplicity, diligence speed, and reliability that AI may provide at a cheaper cost. The improvement in technology from healthcare automation can also get control of more difficulties when technocrats successfully constructed AI systems to complete particular duties. AI, for instance, can greatly improve patient care while also reducing healthcare costs (Pahari, 2023). AI-based medical equipment has already debuted in the field, cancer-detecting nano chips in patients is a case in point for one. There are many others in their nascency but in advent nonetheless. The scope of AI has therefore positive connotations with respect to the medical and healthcare domain. Automated surgery is going far and wide by the day, and it is expected that same shall encompass other sub or allied spheres in health/medical system. Let us hope that the world of AI takes life a notch up in terms of progress and harmony and not the other way around. Over the following 10 years, there will likely be a major increase in the application of AI in healthcare. Grand View Research predicts that the market for AI in healthcare will be worth $208.2 billion in 2030, which is a significant increase from the $15.4 billion market size in 2022 (Grand View Research, 2022). This chapter revolves around the basic applications of AI in healthcare.

Technological advancement

Artificial intelligence is influencing practically every sector of human endeavor. With tools like ChatGPT and AI art generators gaining widespread attention, it is already the primary force behind developing technologies like big data, automation, and nanotechnology. It will continue to be a technical innovator for many years to come. A noteworthy point here is that AI has shown a great impact on the rapid growth of AI -based tools and technologies, in the field of healthcare, which has been made possible by an ideal triangulation of higher computational speed, expanded gathering of data libraries, and a vast AI capacity pool (Erickson et al., 2017). For instance, the use of AI in radiography treatment is one of the most noticeable trends. Radiologists can use AI algorithms to help them with tasks like image comprehension, tumor diagnosis, and outcome prediction. Massive volumes of medical imaging data can now be analyzed by AI systems using ML and deep computing methodologies, which help radiologists, make more precise diagnoses and treatment choices (Hu et al., 2019). Furthermore, ML algorithms can spot characteristics in medical pictures that might be invisible to the naked eye, leading to more precise diagnoses than could be achieved using conventional techniques (Zhang et al., 2023). Thus, the technological improvements in AI will anticipate leading to a decrease in hospital stays, medical visits, and procedures. Through constant monitoring and coaching, AI -powered technology will play a significant part in assisting people in maintaining their health by ensuring earlier diagnosis, personalized therapies, and more effective subsequent examinations (Erickson et al., 2017). As technology automates processes like reading radiologic images, some research suggests that AI could result in large employment losses. However, some experts think this is unlikely to happen. In fact, AI has the potential to increase jobs (The White House, 2022).

How is AI used in healthcare?

The world as we know it has already undergone significant change as a result of AI, from the automation of systems to the enhancement of our decision-making processes. This includes the use of decision support systems for management and monitoring (Hamadani and Ganai, 2022). However, the areas of healthcare where AI is being used to diagnose, develop individualized treatment plans, and even predict patient survival rates may be the most significant and personally relevant ones. Computers and automated processes are used in AI to mimic human intelligence and carry out challenging computerized jobs. AI-enabled robots may surpass the human intellect in a variety of ways while also attempting to mimic them (Dong et al., 2020). This is especially true when it comes to swiftly sorting through massive amounts of big data in order to find patterns, anomalies, and trends. Unsurprisingly, AI offers a plethora of prospects in the field of healthcare, where it can be used to improve a number of standard medical procedures, from disease diagnosis to determining the best course of action for patients with serious illnesses like cancer (Bhinder et al., 2021; Friedman et al., 2015). AI-enhanced robotic surgical equipment can improve surgical outcomes by reducing physical fluctuations and delivering real-time information to the physician (Hussain et al., 2014). As technology continues to advance, AI has the potential to further transform healthcare delivery, enhance precision medicine, and improve patient outcomes by augmenting human expertise and providing valuable insights and support to healthcare professionals. AI has already made significant contributions to healthcare by revolutionizing various aspects of the industry and has the potential for much more in the near future. AI is a general word that refers to several different, yet connected, processes. Various typical applications of healthcare are listed below and are also briefly described in Fig. 5.1. A quick overview of how AI is helping to improve healthcare is given in Fig. 5.2.

Machine learning: MLis the process of training algorithms utilizing large data sets, including those from health records, to build models that can do things like categorize data or predictions. Deep learning uses larger datasets, longer training periods, and several layers of ML algorithms to create neural networks that can handle increasingly difficult applications and is rapidly becoming very popular. ML algorithms enable computers to analyze large datasets and identify patterns, trends, and correlations. In healthcare, ML is used for tasks such as disease diagnosis, risk prediction, treatment planning, and outcome forecasting. ML models can learn from vast amounts of patient data and provide insights to support clinical decision-making. All this is possible today because of the availability of big data, which is being generated in every sphere of life (Hamadani et al., 2022a,c).

Neural language processing: Neural language processing (NLP) is the application of ML to comprehend spoken or written human language. NLP is used in the healthcare industry to interpret written materials like reports, notes, and published research. In healthcare, NLP algorithms can extract relevant information from medical records, clinical notes, and scientific literature, helping to automate administrative tasks, extract patient data, and support clinical research. NLP also enables chatbots and

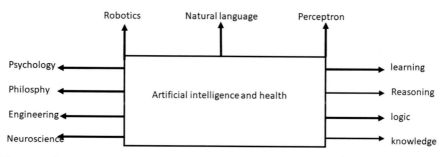

FIGURE 5.1

Artificial intelligence and health map.

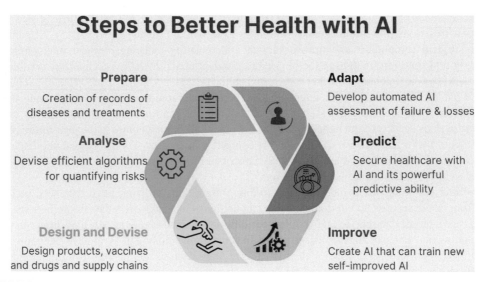

FIGURE 5.2

Steps to improve healthcare with artificial intelligence (AI).

virtual assistants to understand and respond to natural language queries from patients and healthcare professionals.

Computer vision: Computer vision involves teaching computers to understand and interpret visual information from images and videos (Hamadani et al., 2023). In healthcare, computer vision algorithms can analyze medical images, such as X-rays, computed tomography (CT) scans, and magnetic resonance imagings (MRIs), to assist in disease detection, classification, and treatment planning. It can help radiologists identify abnormalities or assist in surgery by providing real-time guidance and enhancing visualization.

Robotics and automation: Robotic process automation (RPA) is the practice of automating administrative and clinical operations using AI in computer programs. RPA is used by some healthcare organizations to enhance the patient experience and regular operations of their facilities. AI-powered robotics and automation technologies have the potential to transform various healthcare workflows. Surgical robots can assist surgeons during complex procedures, offering enhanced precision and control. Robots and automation can also be used for tasks such as medication dispensing, patient monitoring, and logistics management in hospitals, contributing to improved efficiency and patient care. Robots can also be used for cleaning and maintaining workplaces to reduce labor requirements (Hamadani and Khan, 2015).

Expert systems: Expert systems are AI programs designed to replicate the expertise and decision-making capabilities of human experts in specific domains. These provide appropriate decision support (Hamadani and Ganai, 2022). In healthcare, expert systems can assist in clinical diagnosis by providing recommendations based on patient symptoms, medical history, and diagnostic test results. These systems can serve as decision support tools for healthcare professionals, especially in complex and rare cases.

Predictive analytics: Predictive analytics uses historical data and statistical models to forecast future events or outcomes (Hamadani et al., 2019, 2020, 2021, 2022b). In healthcare, predictive analytics can be applied to anticipate disease outbreaks, optimize hospital resource allocation, predict patient readmissions, and identify individuals at high risk for certain conditions. These insights enable proactive interventions and resource planning to improve patient outcomes.

Virtual assistants: Virtual assistants, powered by AI, offer personalized and interactive support to patients and healthcare professionals. They can answer medical queries, provide health education, offer medication reminders, and even triage patient symptoms. Virtual assistants enhance patient engagement, deliver healthcare information, and facilitate access to care, especially in remote or underserved areas (Curtis et al., 2021).

Applications of artificial intelligence in healthcare

The applications of AI in healthcare and pharmacy are given in Table 5.1, and a few are described in more detail below.

Table 5.1 Various realms of artificial intelligence (AI) in healthcare.

AI types	Schemes used in different types of AI	Application area
Robotics	Improvement of accuracy and precision of surgical procedures by providing high-quality treatment.	Medical devices, health IT
Digital assistants	You can choose the best moment for intervention by continuously checking the patient's condition indicators and updating the nurse as necessary.	Health IT
Deep learning	Identify and analyze patterns based on information affecting treatment outcomes. The level of ambiguity in the choice of medical therapy can be reduced by self-learning through the analysis of a large number of diagnostic medical images.	Diagnostic studies
Digital signal processing	Analysis of digital data and then using that analysis outcome to determine whether the results are positive or negative.	Data analysis for medical diagnostics
Language processing	Make massive unstructured text data, such as those seen in medical records, easier to extract and comprehend.	Medical devices
Recognition of signals like voice	Record the signals of the patient for future use in electronic medical records.	Record keeping
Statistical techniques	Using quick analysis of a lot of patient health record data, forecast patient treatment outcomes.	Health
Data analytics	Process the vast volumes of data kept by healthcare organizations to give patients and treatments personalized recommendations.	Record keeping for health use
Predictive modeling	Apply mathematical models to forecast treatment outcomes, such as forecasting dangerous diseases.	Medicine, health IT

Wearable sensors

Wearable biosensing technologies have attracted a lot of scholarly and commercial attention as a result of the growing need for a healthy lifestyle and wellness. These portable biosensing devices can be put on an individual's skin surface, where they can collect information about their everyday and athletic activity to give them useful health status insight (Ting et al., 2019). The competent Medicaid inventory, which is frequently risky due to the deteriorating neurological health of the elderly population in some nations, is a significant concern. The use of a health-monitoring system is required for conditions including high cholesterol, high blood pressure, memory loss, Alzheimer's, and other health-related problems among the elderly (Patel et al., 2016). However, obtaining relevant information from wearable biosensor networks' data is typically challenging for users without medical understanding since it frequently has a large dimensional and extremely erratic signal disturbance. Because wearable biosensors have excellent denoising, pattern recognition, and abnormality detection capabilities, AI-based technology is increasingly being integrated with them. The adoption of AI creates a chance to address the aforementioned problem and transforms the present healthcare system into a more responsive one where patients receive more individualized care (Ting et al., 2017).

Radiology

Recent AI innovations use data analytics, ML software, and robots to track, identify, and evaluate risks and advantages in the healthcare sector (Pahari, 2023). Radiology plays a prominent role in the diagnosis and treatment of numerous health issues. Technology breakthroughs have increased the level of accuracy and precision of medical imaging. Medical envision analysis is a laborious and frequently subjective process, though, because it depends on the skill and expertise of the interpreter. Recently, algorithms based on AI have been developed to aid in the analysis and interpretation of data from medical imaging (Bohr and Memarzadeh, 2020). The development of radiology will also be influenced by improvements in imaging hardware and software. To enhance diagnostic capabilities, new imaging modalities such as molecular visualization and mixed imaging systems are being developed. Techniques for molecular imaging make it possible to see how cells and molecules function, which helps with the early identification and description of diseases (Hu et al., 2019).

Medical image analysis

Beyond radio diagnosis and treatment, ML algorithms are applied to medical image processing and also the majority of medical specialities that employ images, including the study of pathogens, skin-related studies, the study of heart, gastrointestinal, and the science of ophthalmology. ML algorithms employ data from CT, MRI, ultrasound, pathology image, fundus image, and endoscope for diagnosis and categorization of the severity of the diseases (Currie et al., 2019; Gulshan et al., 2016; Kim et al., 2019; Park et al., 2019; Ting et al., 2017). The accuracy of the diagnosis was 94%, and the negative predictive value was 96% in the analysis of 466 microscopic polyps when the AI-based algorithm was applied to the real-time colonoscopy (Fischer et al., 2020). The convolutional neural network technique, one of many deep learning algorithms, has been shown to be useful in medical image analysis with complicated patterns because it maintains excellent performance in visual pattern analysis (Komura and Ishikawa, 2019). For the benefit of chest CT diagnosis, Siemens Healthineers created AI-based AI-Rad Companion Chest CT software (Hannun et al., 2019). GE Healthcare is also developing

AI-based medical image analysis technology. Additionally, Philips Healthcare is attempting to commercialize its IntelliSite Pathology Solution in the field of digital pathology diagnosis and has created IntelliSpace Discovery, an open platform for AI development and implementation (Vahlsing et al., 2018). For its heart-related AI, hepatic AI, and respiratory AI software to create a platform called the Medical Imaging Cloud AI platform, Arterys has acquired FDA clearance (Fernández-Caramés et al., 2019).

Electronic health records and analytics

Medical records are essential to healthcare because it enables the analysis of information from the beginning of time to the present, which enhances various treatment modalities; AI can be used to decipher the records and give the doctors correct information in a precise manner (Ali et al., 2023). Not only can AI improve patient flow in hospitals, but it can also be used to produce drugs, maintain and analyze patient records, and even assist in the detection of diseases like malignancy.

For example, AI has been successful in detecting COVID-19-related diseases. Early in 2020, the medical community declared coronavirus a global pandemic due to its uncontrolled transmission. So far, COVID-19 has infected more than 82 million people and killed more than 1.8 million people (WHO, 2021). The respiratory organs of the human body are known to be severely impacted by this novel coronavirus and frequent respiratory illnesses, particularly in people who also have comorbid conditions such as prolonged heart disease, diabetes, etc. Medical imaging-based prognostic and diagnostic techniques have grown in popularity in clinical settings over time as the rise of viral infection cases has remained persistent and new variants have become common. CT, X-rays, and USG have been used to detect various lung diseases while fighting COVID. It is interesting to note that AI-based systems have the potential to optimize image interpretation by enhancing efficiency in processes involving detection, description, and measurement, particularly in the context of chest imaging. They also make it easier to automate procedures, which lower diagnostic discrepancies between and within reader diagnoses. In order to help radiologists and physicians diagnose and treat COVID-19 patients, ML algorithms based on CT images show promise. ML algorithms are trained on generated data to generate insights, decision-making enhancement, and health improvement.

Precision medicine

Precision medicine, also known as personalized medicine, is an innovative approach to healthcare that takes into account individual variability in genes, environment, and lifestyle for the prevention, diagnosis, and treatment of diseases. By tailoring medical decisions and interventions to the specific characteristics of each patient, precision medicine aims to improve healthcare outcomes and reduce healthcare costs. While precision medicine holds great promise, its widespread implementation faces challenges such as data privacy concerns, the need for large-scale infrastructure and resources, and ensuring equitable access to these technologies and therapies. However, ongoing advancements in technology, decreasing costs of genomic sequencing, and increasing knowledge about the role of genetics in disease will likely continue to drive the integration of precision medicine into mainstream healthcare, leading to improved patient outcomes and a more personalized approach to medicine (Ashley, 2016).

Targeted therapies: Precision medicine enables the identification of specific genetic or molecular alterations in patients, allowing for targeted therapies that are tailored to their individual characteristics. This approach has been particularly successful in cancer treatment, where genetic testing can guide the selection of drugs that are most likely to be effective for a particular patient's tumor.

Pharmacogenomics: Genetic variations can influence an individual's response to medications. Pharmacogenomic testing can identify these variations, helping to determine the most appropriate and effective drug and dosage for a patient. This can minimize adverse reactions, improve drug efficacy, and reduce trial-and-error prescribing.

Early disease detection and prevention: Precision medicine emphasizes proactive screening and early detection of diseases. Advances in genomics and biomarker research enable the identification of individuals at high risk for certain conditions, allowing for targeted interventions and preventive measures to be implemented. This approach can lead to better health outcomes and a reduction in healthcare costs associated with late-stage disease treatments.

Improved diagnostic accuracy: Precision medicine utilizes advanced diagnostic techniques, such as genomic profiling and molecular diagnostics, to improve the accuracy and speed of disease diagnosis. By analyzing an individual's genetic makeup or specific molecular markers, healthcare providers can make more precise diagnoses, leading to timely and targeted treatment decisions.

Enhanced risk assessment: By combining genetic information, environmental factors, and lifestyle data, precision medicine can provide a more comprehensive risk assessment for individuals. This can help healthcare professionals develop personalized prevention strategies and interventions, including lifestyle modifications, early screening, and targeted interventions.

Patient empowerment and engagement: Precision medicine emphasizes patient-centered care and encourages active patient participation in healthcare decision-making. Patients gain a better understanding of their genetic predispositions and health risks, allowing them to make informed choices about their health and participate in shared decision-making with their healthcare providers.

Research and development: Precision medicine relies on extensive research and collaborations between various disciplines, including genomics, bioinformatics, and clinical medicine. These collaborations foster innovation and accelerate the development of new diagnostic tools, therapies, and treatments, ultimately benefiting healthcare as a whole.

Drug discovery and development: AI techniques, such as ML and data mining, are being used to accelerate the process of drug discovery. AI models can analyze large datasets, identify patterns, and predict the effectiveness of potential drug candidates, potentially reducing the time and cost involved in bringing new drugs to market.

Smart internet of medical things and diagnostic analysis

The development and commercialization of devices and services that can keep track of the health of people by collecting all the information regarding their health using the Internet of medical things (IoMT) and wearable devices are the focus of competition between numerous technology behemoths like IBM, Google, Apple, and Samsung. In 2017, after receiving the FDA approval, Apple included an AI-based system that can identify atrial fibrillation in its smartwatch. These devices using smart technologies can even detect the vital signs of patients like heart rate and blood pressure during rest as well as an activity using the photoplethysmography and accelerometer sensors and delivers a warning notice if there is a considerable departure from the anticipated values.

The electrocardiogram (ECG) analysis capabilities of the deep learning system are also very accurate. A deep learning algorithm was used to analyze 91,232 single-lead ECGs in a recent study (Hannun et al., 2019), and thus, these AI-based mathematical models somehow beat the race conducted by humans. Patients with severe heart diseases and chronic illnesses would gain the most from the application of smart IoMT devices.

Other cutting-edge medical devices, such as noninvasive glucose meters, are also being developed by international corporations (Dankwa-Mullan et al., 2019). These meters can be used for the management of diseases and treatment in various hospitals in addition to real-time data monitoring and hence are installed in all the emergency service areas in hospitals. One excellent illustration of this is the partnership between Medtronic and IBM, which resulted in the creation of Sugar.IQ by fusing IBM's AI Watson and Medtronic's continuous glucose monitoring.

By lessening the frequency of symptoms associated with hypoglycemia and hyperglycemia, sugar IQ has been shown to directly benefit diabetic patients (Ahmed et al., 2019). Additionally, by continuously monitoring patient parameters from wearable biosensors and spotting minor changes in those variables, PhysiQ's pinpointIQ device, which recently obtained FDA approval, aids in planning for abrupt and deadly scenarios (Kim et al., 2019). Additionally, Philips has made its Connected Care Solutions available for purchase. This IT solution enables medical personnel to keep tabs on patients' health from anywhere, including the emergency department, recovery area, and nurses' station, using tablet computers and cell phones (Beccaria et al., 2018).

Machine learning has also been applied to the in-vitro diagnosis of conditions including dementia and tuberculosis (TB). In the interim, the classification of cancer-related biomarkers in real-world situations has been made possible by the findings of studies that suggested an increase in diagnostic sensitivity and specificity by incorporating ML into the multibiomarker analysis (Long et al., 2017).

AI can automate various administrative tasks in healthcare, including appointment scheduling, billing, and coding. This automation can streamline processes, reduce errors, and free up healthcare professionals' time to focus more on patient care.

Conclusion

In conclusion, AI has emerged as a transformative force in healthcare, offering immense potential to improve patient care, enhance diagnostic accuracy, streamline processes, and facilitate medical research. The various types of AI, including ML, NLP, computer vision, robotics, expert systems, predictive analytics, and virtual assistants, are revolutionizing healthcare delivery in numerous ways. By harnessing the power of AI, healthcare professionals can leverage vast amounts of data to gain insights, make more accurate diagnoses, and personalize treatment plans based on individual patient characteristics. AI algorithms can process and analyze medical images, detect patterns, and assist in surgical procedures, ultimately enhancing patient outcomes and reducing medical errors. Moreover, AI technologies enable the automation of administrative tasks, enabling healthcare providers to focus more on patient care. Virtual assistants and chatbots enhance patient engagement, deliver health information, and offer support, particularly in remote or underserved areas. However, it is important to recognize that the adoption of AI in healthcare is not without challenges. Privacy and security concerns, regulatory frameworks, data interoperability, ethical considerations, and ensuring equitable access to AI-driven healthcare solutions are critical aspects that need to be addressed. Nevertheless, the ongoing advancements in AI, coupled with the increasing availability and quality of healthcare data, are propelling the integration of AI into mainstream healthcare systems. As AI continues to evolve and mature, it holds the potential to revolutionize healthcare, making it more personalized, efficient, and effective in meeting the evolving needs of patients and healthcare providers.

References

Ahmed, Md R., Zhang, Y., Feng, Z., Lo, B., Inan, O.T., Liao, H., 2019. Neuroimaging and machine learning for dementia diagnosis: recent advancements and future prospects. IEEE Reviews in Biomedical Engineering 12, 19−33. https://doi.org/10.1109/rbme.2018.2886237.

Ali, O., Abdelbaki, W., Shrestha, A., Elbasi, E., Alryalat, M.A.A., Dwivedi, Y.K., 2023. A systematic literature review of artificial intelligence in the healthcare sector: benefits, challenges, methodologies, and functionalities. Journal of Innovation & Knowledge 8 (1), 100333. https://doi.org/10.1016/j.jik.2023.100333.

Ashley, E.A., 2016. Towards precision medicine. Nature Reviews Genetics 17 (9), 507−522. https://doi.org/10.1038/nrg.2016.86.

Beccaria, M., Mellors, T.R., Petion, J.S., Rees, C.A., Nasir, M., Systrom, H.K., Sairistil, J.W., Jean-Juste, M.A., Rivera, V., Lavoile, K., Severe, P., Pape, J.W., Wright, P.F., Hill, J.E., 2018. Preliminary investigation of human exhaled breath for tuberculosis diagnosis by multidimensional gas chromatography − time of flight mass spectrometry and machine learning. Journal of Chromatography, B: Analytical Technologies in the Biomedical and Life Sciences 1074−1075, 46−50. https://doi.org/10.1016/j.jchromb.2018.01.004.

Bhinder, B., Gilvary, C., Madhukar, N.S., Elemento, O., 2021. Artificial intelligence in cancer research and precision medicine. Cancer Discovery 11 (4), 900−915. https://doi.org/10.1158/2159-8290.cd-21-0090.

Bohr, A., Memarzadeh, K., 2020. The rise of artificial intelligence in healthcare applications. Artificial Intelligence in Healthcare 25−60. https://doi.org/10.1016/B978-0-12-818438-7.00002-2.

Currie, G., Hawk, K.E., Rohren, E., Vial, A., Klein, R., 2019. Machine learning and deep learning in medical imaging: intelligent imaging. Journal of Medical Imaging and Radiation Sciences 50 (4), 477−487. https://doi.org/10.1016/j.jmir.2019.09.005.

Curtis, R.G., Bartel, B., Ferguson, T., Blake, H.T., Northcott, C., Virgara, R., Maher, C.A., 2021. Improving user experience of virtual health assistants: scoping review. Journal of Medical Internet Research 23 (12). https://doi.org/10.2196/31737.

Dankwa-Mullan, I., Rivo, M., Sepulveda, M., Park, Y., Snowdon, J., Rhee, K., 2019. Transforming diabetes care through artificial intelligence: the future is here. Population Health Management 22 (3), 229−242. https://doi.org/10.1089/pop.2018.0129.

Dong, Y., Hou, J., Zhang, N., Zhang, M., Rao, R., 2020. Research on how human intelligence, consciousness, and cognitive computing affect the development of artificial intelligence. Complexity 2020, 1−10. https://doi.org/10.1155/2020/1680845.

Erickson, B.J., Korfiatis, P., Akkus, Z., Kline, T.L., 2017. Machine learning for medical imaging. RadioGraphics 37 (2), 505−515. https://doi.org/10.1148/rg.2017160130.

Fernández-Caramés, T.M., Froiz-Míguez, I., Blanco-Novoa, O., Fraga-Lamas, P., 2019. Enabling the Internet of mobile crowdsourcing health things: a mobile fog computing, blockchain and IoT based continuous glucose monitoring system for diabetes mellitus research and care. Sensors 19 (15), 3319. https://doi.org/10.3390/s19153319.

Fischer, A.M., Varga-Szemes, A., Martin, S.S., Sperl, J.I., Sahbaee, P., Neumann, D., Gawlitza, J., Henzler, T., Johnson, C.M., Nance, J.W., Schoenberg, S.O., Schoepf, U.J., 2020. Artificial intelligence-based fully automated per lobe segmentation and emphysema-quantification based on chest computed tomography compared with global initiative for chronic obstructive lung disease severity of smokers. Journal of Thoracic Imaging 35, S28−S34. https://doi.org/10.1097/RTI.0000000000000500.

Friedman, A.A., Letai, A., Fisher, D.E., Flaherty, K.T., 2015. Precision medicine for cancer with next-generation functional diagnostics. Nature Reviews Cancer 15 (12), 747−756. https://doi.org/10.1038/nrc4015.

Grand View Research, 2022. Artificial Intelligence in Healthcare Market Worth $208.2 Billion by 2030. https://www.grandviewresearch.com/press-release/global-artificial-intelligence-healthcare-market.

Gulshan, V., Peng, L., Coram, M., Stumpe, M.C., Wu, D., Narayanaswamy, A., Venugopalan, S., Widner, K., Madams, T., Cuadros, J., Kim, R., Raman, R., Nelson, P.C., Mega, J.L., Webster, D.R., 2016. Development and validation of a deep learning algorithm for detection of diabetic retinopathy in retinal fundus photographs. JAMA, the Journal of the American Medical Association 316 (22), 2402−2410. https://doi.org/10.1001/jama.2016.17216.

Hamadani, A., Ganai, N.A., 2022. Development of a multi-use decision support system for scientific management and breeding of sheep. Scientific Reports 12 (1). https://doi.org/10.1038/s41598-022-24091-y.

Hamadani, H., Khan, A.A., 2015. Automation in livestock farming − a technological revolution. International Journal of Advanced Research 3, 1335−1344.

Hamadani, A., Ganai, N.A., Khan, N.N., Shanaz, S., Ahmad, T., 2019. Estimation of genetic, heritability, and phenotypic trends for weight and wool traits in Rambouillet sheep. Small Ruminant Research 177, 133−140. https://doi.org/10.1016/j.smallrumres.2019.06.024.

Hamadani, A., Ganai, N.A., Rather, M.A., Raja, T.A., Shabir, N., Ahmad, T., Shanaz, S., Aalam, S., Shabir, M., 2020. Estimation of genetic and phenotypic trends for wool traits in Kashmir Merino sheep. Indian Journal of Animal Sciences 90 (6), 893−897. http://epubs.icar.org.in/ejournal/index.php/IJAnS/article/view/104998.

Hamadani, A., Ganai, N.A., Rather, M.A., 2021. Genetic, phenotypic and heritability trends for body weights in Kashmir Merino Sheep. Small Ruminant Research 205. https://doi.org/10.1016/j.smallrumres.2021.106542.

Hamadani, A., Ganai, N.A., Farooq, S.F., Bhat, B.A., 2022a. Big data management: from hard drives to DNA drives. Indian Journal of Animal Sciences 90 (2), 134−140. https://doi.org/10.56093/ijans.v90i2.98761.

Hamadani, A., Ganai, N.A., Mudasir, S., Shanaz, S., Alam, S., Hussain, I., 2022b. Comparison of artificial intelligence algorithms and their ranking for the prediction of genetic merit in sheep. Scientific Reports 12 (1). https://doi.org/10.1038/s41598-022-23499-w.

Hamadani, A., Ganai, N.A., Alam, S., Mudasir, S., Raja, T.A., Hussain, I., Ahmad, H.A., 2022c. Artificial intelligence techniques for the prediction of body weights in sheep. Indian Journal of Animal Research. https://doi.org/10.18805/ijar.b-4831.

Hamadani, A., Ganai, N.A., Gupta, A.R., Farooq, F., Shah, M.S.U., Jindal, K., Bashir, J., 2023. Comparison of Multiple Deep Convolutional Neural Networks for Image Classification.

Hannun, A.Y., Rajpurkar, P., Haghpanahi, M., Tison, G.H., Bourn, C., Turakhia, M.P., Ng, A.Y., 2019. Cardiologist-level arrhythmia detection and classification in ambulatory electrocardiograms using a deep neural network. Nature Medicine 25 (1), 65−69. https://doi.org/10.1038/s41591-018-0268-3.

Hu, W., Cai, B., Zhang, A., Calhoun, V.D., Wang, Y.P., 2019. Deep collaborative learning with application to the study of multimodal brain development. IEEE Transactions on Biomedical Engineering 66 (12), 3346−3359. https://doi.org/10.1109/TBME.2019.2904301.

Hussain, A., Malik, A., Halim, M.U., Ali, A.M., 2014. The use of robotics in surgery: a review. International Journal of Clinical Practice 68 (11), 1376−1382. https://doi.org/10.1111/ijcp.12492.

Kim, J.P., Kim, J., Park, Y.H., Park, S.B., Lee, J.S., Yoo, S., Kim, E.J., Kim, H.J., Na, D.L., Brown, J.A., Lockhart, S.N., Seo, S.W., Seong, J.K., 2019. Machine learning based hierarchical classification of fronto-temporal dementia and Alzheimer's disease. NeuroImage: Clinical 23. https://doi.org/10.1016/j.nicl.2019.101811.

Komura, D., Ishikawa, S., 2019. Machine learning approaches for pathologic diagnosis. Virchows Archiv 475 (2), 131−138. https://doi.org/10.1007/s00428-019-02594-w.

Long, N.P., Jung, K.H., Yoon, S.J., Anh, N.H., Nghi, T.D., Kang, Y.P., Yan, H.H., Min, J.E., Hong, S.S., Kwon, S.W., 2017. Systematic assessment of cervical cancer initiation and progression uncovers genetic panels for deep learning-based early diagnosis and proposes novel diagnostic and prognostic biomarkers. Oncotarget 8 (65), 109436–109456. https://doi.org/10.18632/oncotarget.22689.

Pahari, E., 2023. Frankfurt University of Applied Sciences—Faculty of Computer Science and Engineering—The Role of AI in Healthcare.

Park, H.J., Kim, S.M., La Yun, B., Jang, M., Kim, B., Jang, J.Y., Lee, J.Y., Lee, S.H., 2019. A computer-aided diagnosis system using artificial intelligence for the diagnosis and characterization of breast masses on ultrasound. Medicine 98 (3), e14146. https://doi.org/10.1097/md.0000000000014146.

Patel, V., Armstrong, D., Ganguli, M., Roopra, S., Kantipudi, N., Albashir, S., Kamath, M.V., 2016. Deep learning in gastrointestinal endoscopy. Critical Reviews in Biomedical Engineering 44 (6), 493–504. https://doi.org/10.1615/CritRevBiomedEng.2017025035.

The White House, 2022. The Impact of Artificial Intelligence on the Future of Workforces in the European Union and the United States of America. https://www.whitehouse.gov/wp-content/uploads/2022/12/TTC-EC-CEA-AI-Report-12052022-1.pdf.

Ting, D.S.W., Cheung, C.Y.L., Lim, G., Tan, G.S.W., Quang, N.D., Gan, A., Hamzah, H., Garcia-Franco, R., Yeo, I.Y.S., Lee, S.Y., Wong, E.Y.M., Sabanayagam, C., Baskaran, M., Ibrahim, F., Tan, N.C., Finkelstein, E.A., Lamoureux, E.L., Wong, I.Y., Bressler, N.M., Sivaprasad, S., Varma, R., Jonas, J.B., He, M.G., Cheng, C.Y., Cheung, G.C.M., Aung, T., Hsu, W., Lee, M.L., Wong, T.Y., 2017. Development and validation of a deep learning system for diabetic retinopathy and related eye diseases using retinal images from multiethnic populations with diabetes. JAMA, the Journal of the American Medical Association 318 (22), 2211–2223. https://doi.org/10.1001/jama.2017.18152.

Ting, D.S.W., Pasquale, L.R., Peng, L., Campbell, J.P., Lee, A.Y., Raman, R., Tan, G.S.W., Schmetterer, L., Keane, P.A., Wong, T.Y., 2019. Artificial intelligence and deep learning in ophthalmology. British Journal of Ophthalmology 103 (2), 167–175. https://doi.org/10.1136/bjophthalmol-2018-313173.

Vahlsing, T., Delbeck, S., Leonhardt, S., Heise, H.M., 2018. Noninvasive monitoring of blood glucose using color-coded photoplethysmographic images of the illuminated fingertip within the visible and near-infrared range: opportunities and questions. Journal of Diabetes Science and Technology 12 (6), 1169–1177. https://doi.org/10.1177/1932296818798347.

WHO, 2021. The Impact of COVID-19 on Global Health Goals. https://www.who.int/news-room/spotlight/the-impact-of-covid-19-on-global-health-goals.

Zhang, Y., Hu, Y., Jiang, N., Yetisen, A.K., 2023. Wearable artificial intelligence biosensor networks. Biosensors and Bioelectronics 219, 114825. https://doi.org/10.1016/j.bios.2022.114825.

Reimagining occupational health and safety in the era of AI

Immad A. Shah[1] and SukhDev Mishra[2]

[1]*ICMR-National Institute of Occupational Health, Ahmedabad, Gujarat, India;* [2]*Department of Biostatistics, ICMR-National Institute of Occupational Health, Ahmedabad, Gujarat, India*

Introduction

Every year, a significant number of workers globally experience injuries or illnesses directly related to their jobs. According to the International Labor Organization (ILO), an estimated 2.3 million individuals worldwide, both women and men, lose their lives due to work-related accidents or diseases each year. Shockingly, this corresponds to an alarming average of over 6000 deaths per day. Furthermore, on a global scale, approximately 340 million occupational accidents occur annually, resulting in around 160 million individuals being affected by work-related illnesses. These staggering figures underline the urgent need for comprehensive efforts to address and mitigate workplace risks and ensure the health and safety of workers worldwide as the consequences of these incidents are far-reaching, affecting workers' health, productivity, and financial stability. The field of occupational health is committed to assisting businesses in creating safe and healthy workplaces. Occupational health professionals play a vital role in identifying and mitigating workplace hazards, imparting training on safe work practices, and delivering preventive care to employees. Occupational health is a crucial component of public health aimed at promoting and sustaining the highest level of physical, mental, and social well-being among workers across all occupations. The objectives of occupational health encompass the below mentioned:

Maintain and promote workers' health and working capacity: This includes preventing occupational diseases and injuries and providing workers with the resources they need to stay healthy and productive.

Improve working conditions and create a safe and healthy work environment: This includes identifying and controlling hazards, providing adequate training, and creating a culture of safety and health.

Develop work organizations and cultures that align with the core values of the organization: This includes implementing effective management systems, personnel policies, and participation principles.

NLP in occupational health and safety

The field of occupational health is an amalgam of a range of disciplines, including occupational medicine, nursing, ergonomics, psychology, hygiene, safety, and others. These disciplines work together to ensure the well-being of workers and the implementation of comprehensive health strategies. On a global scale, the World Health Assembly (WHA), which is a decision-making body of the World Health Organization (WHO), comprising 194 member countries urges to develop of national policies and action plans while establishing institutional capacities in the field of occupational health. It emphasizes on the expansion of the coverage of essential interventions to prevent and control occupational and work-related diseases, injuries, and occupational health services. Collaboration with other pertinent national health programs, such as those addressing communicable and non-communicable diseases, injury prevention, health promotion, mental health, environmental health, and health systems development is among the mandates of the WHA. By following these recommendations, countries can strengthen their occupational health efforts and contribute to the overall well-being of their workforce (Fig. 6.1).

Managing occupational health and safety (OHS) presents a significant challenge for managers and human resources departments alike. It is crucial to acknowledge that accidents and unfortunate incidents can happen to anyone, regardless of their position or role within the organization. However, it is the responsibility of companies to proactively implement well-informed OHS management practices. Data-driven decision-making in healthcare utilizing health data may help inform and guide various aspects of decision-making. This approach leverages the wealth of available data to gain insights, make informed choices in order to minimize risks, and prioritize the safety of their employees. The proliferation and intricacy of data within the field of health sciences have led to a growing adoption of artificial intelligence (AI) methodologies.

Noteworthy AI technologies encompass machine learning (ML) and deep learning (DL), natural language processing (NLP), and rule-based expert systems (RBES). These AI technologies have also made their way into occupational health, facilitating the examination of both structured and unstructured data. The depiction of AI playing a role at a monitored facility is depicted in Fig. 6.2.

FIGURE 6.1

Artificial intelligence (AI) integration—driven benefits in the OSH domain.

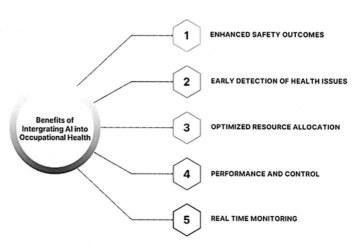

Benefits of Intergrating AI into Occupational Health

1 ENHANCED SAFETY OUTCOMES

2 EARLY DETECTION OF HEALTH ISSUES

3 OPTIMIZED RESOURCE ALLOCATION

4 PERFORMANCE AND CONTROL

5 REAL TIME MONITORING

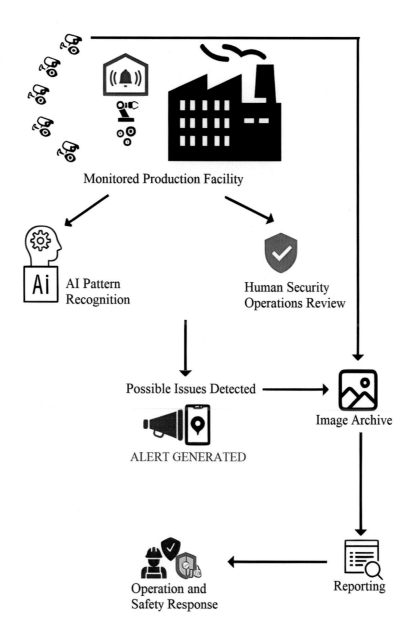

Monitored Production Facility

AI Pattern Recognition

Human Security Operations Review

Possible Issues Detected

Image Archive

ALERT GENERATED

Operation and Safety Response

Reporting

FIGURE 6.2

Schematic map for artificial intelligence (AI) monitored facility.

Their applications extend to various areas, such as job codings (utilizing NLP), exposure assessment (employing NLP and RBES), data analysis (utilizing ML, DL, and neural networks), as well as risk assessment (utilizing RBES).

Effective health, safety, and environment (HSE) management programs are crucial for ensuring the safety of employees and the overall performance of organizations. However, the implementation of new projects, sites, and equipment can introduce a variety of safety risks that require additional resources from HSE teams. HSE teams face challenges in managing the growing operational activities, including employee training, site audits, incident investigations, and implementing corrective and preventive measures. Another significant challenge is extracting actionable insights from historical incident data. Historical incident data are often collected and analyzed using manual, form-based data entry methods. Safety supervisors manually analyze these data to identify risks and develop new regulations, best practice guidelines, and safety training programs. However, this manual process is time-consuming and may not capture emerging or unknown risks adequately. NLP, a branch of AI, can deliver significant value in improving HSE management. It is a specialized branch of AI that focuses on the interaction between computers and human language. It encompasses the ability of machines to understand, interpret, and generate natural language, enabling them to communicate with humans in a more intuitive and human-like manner.

One specific application of NLP is that it plays a crucial role in enhancing safety supervision. NLP helps safety supervisors gain a better understanding of past incidents and their causes. By analyzing textual data, NLP can identify patterns and extract valuable information that sheds light on the underlying factors contributing to these incidents. Furthermore, NLP can also be used to identify leading indicators of potentially hazardous conditions. By analyzing various data sources, such as incident reports, maintenance records, and sensor data, NLP algorithms can identify early warning signs of developing hazardous situations. This empowers safety supervisors to proactively anticipate and mitigate incidents before they occur, thereby enhancing overall safety in the workplace. By leveraging AI technologies like NLP, industrial organizations can harness the power of their data, enhance decision-making processes, and improve safety outcomes. The ability to extract insights from large volumes of data and identify potential risks in advance enables safety supervisors to take proactive measures, thereby mitigating potential incidents and fostering a safer working environment.

Natural language processing can be employed to automate and improve the job coding process. Job coding is an essential process in many organizations that involves assigning specific codes or labels to job positions or roles. These codes help in categorizing and organizing job-related information, such as job descriptions, qualifications, and responsibilities. Job coding plays a crucial role in workplace safety by providing a standardized framework for identifying and addressing potential hazards and risks associated with specific job roles. It allows organizations to systematically identify the inherent risks and hazards associated with each job role. By assigning specific codes or labels to job positions, it becomes easier to recognize the potential dangers involved in performing those roles. This information is vital for developing appropriate safety protocols, training programs, and risk mitigation strategies. It helps in appropriately tailoring safety measures and protocols based on the specific requirements of each job role.

RBES in occupational health and safety

In modern monitoring systems, the process of monitoring goes beyond simple observation of objects. It involves obtaining, accumulating, storing, and reporting data related to the monitoring objects. Additionally, there is a need for data analysis to evaluate the collected information and derive meaningful conclusions about the monitored situations. This expanded approach to monitoring is often referred to as information monitoring. To effectively monitor and make informed decisions, monitoring systems now incorporate units or components that are responsible for data analysis. These units analyze the collected data, record the most important parameters, and provide valuable information to decision-makers. By analyzing the data, these units can identify patterns, trends, anomalies, and other relevant insights that aid in understanding the current state of the monitored objects or systems. Expert systems (ES) based on the production model of knowledge are more suitable for monitoring analysis, since they allow to description of situations arising in the systems for which the monitoring is carried out (Abbasov and Shahbazova, 2014; Russell and Norvig, 2003).

While an ES cannot fully replace the problem-solving capabilities of a knowledgeable worker, it can significantly reduce their workload and preserve their creative and innovative contributions. Some of the potential organizational benefits of ES include:

- **Accelerated task completion:** ESs can perform their assigned tasks much faster compared to human experts.
- **Low error rate:** Successful ESs often demonstrate a lower error rate than humans when performing the same tasks.
- **Consistent recommendations:** ESs provide reliable and consistent recommendations based on predefined rules and knowledge.
- **Facilitating difficult-to-use knowledge:** ESs serve as a convenient platform to apply complex and challenging sources of knowledge in practical scenarios.
- **Capture of scarce expertise:** ESs can preserve the valuable expertise of exceptionally qualified individuals.
- **Accumulation of organizational knowledge:** ESs contribute to the development of organizational knowledge rather than relying solely on individual knowledge.
- **Enhanced learning for novices:** ESs used as training tools can expedite the learning process for inexperienced individuals.
- **Safe operation in hazardous environments:** Companies can deploy ESs in environments that pose risks to human safety.

ML for occupational health and safety

The core concepts of industrial hygiene revolve around identifying, assessing, and managing hazards present in the workplace. Occupational safety and health experts, such as industrial hygienists, undertake the responsibility of evaluating a multitude of intricate factors. However, due to the unavailability of industrial hygienists in certain scenarios, an ES called the workplace exposure assessment systems, for example, WORKSPERT have been developed to aid in conducting workplace exposure assessments (WEAs). These ESs specifically assess hazardous substances, workplace conditions, and worker exposures within designated homogeneous exposure groups (HEGs). The utilization of WORKSPERT or similar ESs should always be subordinate to the expertise and judgment of occupational safety and health professionals. Nevertheless, when employed by competent and knowledgeable technical experts (such as safety and health specialists, chemists, engineers, and toxicologists) who possess a comprehensive understanding of the relevant substances, workplace conditions, and exposure factors associated with designated homogeneous exposure groups (HEGs), ESs can serve as a valuable tool. Employing ESs for conducting workplace exposure assessments (WEAs) contributes to safeguarding workers against hazardous substances and aids in ensuring compliance with occupational safety and health regulations. Additionally, these systems facilitate the concise communication of substance hazards, workplace controls, and worker exposures.

Across various industries like factories, construction sites, and warehouses, accidents have been an ongoing fatal concern. With the added challenges of the pandemic and the increasing frequency of natural disasters, ensuring the safety of employees and citizens becomes more complex. However leveraging ML techniques to reduce workplace accidents, enable the detection of potentially ill employees upon arrival at the worksite, and aid organizations in managing natural disasters enhance workplace safety through targeted applications with clear benefits. With the expertise of technical experts in risk management, computer vision, and asset management, supported by advanced ML tools, industrial workplaces aim to maximize operational efficiency.

ML techniques offer various capabilities that can be leveraged to enhance workplace safety and health including accident prediction, prevention, and avoidance. ML algorithms can analyze historical data and identify patterns or indicators that precede accidents, enabling proactive measures to prevent similar incidents in the future. Such algorithms help realize consistent enforcement of health and safety procedures. Automated job risk analysis and H&S audits: ML algorithms can assess job-related risks by analyzing relevant data, such as task characteristics, environmental conditions, and worker attributes. This automation streamlines the process of risk analysis and health and safety audits. ML-powered monitoring systems can provide centralized and remote oversight of multiple locations, reducing the need for extensive travel to different sites. This enhances efficiency while maintaining a comprehensive view of safety measures. By automating routine tasks and providing insights, ML enables human resources to focus on more complex problem-solving activities related to workplace safety and health. It frees up time and resources for strategic decision-making and developing innovative solutions.

DL for occupational health and safety

Workforce safety concerns within job site safety management (JSM) pose significant challenges for several industries in Architecture, Engineering, and Construction (AEC) industry (Fang et al., 2020). A striking example of these challenges is reflected in the 2018 data from the US Occupational Health and Safety Agency (OSHA), which reported a disturbingly high fatality count of 1008 construction workers. These fatalities were primarily attributed to common accidents on construction sites, such as falling from heights or being struck by falling objects (Prasanna et al., 2012; Sousa et al., 2014). Traditional approaches to onsite safety monitoring rely on site patrols and surveillance, but the dynamic nature of construction sites makes it difficult to ensure comprehensive and proactive safety monitoring (Luo et al., 2019; Wang et al., 2019). Moreover, accurately assessing the fatigue levels of workers remains a challenge.

In recent years, researchers have been exploring the application of MLin various fields, including structural health monitoring (SHM) (Oliveira and Correia, 2008) and job site safety management (JSM) (Ding et al., 2018). ML techniques have shown promise in detecting structural damage and monitoring worker safety onsite. The advancement of graphics processing units (GPUs) has significantly enhanced the computational capabilities required for ML algorithms. This has paved the way for the emergence of DL applications that leverage improved GPU performance. Notably, convolutional neural networks (CNNs), a type of DL algorithm, achieved remarkable results in the ImageNET Large Scale Visual Recognition Challenge (ILSVRC), 2012, a renowned benchmark for object classification and detection across numerous object classes and vast image datasets (Krizhevsky et al., 2017). DL has now surpassed many conventional algorithms in various domains. Consequently, an increasing number of DL applications are being developed and implemented to address challenges in image classification, data augmentation, and object detection (Qian et al., 2020; Qiu et al., 2019).

In the domain of Jobsite Safety Management (JSM), monitoring fatigue is a critical aspect to consider. With the increasing complexity of on-site dynamics, the role of heavy equipment operators, along with their operations and decision-making, becomes crucial in maintaining both safety and productivity at construction sites. However, as work intensity rises, their cognitive awareness may become compromised, posing a safety hazard. Notably, research by Tam and Fung (2011) revealed that approximately 60.5% of crane operators continue working even when feeling fatigued due to long working hours and tight construction schedules (Tam and Fung, 2011). Furthermore, around 52.6% of crane operators reported a lack of breaks, as frequent movement within narrow workspaces was inconvenient. To address these challenges, automated fatigue monitoring systems with timely warnings can offer valuable support to this group of operators. Facial expressions serve as a readily detectable indicator of fatigue among heavy equipment operators and workers. Real-time monitoring of their facial expressions through cameras proves to be an effective, feasible, and noninvasive method for identifying drowsiness and preventing accidents (Shi et al., 2017). CNNs have been applied in real-time fatigue monitoring for on-road driving (Dwivedi et al., 2014; Zhang et al., 2015, 2017). Some researchers have utilized CNNs to extract video-level features and integrate them into long short-term memory (LSTM) models to analyze temporal information for fatigue identification (Guo and Markoni, 2019; Lyu et al., 2018).

Understanding the application of AI/ML in workplace safety through vision algorithms

There have been multiple applications of computer vision algorithms to improve safety in the workplace. One such advancement has been made by an Australia-based computer vision company and involved in developing applications for improving workplace safety by harnessing the power of ML. Their innovative solutions aim to minimize workplace accidents, enable companies to identify potentially unwell employees upon arrival, and assist organizations in effectively managing natural disasters from an operational standpoint. In response to the increased emphasis on OHS, organizations have intensified their efforts to raise employee awareness about workplace hazards and implement enhanced safety measures. While these initiatives have proven beneficial, accidents can still happen. To address this challenge, their team has developed Warny which is claimed as an innovative solution aiming to improve workplace safety and minimize accidents. The Warny system consists of three key applications that work together to provide a comprehensive safety solution. Firstly, there is vehicle collision avoidance, which employs advanced technologies to prevent accidents involving vehicles. Secondly, safety zone alerting enhances situational awareness by notifying individuals when they enter restricted or hazardous areas. Lastly, Warny incorporates thermal analysis capabilities to monitor both people and industrial systems on the factory floor, enabling the early detection of anomalies that may pose safety risks. Leveraging advanced computer vision algorithms, Warny serves as a tool for safeguarding individuals working in proximity to hazardous machinery, including forklifts, trucks, and manufacturing equipment. It detects critical events such as spontaneous combustion of materials, equipment overheating, and fires within the workplace. Moreover, it can perform real-time analysis, generate reports, and promptly alert machine operators regarding unforeseen incidents, even in cases where the operator's line of sight does not directly cover an individual being present in an unsafe area Fig. 6.3.

This platform uses a combination of edge-based software and cloud-based services, the platform receives data streaming from IoT sensors. These data include images and object detection information, which undergo analysis within the platform. Through their developmental work, they have come up with a neural network capable of achieving nearly 100% accuracy in recognizing both equipment and people. An essential aspect of the solution involves enabling depth of field capabilities on standard CCTV cameras. This feature provides valuable depth information, allowing for the determination of object positions, distances between objects, and the speed at which they are moving. By auditing the environment and understanding the movements of individuals and forklifts, the system utilizes

FIGURE 6.3

Illustration of a vision algorithm-based sensor-generated artificial intelligence (AI) system for raising an alert based on proximity breach at a workplace.

trajectory data derived from depth, position, time, and distance to predict the near real-time path of objects using ML algorithms. In the event of an imminent collision, the platform promptly sends an alert to the employee's wearable device, leveraging AWS IoT Greengrass to achieve minimal latency in the range of milliseconds. This swift notification ensures timely response and intervention. Furthermore, the collected data are consolidated in user-friendly dashboards, enabling organizations to analyze their workplace safety practices and make informed improvements.

Traditional manual temperature screening processes can hinder productivity by causing congestion at entry and exit points, and they often fail to identify employees who develop symptoms during the workday. This company addresses these challenges with their prescreening solution called Thermy. It utilizes thermal imaging technology to swiftly and accurately detect elevated temperatures of individuals in real time, operating at scale by scanning 30 people per second or 500–600 people per minute. It also conducts 8.3 scans per second to validate temperature readings. This solution can be deployed at various locations within an organization, including building entrances, factory floors, cafeterias, breakrooms, washrooms, and other areas where employees frequent. Built upon the Warny platform and leveraging advanced analytics and ML, Thermy combines thermal cameras with computer vision technology. Unlike other thermal solutions that solely measure skin temperature, the Thermy platform calculates an accurate representation of a person's core body temperature (CBT) depicted in Fig. 6.4.

By utilizing data from both a thermal camera and an optical camera, it isolates the subject's head and captures skin temperature even when the individual has facial hair, glasses, a hard hat, or other distinguishing features. An ML algorithm then calculates an estimation of the person's core body temperature, enabling the determination of whether the individual has an elevated temperature. Real-time information is presented through dashboards hosted on AWS, facilitating remote viewing and

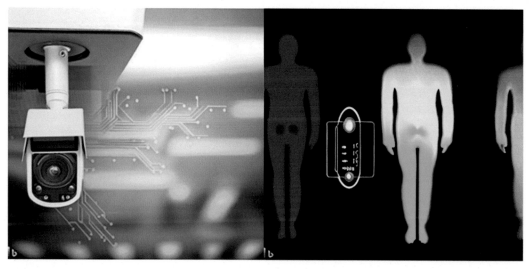

FIGURE 6.4

Thermal imaging camera and sensor generated core body temperature (CBT) image.

trend analysis. This comprehensive approach to temperature screening offers an enhanced and accurate assessment of individuals' core body temperature, empowering organizations to take proactive measures in ensuring a safe and healthy work environment. Some of the advanced technological development using AI is being done by companies such as Bigmate, GOARC, Intenseye, HGS, Mirrag AI, and CronJ AI among others in the EHS domain.

Workplace exposure assessment of toxic gases using AI techniques

With the rapid advancements in industrialization and the automation of chemical plants, gas leakage has become a prevalent issue. Industrial accidents can lead to explosions, fires, spills, leaks, and emissions of hazardous substances (Zhou et al., 2016; Trivedi et al., 2014). Additionally, fume leakages caused by residential cooking and improper waste disposal contribute to unnecessary air pollution. Shockingly, an article in the media highlighted that the burning of wood, biomass, and dung accounted for approximately 326,000 out of the estimated 645,000 premature deaths caused by outdoor air pollution. This alarming figure represents around 50% of the total deaths attributed to outdoor pollution (Lelieveld et al., 2015). Harmful gases, including liquid petroleum gas (LPG), compressed natural gas (CNG), methane, propane, and other flammable and toxic gases, pose significant risks if not handled with care and caution. Failure to use these gases properly can result in accidents and, in some cases, catastrophic consequences. A gas leak occurs when an unintended crack, hole, or porosity in a joint or machinery allows the escape of a closed medium, comprising various fluids and gases. Therefore, performing a gas leak test is a critical quality control step that must be undertaken before the installation of any device in a plant or industrial setting. As a preventive measure, gas sensors are installed near equipment prone to leakage. However, these sensors may not effectively detect gas in mixed gas environments, and their performance is limited by their operating characteristics. The possible role of AI in exposure assessment is described in Fig. 6.5.

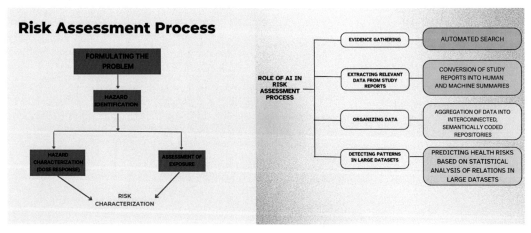

FIGURE 6.5

Descriptive role of artificial intelligence (AI) in risk assessment process.

Human intervention is often impractical in situations involving gas leaks due to the hazardous nature of these gases. Smoke emissions during leaks can obscure visibility, making it difficult to navigate and causing immediate evacuation challenges, especially for individuals with mobility issues. Inhaling these toxic fumes can result in dizziness, and loss of consciousness, and potentially lead to mass disasters if not promptly addressed. In chemical factories, gas leaks can even trigger explosions, emphasizing the criticality of detecting and responding to such incidents within a short timeframe. To achieve early and accurate detection of gas leaks, advanced techniques and assistive technology solutions are indispensable. Detecting specific gases or identifying different gases in gas mixtures poses a significant challenge that requires focused technological advancements. In addition to traditional chemical detection methods and interdisciplinary technologies, AI-based techniques have emerged as a viable approach. Various ML algorithms, including logistic regression (LR), random forest (RF), and support vector machines (SVMs), have been proposed in the literature for gas detection (Yin et al., 2016). However, these methods often require extensive hyperparameter tuning and statistical calculations to achieve accurate and robust gas classification. This not only increases processing time but also consumes more power and computational resources (Brahim-Belhouari et al., 2005).

Different research studies have investigated various strategies for gas detection and classification. Khalaf (2012) proposed the use of an electronic nose system that employs least square regression for estimating gas concentration (Khalaf, 2012). To enhance the accuracy of gas classification, deep CNNs have been utilized to learn features from gas sensor measurements (Peng et al., 2018; Bilgera et al., 2018) presented a fusion of AI models for gas source localization, which involves multiple gas sensors to identify the point of leakage in the ground. Pan et al. (2019) introduced a hybrid framework that combines CNNs and LSTM to extract sequential information from transient response curves. Another noteworthy development is the fast gas recognition algorithm, which employs a hybrid CNN and recurrent neural network (RNN) and demonstrates superior performance compared to models like SVM, RF, and k-nearest neighbors (Brahim-Belhouari et al., 2005; Liu et al., 2015) utilized deep belief networks and stacked autoencoders to extract abstract gas features from an electronic nose, followed by constructing Softmax classifiers using these features. This research approaches primarily focus on utilizing gas sensor data directly to develop sequential methods for gas analysis.

Workplace exposure assessment of hazardous chemicals using AI techniques

Quantitative structure—activity relationship (QSAR) models are powerful tools in the field of computational chemistry and toxicology. They are employed to uncover complex relationships between various chemical features, such as molecular structures, physicochemical properties, and toxicity values. These models provide a quantitative understanding of how different chemical characteristics influence the biological activity or toxicity of a compound. In the absence of sufficient experimental data, the QSAR/QSTR approach represents a convincing substitute to predict the possible hazards, from their chemical structure information (Cassani and Gramatica, 2015). By analyzing large datasets of chemical compounds with known activities or toxicities, QSAR models can identify patterns and correlations between specific chemical features and their corresponding effects on biological systems. This allows researchers to make predictions and prioritize the screening of new compounds based on their potential activity or toxicity profiles. The development of QSAR models involves the application of statistical and ML techniques to extract meaningful information from chemical data. These models can be used to assess the safety and potential risks associated with exposure to various chemicals, aiding in the design and optimization of safer and more effective compounds in industries such as pharmaceuticals, environmental sciences, and chemical engineering. QSAR, being a nonexperimental approach, is significant for reducing the time, animal testing, and cost associated with predicting the toxicity of pharmaceutical compounds (Roy et al., 2015). OPERA is an open-source and free suite of QSAR models that have been developed to serve various research and regulatory purposes. It offers a wide range of models, including those predicting physicochemical properties, environmental fate properties, and important absorption, distribution, metabolism, and excretion (ADME) endpoints. These ADME-related models, which were added in 2020 and 2021, cover essential endpoints for physiologically based pharmacokinetic modeling and in vitro to in vivo extrapolation (IVIVE) studies. They encompass predictions for octanol/water partition and distribution coefficients, acidic dissociation, the fraction of chemical unbound to plasma protein, intrinsic hepatic clearance, and $CaCo_2$ permeability. All OPERA models have been developed using carefully curated datasets, which were divided into training and test sets. The models utilize molecular descriptors derived from standardized QSAR-ready chemical structures. The development of these models adheres to the five principles for QSAR model development as adopted by the Organization for Economic Co-operation and Development (OECD). These principles ensure the creation of scientifically valid and highly accurate models with minimal complexity, allowing for mechanistic interpretation when feasible. The modeling suite, OPERA, is available for download and provides open access to a range of predictive models for assessing the toxicity of various agents, including flame retardants, pesticides, and air pollutants. It offers the capability to identify factors that may contribute to or exacerbate adverse health outcomes. OPERA encompasses predictions for nearly 1 million chemical structures and provides a comprehensive set of resources, including training videos, testing and assessment strategies, computational models, and workflows for analyzing chemical data.

Generative AI models

Generative AI models employ neural networks to recognize patterns and structures present in available data, enabling them to generate fresh and unique content. A notable advancement in generative AI models is their capability to utilize diverse learning methods, such as unsupervised or semi-supervised learning, during the training process. This allows organizations to effectively and swiftly utilize a vast amount of unlabeled data to build foundational models. As the name implies, these foundation models can serve as groundwork for AI systems that are capable of performing multiple tasks. Generative AI has the potential to revolutionize various aspects of the workplace, ranging from employee performance assessment and predictive analysis to the creation of personalized training programs. In today's complex business environment, this technology can act as a vital link that seamlessly integrates different operations, leading to enhanced productivity. The impact of generative AI on HR departments is significant, particularly in terms of redefining their approach to employee performance and productivity evaluation. The fusion of generative AI and management frameworks can be a powerful tool for departments like safety, especially in the context of contractor management. Activities such as evaluating responses, conducting audits, and preparing safety documentation materials (e.g., safety manuals and hazard analyses) align well with the capabilities of AI. By leveraging a vast dataset of previous contractor responses, AI can identify key phrases, patterns, and indicators of safety preparedness, efficiently filtering out those who meet established safety standards and highlighting those who do not. When conducting audits of contractor safety practices, generative AI can analyze a database of past safety incidents, near misses, and relevant data, extracting valuable insights that inform decision-making (an area where AI excels). This analysis can pinpoint areas most likely to present problems, allowing safety departments to focus their audits and enhance overall effectiveness. Moreover, generative AI tools can identify crucial elements and best practices from existing safety documentation, leveraging this information to generate tailored materials that meet the specific needs of the organization. This approach saves time and resources for safety departments while ensuring that the resulting materials are authentic and impactful.

AI for diagnostic and prevention of occupational lung diseases

Among the most prevalent work-related diseases, pneumoconiosis or occupational lung disease is the prime reason for respiratory suffering among workers due to exposure to respirable crystalline silica (RCS). With recent advancements, AI algorithms with DL functionality have been found very useful in lung image processing making a wider impact on the disease diagnosis process (Çallı et al., 2021). AI algorithms can analyze lung images, such as chest X-rays or computed tomography (CT) scans, to detect and diagnose various lung conditions. They can identify abnormalities, nodules, masses, or patterns indicative of diseases like mesothelioma, chronic obstructive pulmonary disease (COPD), and silicosis.

Imaging has a vital major role in the assessment of pulmonary diseases, various AI algorithms have been developed for chest imaging which are recognized by regulatory bodies and are now commercially available in the market (Choe et al., 2022). A chest radiograph is performed in all patients undergoing evaluation for silicosis, which remains an important health issue among silica-exposed workers and causes a potentially significant burden worldwide with a greater burden in low- and middle-income countries (Chen et al., 2022). It is the confirmatory method to assess the presence of nodules in the lungs. Confirmatory diagnosis of silicosis is a complex decision-making and offers a critical challenge to Radiolosgists. AI models can be very effective in analyzing imaging data with high precision and accuracy (Cellina et al., 2022). Advanced AI techniques can help in data augmentation and synthetic data generation, which can generate synthetic lung images that resemble real patient data. This can be valuable for limiting the exposure of the workers to a hazardous work process in dust-exposed industries Fig. 6.6.

Early algorithm-based attempts can be traced back by Kruger et al., (1974) when medical decision-making applied hybrid optical-digital methods involving the optical Fourier transformation for the purpose of screening for diagnosis related to maintenance and compensation for affected individuals. The further investigations were joined by the application of a multiresolution SVM-based algorithm and advanced by the application of CNNs and DL algorithms for chest X-ray (CXR) image analysis (Kruger et al., 1974; Litjens et al., 2017; Sundararajan et al., 2010). The AI-based tools offer huge potential gains from algorithms-based CXR analysis for developing countries where there could be a delay in the availability of radiologists due to the super-speciality nature of the clinical branch and can play a crucial role in assessing the eligibility of dust-exposed workers for compensation claims in Pneumoconiosis Board (Sishodiya, 2022), which may get delayed due to nonconfirmatory diagnosis of various pneumoconiosis conditions including silicosis.

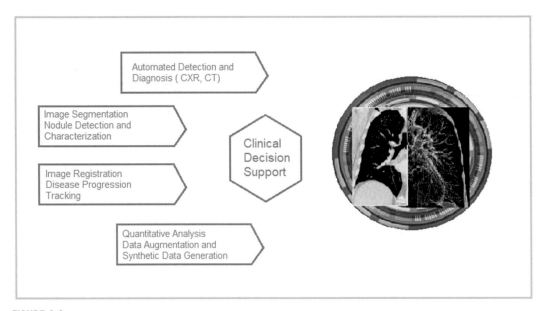

FIGURE 6.6

Working with artificial intelligence (AI) models in detection of occupational lung diseases.

NLP utility for workplace health education and awareness

Workers' education is a paramount aspect to make them understand the social, economic, and workplace environment, their rights to protect their health and compensation at the workplace. Education and training in vernacular or local languages could be beneficial to improve learning and learning outcomes (i.e., safety behavior) (UNESCO, 2023), which can affect the worker in multiple ways including mental, physical, environmental, experiential, technological, and social factors. NLP techniques can be used to automatically translate OHS training materials and resources into different languages, making them accessible to a diverse workforce. Localization of training content ensures that cultural nuances and language variations are taken into account, improving comprehension and engagement. Countries with diverse linguistic background face challenges in the education of workers in vernacular language. The impact of linguistic barriers during training has been widely discussed in a report by Rosetta Stone and Forbes Insights (Stone and insights, 2011). In India, there are 122 officially recognized languages other than English, which are used by only approximately 11% of the population (Webpage, 2011); hence, training workers on health and safety parameters is a challenging task. The use of AI-based voice automation platforms could play a vital role in this area, as the technologies such as Vernacular Intelligent Virtual Assistant (VIVA), a next-gen multilingual voice AI platform, which is designed to have natural conversations (Linares-Garcia et al., 2022) not only accelerate engagement strategy with user but also utilizes speech recognition and natural language understanding (NLU), which is crucial for a training environment used for worker at the time of health and safety trainings. NLP can power chatbots or virtual assistants that interact with employees, answering their questions, providing guidance, and delivering on-demand training support. These conversational interfaces can offer personalized training recommendations, explanations, and step-by-step instructions, facilitating self-paced learning. The introduction of technologies such as virtual and augmented reality (VR/AR) has been reviewed for application to minimize the risk of workplace fatalities and injuries by increasing their educative role in building up risk-preventive knowledge (Li et al., 2018). The integration of VR/AR with voice-based intelligent virtual agents can be very effective for the education of workers to make them aware of workplace hazards and safety measures requirements.

Workers who are engaged in high-risk jobs, such as construction, chemical handling, etc., usually come from lower socioeconomic status, less education, ethnically and linguistically diverse, hard-to-reach populations of workers with a high burden of occupational injury and illness (Anderson et al., 2022). It has been reported that linguistic minority status was associated with longer time loss and higher median medical costs. The challenge of language by workers is compounded when they have to go through a compensation claim process, which is usually not in their vernacular language and leaves them to rely on information provided by others. AI-driven platforms can not only interact with these workers on needed information but also make them empowered with legal aid available. Not all but the least, AI offers a promising future to the linguistically diverse and challenged by multiple risk factors at the workplace compounding the vulnerability to poor understanding of workplace risks. AI technologies can be used to understand and address the need of various occupational groups better inform policy and help formulate strategies to improve workers' health and safety environment.

References

Abbasov, A.M., Shahbazova, S.N., 2014. Model of the applicability of expert system based on neural networks technology and hybrid systems for decision making. Studies in Fuzziness and Soft Computing 317, 3–18. https://doi.org/10.1007/978-3-319-06323-2_1.

Anderson, N.J., Smith, C.K., Foley, M.P., 2022. Work-related injury burden, workers' compensation claim filing, and barriers: results from a statewide survey of janitors. American Journal of Industrial Medicine 65 (3), 173–195. https://doi.org/10.1002/ajim.23319. http://onlinelibrary.wiley.com/journal/10.1002/(ISSN)1097-0274.

Bilgera, C., Yamamoto, A., Sawano, M., Matsukura, H., Ishida, H., 2018. Application of convolutional long short-term memory neural networks to signals collected from a sensor network for autonomous gas source localization in outdoor environments. Sensors 18 (12), 4484. https://doi.org/10.3390/s18124484.

Brahim-Belhouari, S., Bermak, A., Shi, M., Chan, P.C.H., 2005. Fast and Robust gas identification system using an integrated gas sensor technology and Gaussian mixture models. IEEE Sensors Journal 5 (6), 1433–1444. https://doi.org/10.1109/JSEN.2005.858926.

Çallı, Erdi, Sogancioglu, Ecem, Ginneken, Bram van, Leeuwen, Kicky G. van, Murphy, Keelin, 2021. Deep learning for chest X-ray analysis: A survey. Medical Image Analysis 72, 102125. https://doi.org/10.1016/j.media.2021.102125.

Cassani, S., Gramatica, P., 2015. Identification of potential PBT behavior of personal care products by structural approaches. Sustainable Chemistry and Pharmacy 1, 19–27. https://doi.org/10.1016/j.scp.2015.10.002. http://www.journals.elsevier.com/sustainable-chemistry-and-pharmacy.

Cellina, M., Cè, M., Irmici, G., Ascenti, V., Khenkina, N., Toto-Brocchi, M., Martinenghi, C., Papa, S., Carrafiello, G., 2022. Artificial intelligence in lung cancer imaging: unfolding the future. Diagnostics 12 (11). https://doi.org/10.3390/diagnostics12112644. http://www.mdpi.com/journal/diagnostics/.

Chen, S., Liu, M., Xie, F., 2022. Global and national burden and trends of mortality and disability-adjusted life years for silicosis, from 1990 to 2019: results from the global burden of disease study 2019. BMC Pulmonary Medicine 22 (1). https://doi.org/10.1186/s12890-022-02040-9. http://www.biomedcentral.com/bmcpulmmed/.

Choe, J., Lee, S.M., Hwang, H.J., Lee, S.M., Yun, J., Kim, N., Seo, J.B., 2022. Artificial intelligence in lung imaging. Seminars in Respiratory and Critical Care Medicine 43 (6), 946–960. https://doi.org/10.1055/s-0042-1755571. http://www.thieme-connect.com/ejournals/toc/srccm.

Ding, L., Fang, W., Luo, H., Love, P.E.D., Zhong, B., Ouyang, X., 2018. A deep hybrid learning model to detect unsafe behavior: integrating convolution neural networks and long short-term memory. Automation in Construction 86, 118–124. https://doi.org/10.1016/j.autcon.2017.11.002.

Dwivedi, K., Biswaranjan, K., Sethi, A., 2014. Drowsy driver detection using representation learning. In: Souvenir of the 2014 IEEE International Advance Computing Conference, IACC 2014. IEEE Computer Society, India, pp. 995–999. https://doi.org/10.1109/IAdCC.2014.6779459.

Fang, W., Ding, L., Love, P.E.D., Luo, H., Li, H., Peña-Mora, F., Zhong, B., Zhou, C., 2020. Computer vision applications in construction safety assurance. Automation in Construction 110. https://doi.org/10.1016/j.autcon.2019.103013.

Guo, J.M., Markoni, H., 2019. Driver drowsiness detection using hybrid convolutional neural network and long short-term memory. Multimedia Tools and Applications 78 (20), 29059–29087. https://doi.org/10.1007/s11042-018-6378-6.

Khalaf, W.M.H., 2012. Electronic nose system for safety monitoring at refineries. Journal of Engineering and Sustainable Development 16, 220–228.

Krizhevsky, A., Sutskever, I., Hinton, G.E., 2017. ImageNet classification with deep convolutional neural networks. Communications of the ACM 60 (6), 84–90. https://doi.org/10.1145/3065386. http://www.acm.org/pubs/cacm/.

Kruger, R.P., Thompson, W.B., Turner, A.F., 1974. Computer diagnosis of pneumoconiosis. IEEE Transactions on Systems, Man, and Cybernetics SMC-4 (1), 40–49. https://doi.org/10.1109/TSMC.1974.5408519.

Lelieveld, J., Evans, J.S., Fnais, M., Giannadaki, D., Pozzer, A., 2015. The contribution of outdoor air pollution sources to premature mortality on a global scale. Nature 525 (7569), 367–371. https://doi.org/10.1038/nature15371. http://www.nature.com/nature/index.html.

Li, X., Yi, W., Chi, H.L., Wang, X., Chan, A.P.C., 2018. A critical review of virtual and augmented reality (VR/AR) applications in construction safety. Automation in Construction 86, 150–162. https://doi.org/10.1016/j.autcon.2017.11.003.

Linares-Garcia, D.A., Roofigari-Esfahan, N., Pratt, K., Jeon, M., 2022. Voice-based intelligent virtual agents (VIVA) to support construction worker productivity. Automation in Construction 143, 104554. https://doi.org/10.1016/j.autcon.2022.104554.

Litjens, G., Kooi, T., Bejnordi, B.E., Setio, A.A.A., Ciompi, F., Ghafoorian, M., van der Laak, J.A.W.M., van Ginneken, B., Sánchez, C.I., 2017. A survey on deep learning in medical image analysis. Medical Image Analysis 42, 60–88. https://doi.org/10.1016/j.media.2017.07.005.

Liu, Q., Hu, X., Ye, M., Cheng, X., Li, F., 2015. Gas recognition under sensor drift by using deep learning. International Journal of Intelligent Systems 30 (8), 907–922. https://doi.org/10.1002/int.21731.

Luo, X., Li, H., Yang, X., Yu, Y., Cao, D., 2019. Capturing and understanding workers' activities in far-field surveillance videos with deep action recognition and bayesian nonparametric learning. Computer-Aided Civil and Infrastructure Engineering 34 (4), 333–351. https://doi.org/10.1111/mice.12419.

Lyu, J., Yuan, Z., Chen, D., 2018. Long-term multi-granularity deep framework for driver drowsiness detection. arXiv. https://arxiv.org.

Oliveira, H., Correia, P.L., 2008. Supervised strategies for cracks detection in images of road pavement flexible surfaces. In: European Signal Processing Conference. Portugal.

Pan, X., Zhang, H., Ye, W., Bermak, A., Zhao, X., 2019. A Fast and Robust gas recognition algorithm based on hybrid convolutional and recurrent neural network. IEEE Access 7, 100954–100963. https://doi.org/10.1109/ACCESS.2019.2930804.

Peng, P., Zhao, X., Pan, X., Ye, W., 2018. Gas classification using deep convolutional neural networks. Sensors 18 (2), 157. https://doi.org/10.3390/s18010157.

Prasanna, P., Dana, K., Gucunski, N., Basily, B., 2012. Computer vision based crack detection and analysis. Proceedings of SPIE - The International Society for Optical Engineering 8345. https://doi.org/10.1117/12.915384.

Qian, J., Feng, S., Tao, T., Hu, Y., Li, Y., Chen, Q., Zuo, C., 2020. Deep-learning-enabled geometric constraints and phase unwrapping for single-shot absolute 3D shape measurement. APL Photonics 5 (4). https://doi.org/10.1063/5.0003217.

Qiu, T., Shi, X., Wang, J., Li, Y., Qu, S., Cheng, Q., Cui, T., Sui, S., 2019. Deep learning: a rapid and efficient route to automatic metasurface design. Advanced Science 6 (12), 1900128. https://doi.org/10.1002/advs.201900128.

Roy, K., Kar, S., Das, R.N., 2015. Understanding the Basics of QSAR for Applications in Pharmaceutical Sciences and Risk Assessment. Elsevier Inc, India, pp. 1–479. https://doi.org/10.1016/C2014-0-00286-9. http://www.sciencedirect.com/science/book/9780128015056.

Russell, S., Norvig, P., 2003. Artificial Intelligence: A Modern Approach. Pearson.

Shi, S.-Y., Tang, W.-Z., Wang, Y.-Y., Long, L., Li, Y., Li, X., Dai, Y., Yang, H., 2017. A review on fatigue driving detection. ITM Web of Conferences 12, 01019. https://doi.org/10.1051/itmconf/20171201019.

Sishodiya, P., 2022. Silicosis-An ancient disease: providing succour to silicosis victims, lessons from Rajasthan model. Indian Journal of Occupational and Environmental Medicine 26 (2), 57–61. https://doi.org/10.4103/ijoem.ijoem_160_22.

Sousa, V., Almeida, N.M., Dias, L.A., 2014. Risk-based management of occupational safety and health in the construction industry - Part 1: background knowledge. Safety Science 66, 75–86. https://doi.org/10.1016/j.ssci.2014.02.008.

Stone, R., insights, F., 2011. Reducing the Impact of Language Barriers.

Sundararajan, R., Xu, H., Annangi, P., Tao, X., Sun, X.W., Mao, L., 2010. A multiresolution support vector machine based algorithm for pneumoconiosis detection from chest radiographs. In: 2010 7th IEEE International Symposium on Biomedical Imaging: From Nano to Macro, ISBI 2010 - Proceedings, pp. 1317–1320. https://doi.org/10.1109/ISBI.2010.5490239. India.

Tam, V.W.Y., Fung, I.W.H., 2011. Tower crane safety in the construction industry: a Hong Kong study. Safety Science 49 (2), 208–215. https://doi.org/10.1016/j.ssci.2010.08.001.

Trivedi, P., Purohit, D., Soju, A., Tiwari, R.R., ENVIS-NIOH, 2014. Major industrial disasters in India an official newsletter of ENVIS-NIOH. https://niohenvis.nic.in/newsletters/vol9_no4_Indian%20Industrial%20Disasters.pdf. (Accessed 19 June 2023).

UNESCO, 2023. What You Need to Know about Languages in Education. https://www.unesco.org/en/languages-education/need-know. (Accessed 20 June 2023).

Wang, M., Wong, P., Luo, H., Kumar, S., Delhi, V., Cheng, J., 2019. Predicting safety hazards among construction workers and equipment using computer vision and deep learning techniques. In: Proceedings of the 36th International Symposium on Automation and Robotics in Construction, ISARC 2019. International Association for Automation and Robotics in Construction I.A.A.R.C), Hong Kong, pp. 399–406. https://doi.org/10.22260/isarc2019/0054.

Webpage, 2011. Multilingualism in India: First, Second, and Third Languages by Number of Speakers in India.

Yin, X., Zhang, L., Tian, F., Zhang, D., 2016. Temperature modulated gas sensing E-nose system for low-cost and fast detection. IEEE Sensors Journal 16 (2), 464–474. https://doi.org/10.1109/JSEN.2015.2483901.

Zhang, W., Murphey, Y.L., Wang, T., Xu, Q., 2015. Driver yawning detection based on deep convolutional neural learning and robust nose tracking. In: Proceedings of the International Joint Conference on Neural Networks. Institute of Electrical and Electronics Engineers Inc., China https://doi.org/10.1109/IJCNN.2015.7280566.

Zhang, F., Su, J., Geng, L., Xiao, Z., 2017. Driver fatigue detection based on eye state recognition. In: Proceedings - 2017 International Conference on Machine Vision and Information Technology, CMVIT 2017. Institute of Electrical and Electronics Engineers Inc., China, pp. 105–110. https://doi.org/10.1109/CMVIT.2017.25.

Zhou, Y., Zhao, X., Zhao, J., Chen, D., 2016. Research on fire and explosion accidents of oil depots. Chemical Engineering Transactions 51, 163–168. https://doi.org/10.3303/CET1651028.

From data to insights: Leveraging machine learning for diabetes management

Asra Khanam[1], Faheem Syeed Masoodi[1,a] and Alwi Bamhdi[2]

[1]*University of Kashmir, Srinagar, India;* [2]*Department of Computer Science, Umm AlQura University, Mecca, Saudi Arabia*

Introduction

Metabolic disorders are of significant importance due to their widespread impact on individuals' health and well-being. They can affect various aspects of an individual's life, including physical health, mental health, and overall quality of life. These disorders often result in abnormal metabolic processes, leading to imbalances in essential substances, such as glucose, lipids, and amino acids. If left untreated or unmanaged, metabolic disorders can lead to a range of serious complications, including cardio-vascular disease, organ damage, developmental delays, cognitive impairment, and even premature death. Moreover, metabolic disorders are often chronic conditions that require long-term management, including medication, dietary modifications, and lifestyle changes. By raising awareness, promoting early detection, and implementing appropriate interventions, healthcare professionals can help individuals with metabolic disorders achieve better outcomes, prevent complications, and improve their overall health and quality of life. By assisting with diagnosis, individualized therapy planning, and patient monitoring, artificial intelligence (AI) can significantly contribute to the treatment of metabolic disorders. The following are some applications of AI in the treatment of metabolic disorders.

1. **Diagnosis and early detection:** The application of AI algorithms in diagnosis and early detection presents a remarkable opportunity to extract valuable insights from extensive patient data. By analyzing diverse sources such as medical records, laboratory test results, genetic information, and imaging studies, these algorithms can uncover intricate patterns and indicators that might signify the presence of metabolic disorders. This sophisticated analysis goes beyond human capabilities, enabling healthcare professionals to identify potential risk factors and symptoms at an earlier stage. Consequently, prompt interventions can be initiated, offering patients a higher likelihood of successful treatment outcomes and improved overall health. The integration of AI into the diagnostic process has the potential to revolutionize healthcare by providing comprehensive and proactive care that addresses metabolic disorders before they manifest into more severe conditions (Srikanthan et al., 2016).

[a]The corresponding author would like to acknowledge NIDA ul Islam for their great support with peer review.

2. **Personalized treatment planning:** AI plays a pivotal role in the realm of personalized treatment planning by leveraging patient-specific data to devise tailored and optimized intervention strategies. Through comprehensive analysis, AI algorithms can predict individual responses to various treatments, enabling healthcare providers to develop personalized treatment plans. These plans encompass not only medication dosages but also recommendations for dietary modifications and customized exercise regimens that align with the specific needs and characteristics of each patient. By harnessing the power of AI, healthcare professionals can account for the intricate interplay between patient factors, such as genetics, lifestyle, and medical history, to devise more effective and precise treatment plans. AI algorithms can analyze vast amounts of data from diverse sources, including electronic health records (EHRs), genomic information, wearable devices, and real-time patient monitoring systems. This holistic approach enables healthcare providers to gain a deeper understanding of each patient's unique profile and make informed decisions about the most suitable interventions. Furthermore, AI algorithms continually learn and adapt based on new data and insights, allowing treatment plans to evolve and be refined over time. This dynamic nature of AI-driven personalized treatment planning ensures that patients receive the most up-to-date and relevant care, maximizing the potential for positive outcomes and minimizing the risk of adverse effects. Ultimately, the integration of AI in personalized treatment planning holds tremendous promise in transforming healthcare by providing highly individualized, evidence-based, and precise interventions that address the specific needs of each patient, leading to improved treatment efficacy and patient satisfaction (Ahamed and Farid, 2019).

3. **Drug discovery and development:** AI plays a pivotal role in expediting the complex and time-consuming process of drug discovery and development, particularly in the context of metabolic disorders. By harnessing the power of AI, scientists can analyze immense volumes of scientific literature, molecular databases, and clinical trial data to identify promising targets and compounds for further investigation. Traditionally, drug discovery involves an arduous and costly process of trial and error. However, AI algorithms can rapidly sift through vast amounts of information and identify patterns, relationships, and potential therapeutic candidates. These algorithms can detect subtle molecular interactions, predict drug–target interactions, and assess the safety and efficacy of potential compounds with high accuracy. The utilization of AI in drug discovery not only expedites the identification of potential targets and compounds but also enhances the precision and efficiency of the entire process. AI algorithms can assist in the design of new drug molecules by generating and evaluating numerous chemical structures, optimizing their properties, and predicting their biological activities. Moreover, AI enables researchers to explore novel therapeutic avenues and repurpose existing drugs for metabolic disorders. By uncovering hidden connections and repurposing existing compounds, AI algorithms facilitate the identification of new treatment options with potentially improved effectiveness and reduced development time. The integration of AI in drug discovery and development has the potential to revolutionize the field, leading to faster identification of therapeutic targets, more efficient lead optimization, and improved success rates in clinical trials. Ultimately, AI-driven drug discovery holds the promise of delivering safer, more effective, and personalized medications for individuals affected by metabolic disorders, addressing their specific needs and improving their quality of life (Paul et al., 2021).

4. **Continuous monitoring and predictive analytics:** The integration of AI-powered wearable devices and sensors enables continuous monitoring of a multitude of physiological parameters, including blood glucose levels, heart rate, and activity patterns. This real-time data are invaluable

in detecting anomalies, predicting disease progression, and proactively alerting healthcare providers to potential complications. By leveraging AI algorithms, the collected data from wearable devices can be analyzed in a comprehensive and dynamic manner. These algorithms possess the capability to detect patterns and trends that might indicate deviations from normal physiological states. For instance, in the context of metabolic disorders, AI can identify fluctuations in blood glucose levels that may signal potential complications, such as hyperglycemia or hypoglycemia. Furthermore, AI algorithms can employ predictive analytics to anticipate disease progression based on continuous monitoring data. By discerning subtle changes and correlating them with historical data from similar cases, AI can forecast the likelihood of future complications or deteriorations in a patient's condition. This early warning system allows healthcare providers to intervene promptly, potentially preventing adverse events or enabling timely adjustments to treatment plans. The combination of continuous monitoring and predictive analytics empowers both patients and healthcare providers. Patients benefit from personalized insights into their health status, allowing them to make informed decisions regarding lifestyle modifications or medication adherence. Healthcare providers, on the other hand, gain access to real-time data that assist in identifying high-risk patients, allocating resources efficiently, and providing proactive and personalized care. The implementation of AI-powered continuous monitoring and predictive analytics has the potential to revolutionize healthcare by shifting from a reactive model to a proactive one. Early detection of anomalies and timely interventions can improve patient outcomes, enhance disease management, and reduce healthcare costs associated with complications and hospitalizations. In summary, the integration of AI with wearable devices and sensors enables continuous monitoring of physiological parameters, leveraging real-time data to detect anomalies, predict disease progression, and provide proactive alerts. This transformative approach enhances patient care, empowers individuals to take control of their health, and supports healthcare providers in delivering timely and personalized interventions (Bhatt and Chakraborty, 2021).

5. **Treatment optimization and decision support:** AI plays a crucial role in optimizing treatment plans and providing decision support to healthcare providers by analyzing patient responses to therapies over time. Through the integration of real-time patient data, AI algorithms offer valuable tools that assist healthcare professionals in making informed choices regarding treatment adjustments and interventions. By continuously analyzing patient data, including symptoms, biomarkers, medication adherence, and treatment outcomes, AI algorithms can identify patterns and trends that may not be readily apparent to human observers. This deep analysis allows AI to uncover potential correlations and relationships between various factors and treatment efficacy. As a result, healthcare providers can gain insights into which treatments are most effective for specific patients, helping to personalize and optimize their care. Furthermore, AI algorithms can offer decision support by providing evidence-based recommendations for treatment adjustments. By considering the patient's unique characteristics, medical history, and current health status, AI can suggest potential modifications to medication dosages, changes in treatment protocols, or alternative interventions. These recommendations are based on the analysis of large datasets, clinical guidelines, and the latest research findings, offering healthcare providers valuable insights to guide their decision-making process. The integration of AI in treatment optimization and decision support enhances the quality of patient care by considering a multitude of factors and providing a more holistic perspective. It allows healthcare providers to

leverage real-time patient data, which may include information from wearable devices, EHRs, and patient-reported outcomes, to make well-informed decisions about treatment adjustments and interventions. By utilizing AI-powered decision support tools, healthcare providers can optimize treatment plans, improve patient outcomes, and ensure that care aligns with the most up-to-date evidence-based practices. In summary, AI-driven treatment optimization and decision support enable healthcare providers to leverage real-time patient data and advanced analytical capabilities. This integration empowers them to make informed choices regarding treatment adjustments and interventions, ultimately improving the effectiveness and precision of patient care. AI serves as a valuable ally in the decision-making process, enabling personalized and evidence-based treatments that can lead to better outcomes and enhanced patient satisfaction (Giordano et al., 2021).

Nowadays, the rising metabolic disorder worldwide is diabetes according to the article published in Nature on March 07, 2023 "Diabetes and obesity are rising globally, but some nations are hit harder." Diabetes manifests as a complex metabolic disorder characterized by intricate disruptions within the body's fundamental metabolic processes, primarily involving the regulation of glucose, production of insulin, and utilization of insulin. It encompasses two primary classifications: type 1 diabetes, an autoimmune condition wherein the immune system launches an assault on the insulin-producing cells, leading to inadequate insulin synthesis, and type 2 diabetes, which encompasses insulin resistance and diminished efficacy of insulin. These disturbances culminate in heightened levels of glucose in the bloodstream, triggering the detrimental consequences of hyperglycemia on diverse organs and systems. The holistic consideration of diabetes as a metabolic disorder stems from its profound impact on the intricate interplay of metabolic pathways crucial for maintaining the delicate equilibrium of glucose homeostasis.

Overview of diabetes and its management challenges

Muscles and tissues of the body are made up of cells, and glucose is the essential energy source for both as it is a ubiquitous energy source for cells. Glucose is also a primary source of fuel for the brain, and the ability of the body to process glucose is essential for survival. Diabetes is a sickness that diminishes the capability of the body to process glucose in the blood and, in turn, results in the increase of sugar in the blood, which increases the complications of the body system, including heart diseases, stroke, eye problems, nerve damage (Tu, 2019). 5,37,000,000 (537 million) people suffer from diabetes globally as per International Diabetes Federation (IDF) 2021 report. Ninety million people in Southeast Asia, of these 90 million people, 74.2 million people are from India. Some of the worldwide figures from Diabetes Atlas (2021) are as about 537 million persons between the ages of 20 and 79 now have diabetes. The entire population is anticipated to increase, reaching 643 million (11.3%) by 2030 and 783 million (12.2%) by 2045. More than half of diabetics in Africa, South-East Asia, and the Western Pacific remain undiagnosed. Type 1 diabetes affects more than 1.2 million children and adolescents. 54% of the population is under the age of 15. China, India, and Pakistan are expected to continue to have the highest rates of adults with diabetes aged 20–79 in 2021 and 2045, respectively. In India, there are now 74.2 million diabetics, and the IDF predicts that number will rise to 124.9 million by 2045 (International Diabetes Federation, 2019).

Various forms of diabetes can occur, and how to manage it depends on the type.

Type 1 (juvenile diabetes)

This diabetes is most prevalent in adolescent age, where the body cannot produce insulin. Here, the beta cells of the pancreas responsible for producing insulin are falsely attacked by its immune system. Since the body produces no insulin, people have to depend on artificial insulin. The Eli Lilly Company made the first synthetic insulin. Why the immune system attacks its cells is not known. Symptoms for type 1 are excessive thirst, vomiting, an upset stomach, fatigue, vision is blurry, frequent infections of the skin, urinary tract, or vaginal area, irritability or mood swings, hunger is increased even if a person has eaten, unexplained loss of weight, a person urinates more frequently, difficulty in breathing (Foster et al., 2019). Type 1 diabetes can be a genetic problem or due to some environmental causes (Arneth et al., 2019).

Type 2 (diabetes mellitus)

It is a situation where the body's insulin is not utilized efficiently. Since the body's cells do not get enough glucose, the pituitary gland triggers the beta cells for more insulin production, resulting in cells being unable to cope with this much insulin production demand. Thus, insulin production will eventually decrease, leading to a greater level of sugar in the blood. The symptoms of type 2 are excessive thirst, exhaustion, injuries that do not heal, and loss of weight (Inzucchi et al., 2015). People with this type of diabetes are more open to infections, have numbness in the hands and feet, vision is also blurry, and feel hungry; urination is frequent. The contributing factors are genetics, lack of exercise, being overweight, and environmental resources (Arneth et al., 2019).

Gestational diabetes

This diabetes is not permanent or lifetime diabetes. It is common in pregnant women. Women show a little insensitivity to insulin during pregnancy, due to which sugar levels in the blood rise. No notable indications or symptoms are there for women with gestational diabetes; the only thing is that woman feels more thirsty, which eventually leads to excessive urination. This diabetes disappears after the woman gives birth (Arneth et al., 2019).

Diabetes management is difficult for patients, medical professionals, and society at large. Several of these difficulties include.

1. **Blood sugar monitoring:** To maintain the best degree of control, diabetics must routinely check their blood sugar levels. This often entails utilizing continuous glucose monitoring (CGM) devices or pricking the finger to collect a blood sample. Frequent inspections may be time-consuming and perhaps painful.
2. **Medication and insulin management:** Medication or insulin therapy is frequently necessary for the management of diabetes. It can be difficult to find the ideal medicine mix, modify dosages, and time medication administration to coincide with food planning and physical activity. Since basal and bolus insulin dosages must be precisely balanced, carbohydrate consumption must be tracked, and insulin must be adjusted based on blood sugar levels, managing insulin is extremely complex.
3. **Diet and lifestyle changes:** A good diet and consistent exercise are essential for managing diabetes. However, some people may find it challenging to stick to their exercise routines, maintain a healthy weight, and make long-term dietary modifications. It might be difficult to overcome obstacles including limited dietary options, cultural preferences, and lifestyle restrictions.

4. **Hypoglycemia and hyperglycemia:** Keeping blood sugar levels in check to prevent hypoglycemia (low blood sugar) and hyperglycemia (high blood sugar) is a never-ending task. Hypoglycemia can be fatal; thus, it has to be treated right away using a glucose supply. If hyperglycemia is not treated effectively, long-term consequences may result. It might be challenging to achieve and keep tight glycemic control without having to deal with significant blood sugar swings.

5. **Impact on emotions and mental health:** Having diabetes can have a serious emotional and mental health impact. Diabetes-related sadness, anxiety, depression, and reduced quality of life might result from the need for ongoing self-care, the worry about complications, and the stress associated with blood sugar management. The management of diabetes must include access to mental health services and emotional support.

6. **Long-term complications:** Poorly treated diabetes can result in a number of long-term problems that have an impact on the organ systems. Diabetes has a number of side effects, some of which are cardiovascular disease, diabetic foot ulcers, diabetic retinopathy (DR), diabetic neuropathy, and damage to the kidneys and nerves. Continuous medical attention, routine tests, adherence to medication schedules, and lifestyle modifications are necessary for preventing and managing these problems.

7. **Access to care and resources:** Diabetes management can be difficult in locations where access to healthcare services and resources is restricted. Some people can find it difficult to pay for diabetes education programs, supplies, or prescriptions. Diabetes management results can also be impacted by health disparities and inequality.

Healthcare providers, including doctors, nurses, dietitians, and diabetes educators, play a crucial role in providing education, support, and individualized care plans. Addressing the issues of diabetes management needs a thorough and multidisciplinary approach. The availability and affordability of technological innovations like insulin pumps and CGM devices have revolutionized diabetes care, yet these can also be limiting factors (Masoodi et al., 2023). Public health programs that encourage active living, increase knowledge, and enhance access to diabetes care are also crucial. The key to effective diabetes control is personal empowerment, knowledge, self-monitoring, and self-care. Diabetes patients must take an active role in their care, alter their lifestyles, follow treatment programs, and ask for assistance when necessary. People can enhance their quality of life, lower the risk of complications, and lead more rewarding lives by appropriately managing their diabetes.

Role of machine learning in diabetes management

With its many applications in the medical field (Teli et al., 2023), applications for prediction (Hussain Bukhari et al., 2023), diagnosis, medication optimization, and individualized care, machine learning (ML) plays a crucial role in the management of diabetes. Here is a description of several important functions of ML in the management of diabetes.

1. **Risk prediction and early detection:** ML models can analyze clinical, genetic, and lifestyle data to predict a person's risk of developing diabetes or its complications. These models can help identify high-risk people who may benefit from preventive interventions and targeted screening (Esteva et al., 2019).

2. **Glucose forecasting:** ML algorithms can anticipate future glucose levels by using past glucose data, insulin dose, meal information, physical activity, and other pertinent parameters. This makes it possible for people to actively manage their blood sugar levels and to choose their food, exercise, and insulin dosage after doing their research (Waring et al., 2020).

3. **Systems for making decisions:** ML methods may be used to create systems for making decisions that offer individualized suggestions for the treatment of diabetes. To provide individualized treatment plans, insulin dose recommendations, and lifestyle changes, these systems analyze patient data, including glucose measurements, medication use, and clinical factors (Kovatchev et al., 2020).

4. **Automated insulin delivery:** Automated insulin delivery is possible using closed-loop systems, which are made possible by ML and control algorithms. These devices replicate the functioning of a healthy pancreas and maintain constant glucose levels by controlling insulin infusion using real-time glucose monitoring and ML algorithms (Thompson et al., 2018).

5. **Personalized medicine:** By combining clinical data, genetic information, and patient preferences, ML enables the generation of personalized treatment plans. ML models can help in customizing treatment regimens, choosing the best drugs, and predicting reactions to treatments by taking into account individual characteristics (Kleinberger and Pollin, 2015).

6. **Population health management:** By identifying patterns and trends in massive amounts of diabetic data, ML may assist healthcare professionals and public health organizations learn more about the patterns of illness, risk factors, and treatment results. Decisions on resource allocation, policy, and interventions at the population level can be supported by this knowledge (Luo et al., 2018).

Understanding data collection and preprocessing of diabetes-related data

Collection and preprocessing of diabetes-related data

The process of acquiring pertinent data and getting it ready for analysis is known as collecting and preprocessing diabetes-related data.

Data collection

1. **Clinical data:** Clinical information may be gathered via clinical trials, EHRs (EHRs), or healthcare institutions. It comprises the patient's demographics, medical history, findings from blood tests, prescription history, and other pertinent clinical factors.

2. **Self-reported data:** Data on lifestyle elements, dietary practices, physical activity, and quality of life can be gathered by patient surveys or questionnaires.

3. **Continuous glucose monitoring (CGM):** CGM equipment offers in-the-moment glucose readings. The analysis of glucose patterns, trends, and variability across time may be done using data from CGM devices.

4. **Hereditary information:** Genetic information, such as DNA samples, may be gathered to investigate the hereditary basis of diabetes or to develop personalized medical treatments. Whole-genome sequencing or genotyping may be used for this.

5. **Mobile apps and wearables:** Data from wearables and mobile health applications may be used to track a variety of elements of managing diabetes, such as blood glucose levels, insulin dosages, physical activity levels, and sleep patterns.

Data preprocessing

1. **Data cleaning:** Errors, inconsistencies, and missing values are removed from the dataset during data cleaning. Missing data can be handled using methods like estimate, deletion, or imputation.
2. **Data transformation:** Data transformation may be necessary to guarantee that the variables satisfy the analysis's underlying presumptions. Transformations that are often used include square root, logarithmic, and z-score transformations.
3. **Feature engineering:** The process of feature engineering includes the creation of fresh variables or the extraction of useful features from the raw data. For instance, determining indices of glycemic variability or extracting characteristics from time series data (such as glucose trends or daily patterns).
4. **Data integration:** Data integration is necessary to ensure compatibility and consistency between various sources when using several datasets.
5. **Data scaling and normalization:** When utilizing ML methods that are sensitive to such variations, it may be necessary to scale or normalize variables such that they have comparable ranges and distributions.

The flow of preprocessing of data is shown below in Fig. 7.1.

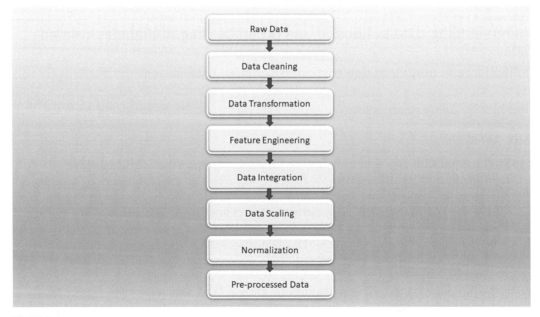

FIGURE 7.1

Flow diagram of data preprocessing.

Standard datasets for diabetes research

For diabetes research, several standard datasets have been extensively utilized in several studies.

1. **Pima Indians diabetes database:** One of the most frequently utilized datasets in diabetes research is the Pima Indians Diabetes Database. Data from research on Pima Indian women in Arizona, USA, are included in the Pima Indians Diabetes Dataset. It includes information on 768 female Pima Indians who are 21 years of age and older, including details about their age, number of pregnancies, blood glucose levels, blood pressure, body mass index (BMI), and the presence or absence of diabetes. When attempting to forecast the onset of diabetes, this dataset is frequently utilized for binary classification tasks (Smith et al., 1988).

2. **NHANES (National Health and Nutrition Examination Survey):** The U.S. Centers for Disease Control and Prevention (CDC) conducts the NHANES survey to gauge the population's nutritional and physical health. Each year, a sample of around 5000 nationally representative people is examined for the survey. 15 of the country's counties, where these people are situated, receive annual visitors. It contains a plethora of data, including information on variables connected to diabetes, including blood sugar levels, insulin levels, glycated hemoglobin (HbA1c), and other health indicators. To examine diabetes prevalence, risk factors, and trends, researchers can acquire data from several NHANES waves.

3. **The Swedish National Diabetes Register (NDR):** It is a comprehensive and long-term database that contains clinical information from people with diabetes in Sweden. It contains data on demographics, test findings, drugs, side effects, and previous treatments (Hallgren Elfgren et al., 2016). Studying diabetes outcomes, treatment efficacy, and healthcare quality have all benefited from the NDR (Celis-Morales et al., 2022).

4. **Electronic Health Records (EHR) datasets:** EHR datasets are a valuable source of longitudinal clinical data for diabetes research. They are gathered from electronic medical record systems. The patient demographics, diagnoses, prescriptions, test findings, and treatment histories are all included in these datasets. Deidentified EHR data are frequently used by researchers to study several facets of managing diabetes, including risk factors, treatment results, and the effects of interventions (Cowie et al., 2017).

5. **Dataset for diabetes 130-US hospital:** The diabetes 130-US hospitals dataset is made up of patient records pertaining to the treatment of diabetes and is obtained from hospital data. This dataset is frequently used to forecast readmissions to hospitals and examine variables related to diabetes care. It is a hospital admission or an inpatient encounter. The dataset represents clinical treatment provided over 10 years (1999–2008) at 130 US hospitals and integrated delivery networks. More than 50 features that represent patient and hospital outcomes are available. For interactions that met the following requirements, data were taken out of the database. It is a diabetic encounter, that is, one in which a diagnosis of diabetes of any kind was made. The visit was at least 1 day long and no longer than 14 days. Laboratory tests were conducted while the meeting was taking place. Medicines were given during the interaction. The data include characteristics like the patient number, race, gender, and age; admission type, length of stay in the hospital; admitting physician's medical speciality; the number of lab tests conducted; the results of the HbA1c test; the diagnosis; the number of medications; the number of diabetic medications; and the number of outpatients, inpatient, and emergency visits in the year before the hospitalization (Strack et al., 2014).

6. **The United Kingdom Prospective Diabetes Study (UKPDS) dataset**: It is a collection of longitudinal information from a clinical experiment that examined how different therapies for type 2 diabetes affected patients. It contains data on demographics, medical evaluations, medication use, and diabetic complications. This dataset is frequently used to examine the course of diabetes, the effectiveness of therapy, and risk factor analysis (effect of intensive blood-glucose control with metformin on complications in overweight patients with type 2 diabetes (UK Prospective Diabetes Study (UKPDS) Group, 1998).

7. **The T1D Exchange Clinic Registry**: It is a large-scale, multicenter registry that gathers thorough information on people with type 1 diabetes. It consists of clinical data, self-reported data, therapy information, and lab test results. This dataset is useful for type 1 diabetes research, including analyses of clinical outcomes, disease management, and personalized treatment (Russel and Norvig, 2012).

Machine learning models for diabetes risk prediction
Logistic regression model

Logistic regression is used whenever a binary dependent variable is present and the relationship between the dependent variable and the independent variable has to be established. Logistic regression uses a sigmoid function, given as:

$$y = \frac{1}{1 + e^{-x}} \tag{7.1}$$

x is the independent variable, e is Euler's constant with a value of 2.718, and y is the output. The sigmoid function converts the independent variable into the probability expression between 0 and 1 concerning the dependent variable. Here, all the predictions are converted into probability ranging between 0 and 1. So, we say the mapping of absolute values here is done in the interval of 0 and 1. Here, we have a decision boundary above which we say it is one class, and below is another class (Russel and Norvig, 2012).

In diabetes risk prediction, based on the features that are provided as input, logistic regression calculates the likelihood that a person belongs to a specific class. Using clinical indicators, lifestyle factors, and genetic data, logistic regression can estimate the likelihood that a person would develop diabetes.

Decision tree model

This supervised learning algorithm uses tree design to solve a problem. It follows the hierarchy of if/else questions to arrive at a particular decision. In a tree structure, intermediate nodes correspond to attributes (questions), and leaf nodes to class labels (answers). The decision tree gets us to the actual answer very quickly. Here if/else questions are called tests. Usually, data are not in the form of binary yes/no features but is represented as continuous features, and the tests run on these features can be "is the feature I greater than J." The algorithm examines all possible tests and finds the one that is closely related to the target variable (Ethem, 2020). To make a decision tree, let us consider three or more features say f1, f2, and f3. Based on these features, we do split and get an output as yes/no. The important thing here is the selection of the feature to split. The more appropriate feature selected for splitting, the more quickly algorithm can reach the leaf node (goal state). For this feature selection, concept of entropy is used.

Entropy

Entropy measures the purity of the split. We can also state as the aim of the decision tree is to get the leaf node quickly, we have to select the right parameters (features) for this, and we calculate the purity split of each node every time, and for this, we use entropy. The node with less entropy is used for further splitting. Entropy is given as

$$H(s) = -(P_+)\log_2(P_+) - (P_-)\log_2(P_-) \qquad (7.2)$$

$P+ =$ Number of positive classes $P- =$ Number of negative classes

If we have a complete impure sub-split (say 3 yes/3 no) for a particular node, then entropy $= 0$. If we have a pure sub-split (say 3 yes/0 no) for a particular node, then entropy $= 1$. Entropy is calculated for each node until we reach the leaf or goal node. This is known as information gain. So, information gain calculates the total entropy gain/value from the top to the leaf node (Kingsford and Salzberg, 2008).

Information gain: (insert proper source)

It calculates an average of the entropy based on a specific split. Information gain is given as

$$\text{Gain}(S, A) = H(s) \sum \frac{|s_v|}{|s|} H(s_v) \qquad (7.3)$$

Gini impurity

It is the same as entropy, but Gini impurity is used more because it is computationally efficient and takes a shorter period of execution time. It is given as

$$GI = 1 - \sum (P^2) \qquad (7.4)$$

$$GI = 1 - \left[(P_+)^2 + (P_-)^2 \right] \qquad (7.5)$$

Decision trees use a tree-like structure to generate choices based on input features in diabetes risk prediction as shown below in Fig. 7.2. Each leaf node represents a class label (such as higher risk of diabetes or lower risk of diabetes), while each internal node represents a characteristic. Decision trees are useful for capturing intricate relationships and providing rules that are easy to understand.

Random forest

Here, many decision trees are built, and each tree does a particular job of predicting the target. The training set is randomly divided into subsets, and decision trees are made from these subsets, and finally, votes from different decision trees are gathered to decide the final prediction. Since the main drawback of decision trees was overfitting the data so, random forests are one way of reducing this problem. In random forest, many decision trees are made that overfit data differently and give the results. Taking the average results from each tree will reduce the over-fitting while retaining the predictive power (Russel and Norvig, 2012).

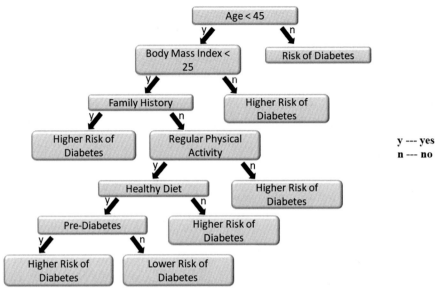

FIGURE 7.2

Decision tree for diabetes risk prediction considering a few features.

A random forest model integrates diabetes risk projections from individual decision trees to provide a more reliable and precise forecast. High-dimensional data can be handled, feature interactions can be recorded, and overfitting can be reduced.

Support vector machine

Support vector machine (SVM) is a powerful supervised ML algorithm extensively employed for classification and regression tasks. Its inherent strength lies in its exceptional performance in binary classification problems, but it can also be extended to handle multiclass classification scenarios.

At its core, SVM strives to discover an optimal hyperplane within a high-dimensional feature space that effectively segregates data points across different classes. This hyperplane serves as a decisive boundary, demarcating data points into their respective classes. The key objective of SVM is to maximize the margin between the hyperplane and the nearest data points for each class, thereby enhancing the model's ability to generalize and make accurate predictions.

The term "support vector machines" originates from the pivotal role played by support vectors, which are the data points in proximity to the hyperplane. These support vectors significantly influence the positioning and orientation of the hyperplane, ultimately shaping its ability to effectively separate the classes.

Thus, SVM harnesses the power of an optimal hyperplane to discern intricate patterns within high-dimensional data, enabling precise classification and regression. By maximizing the margin and leveraging the support vectors, SVM unleashes its potential as a robust and versatile ML algorithm for various complex classification tasks (Russel and Norvig, 2012).

SVM can be used to categorize people into diabetes or nondiabetic groups based on input characteristics in diabetes risk prediction. SVM seeks to maximize the distance between each class's closest data points and the hyperplane. It is efficient when the data cannot be separated linearly and can handle high-dimensional data by converting it into a higher-dimensional feature space using a kernel function (Barakat et al., 2010).

K-means clustering

K-means clustering stands out as a highly prevalent unsupervised ML algorithm renowned for its ability to organize data points into clusters by exploiting their inherent similarities. The principal objective of this algorithm revolves around the minimization of within-cluster variance, which corresponds to the sum of squared distances existing between individual data points and their corresponding cluster centroids. By means of iteratively assigning data points to the nearest centroids and subsequently updating the centroids based on the computed means, K-means clustering refines the clustering arrangement until either convergence is achieved—wherein the centroids cease to undergo significant changes—or a predetermined maximum number of iterations is reached. Consequently, K-means clustering produces a set of K clusters, where each cluster is represented by its centroid, furnishing valuable insights for subsequent analysis and classification endeavors. It is imperative to acknowledge that the algorithm's outcomes can be heavily influenced by the initial placement of centroids, thereby warranting the execution of multiple runs with diverse initializations to ascertain the clustering result that exhibits the lowest within-cluster variance. With its broad array of applications encompassing customer segmentation, image compression, document categorization, anomaly detection, and beyond, K-means clustering exemplifies an indispensable tool in unsupervised learning; however, it is also crucial to acknowledge its limitations, such as its sensitivity to the number of clusters and its underlying assumption of spherical-shaped clusters with akin variances. Overall, K-means clustering epitomizes a highly acclaimed algorithm that seamlessly and effectively enables the grouping of data points into clusters based on their shared characteristics, thus unlocking a multitude of possibilities for insightful data analysis (Russel and Norvig, 2012).

K-means clustering, although not designed specifically for diabetes detection, serves as a versatile clustering technique utilized to amalgamate-related data points based on their shared features. However, when employed as part of a comprehensive analysis pipeline, it can provide valuable insights into patterns within a dataset that pertain to diabetes (Hassan et al., 2022).

In the context of diabetes detection, the application of k-means clustering unfolds as follows.

1. **Dataset preparation:** Assemble a comprehensive dataset containing pertinent features associated with diabetes, encompassing parameters such as glucose level, BMI, blood pressure, age, and others.
2. **Feature selection:** Discern a subset of features from the dataset deemed informative for diabetes detection, considering domain expertise, and consulting medical professionals to identify the most relevant attributes.
3. **Data preprocessing:** Conduct necessary preprocessing steps on the dataset, encompassing handling missing values, normalizing or scaling features, and potentially removing outliers to ensure data quality.

4. **Implementation of K-means clustering:** Employ the preprocessed dataset in the K-means clustering algorithm. The primary objective of k-means clustering is to partition the data into k clusters, with each cluster's centroid (representing the average of its constituent data points) acting as a representative for that cluster.

5. **Determining optimal k:** Determine the suitable value of k, indicating the desired number of clusters. Several techniques, such as the elbow method, silhouette score, or domain knowledge, can be employed to ascertain the optimal value of k.

6. **Interpretation of clusters:** Analyze the resulting clusters after the clustering process. Scrutinize the intricacies of each cluster, including the distribution of data points within them and the ranges of feature values they encapsulate.

7. **Diabetes correlation:** Investigate whether specific clusters exhibit patterns or characteristics associated with diabetes. This analysis may involve assessing the average feature values within each cluster, comparing them to established diabetes indicators, or conducting statistical tests to identify significant associations.

By leveraging the potential of k-means clustering in this manner, the objective is to unearth intricate patterns or subgroups within the dataset that may exhibit correlations with diabetes. This analytical approach aids in the identification of potential relationships and insights that can guide subsequent analyses or feature selections for effective diabetes detection.

Neural networks

Neural networks, which is a subset of ML and are at the core of deep learning algorithms, are also known as artificial neural networks (ANNs) or simulated neural networks (SNNs). In order to mirror the way that organic neurons communicate with one another, their name and structure are both derived from the human brain.

A node layer of an ANN consists of an input layer, one or more hidden layers, and an output layer. Each node, or artificial neuron, is connected to others and has a weight and threshold that go along with it. Any node whose output exceeds the defined threshold value is activated and begins providing data to the network's uppermost layer. Otherwise, no data are transmitted to the network's next tier.

Training data are essential for neural networks to develop and enhance their accuracy over time. However, these learning algorithms become effective tools in computer science and AI once they are adjusted for accuracy, enabling us to quickly classify and cluster data.

With numerous layers of interconnected nodes (neurons), neural networks can capture complicated linkages in diabetes risk prediction. These models have the ability to learn hierarchical representations of input features and generate predictions using patterns they have discovered (Russel and Norvig, 2012).

Artificial neural networks can be used for diabetes detection by training a model on a dataset containing relevant features and corresponding diabetes labels (Pradhan et al., 2020). In general, the steps involved in using ANNs for diabetes detection are as follows.

1. **Dataset preparation:** Gather a dataset that includes features related to diabetes, such as glucose level, BMI, blood pressure, age, etc. The dataset should also include corresponding labels indicating whether each individual has diabetes or not.

2. **Data preprocessing:** Preprocess the dataset using the relevant methods, such as handling missing values, normalizing or scaling the features, and creating training and test sets from the dataset.

3. **Architecture design:** It encompasses the meticulous selection of suitable activation functions, determining the optimal number of layers, and fine-tuning the density of neurons within each layer. A conventional architecture typically consists of an input layer, multiple hidden layers, and an output layer, all meticulously arranged to maximize the model's performance.

4. **Model training:** It is a critical phase wherein the neural network undergoes rigorous training using the designated training set. This intricate process entails systematically presenting the training examples to the network, subsequently computing the loss, which represents the disparity between the predicted and actual labels. By leveraging the powerful technique of backpropagation, the network's weights and biases are iteratively adjusted to minimize the loss. The overarching objective of this training process is to refine the network's capabilities, enhancing its proficiency in accurately classifying instances of diabetes.

5. **Model evaluation:** It involves the astute utilization of the designated testing set to gauge the performance of the trained model. This crucial step entails assessing the model's ability to accurately predict instances of diabetes by computing essential metrics such as the F1 score, accuracy, precision, and recall. These metrics provide valuable insights into the model's efficacy and its capacity to generalize well beyond the training data. By meticulously analyzing these evaluation metrics, one can ascertain the effectiveness and overall capabilities of the trained model in the context of diabetes prediction.

6. **Fine-tuning and optimization:** To enhance the performance of the model, experiment with various hyperparameters, optimization strategies, and network designs. This may involve adjusting the learning rate, adding regularization techniques (e.g., dropout), or using different activation functions. Techniques like cross-validation and grid search can also be employed for hyperparameter tuning.

7. **Deployment and prediction:** Once the model performs well on the testing set, it can be deployed to predict diabetes for new, unseen data. The trained model takes in the relevant features of an individual and provides a prediction of whether they have diabetes or not.

The standard and generality of the dataset, the features selected, and the design of the model all have a role in how well ANNs identify diabetes and optimization. The process may require iterations and experimentation to achieve the best results (Pradhan et al., 2020).

Deep learning models

Machine learning, a division of AI, is the foundation of deep learning. Deep learning will operate because neural networks mimic the functioning of the human brain. Nothing is explicitly programmed in deep learning. In essence, it is a class of ML that does feature extraction and transformation using a large number of nonlinear processing units. Each of the next layers uses the output from the one below as its input.

Deep learning models are quite useful in resolving the dimensionality issue since they are able to focus on the accurate features by themselves with just a little programming assistance. Particularly when there are many inputs and outputs, deep learning methods are used.

Since deep learning is a development of ML, which is a branch of AI on its own, and since the goal of AI is to mimic human behavior, the goal of deep learning is to create an algorithm that can do the same.

Deep learning models can process many data sources (such as pictures, time series, and textual information) and catch subtle correlations, improving prediction accuracy in the area of diabetes risk (Rajkomar et al., 2019).

Some deep learning models are as.

1. **Recurrent neural network:** A special kind of ANN called a recurrent neural network (RNN) is made for processing sequential data. RNNs have connections that enable information to flow in loops, in contrast to feedforward neural networks, which process data in a single direction. RNNs can keep an internal memory or context owing to this loop structure, which makes them suited for jobs involving sequential or time-dependent input.

   The ability of RNNs to handle input sequences of any length and capture dependencies between sequence pieces is its key feature. The RNN normally processes each element in the sequence one at a time while retaining a state that contains data from earlier steps. The current input and the prior hidden state are used to update this internal state, also known as the hidden state, at each step.

   The recurrent unit, which executes the same operation at each step, is the fundamental component of an RNN. The **long short-term memory (LSTM)** cell is the most popular type of recurrent unit. For applications like language modeling, machine translation, speech recognition, and sentiment analysis, LSTMs are particularly effective because of their more complicated structure, which enables them to learn and remember long-term connections in the data.

   The **gated recurrent unit (GRU)**, which resembles LSTM but has a more straightforward layout and fewer gates, is another type of recurrent unit. GRUs can be a suitable option for applications with limited computational resources because they are computationally more effective than LSTMs.

   Recurrent neural networks can handle sequential data, which make it possible to use them to forecast diabetes. RNNs can examine temporal patterns and dependencies in a patient's data across time, including blood glucose levels, insulin use, and other pertinent factors, in the context of diabetes prediction (Sharma and Shah, 2021).

2. **Convolution neural network:** Convolutional neural networks (CNNs) are a particular kind of ANN created with the purpose of processing organized input with a grid-like structure, such as photographs. In computer vision applications including image classification, object recognition, and picture segmentation, CNNs are quite effective.

   The convolution procedure, which entails conducting element-wise multiplications and summations while a tiny window (also referred to as a filter or kernel) is slid over the input data, is the fundamental idea of CNNs. The network can now extract regional patterns and spatial hierarchies from the data (Swapna et al., 2018).

   The fundamental elements and ideas behind CNNs are as.

1. **Convolutional layers:** The fundamental components of CNNs are convolutional layers. Multiple learnable filters are included in each layer and are convolved with the input data. The filters extract numerous elements from the input at different spatial scales, producing feature maps that emphasize particular patterns.

2. **Pooling layers:** To cut down on spatial dimensions and manage overfitting, pooling layers are frequently added after convolutional layers. A popular pooling method is called max pooling, which down samples feature maps by choosing the highest value in each pooling region.

3. **Activation functions:** By introducing nonlinearities to the network, activation functions enable the network to learn intricate connections between inputs and outputs. Rectified Linear Unit (ReLU) is frequently employed as the activation function in CNNs since it facilitates training and reduces the issue of vanishing gradients.

4. **Fully connected layers:** High-level representations can be learned due to fully connected layers, which link every neuron in one layer to every neuron in the following layer. To map the retrieved characteristics to particular classes or labels, fully connected layers are typically added at the CNN's end.

5. **Training:** For multiclass classification issues, CNNs are trained using labeled training data and a loss function like categorical cross-entropy. Stochastic gradient descent (SGD) or its variants are gradient-based optimization methods that are used to optimize the network's parameters, including the filter weights and biases. The gradients are calculated and the parameters are updated via backpropagation.

6. **Transfer learning:** It is a method where pretrained CNN models that have been trained on huge datasets, like ImageNet, are used as a starting point for a particular task. The network can learn from the generic feature representations of the pretrained models and adapt them to the current task with less data by utilizing them.

Diabetes identification and management have been aided by the use of CNNs. Although CNNs are most often used for computer vision jobs, they can also be used to analyze medical data, including information about diabetes. CNNs can be used to automatically diagnose and categorize the severity of DR using fundus pictures.

Photographs of the retina, the light-sensitive tissue at the back of the eye, are called fundus pictures. Blindness may result from the retinal damage caused by DR. Blindness can be avoided with the early diagnosis and treatment of DR. CNNs learn to recognize the characteristics in fundus images that are connected to DR. Hemorrhages, hard exudates, and microaneurysms are a few examples of these characteristics. The CNN may be used to categorize fundus images as having no DR, mild DR, moderate DR, severe DR, or proliferative DR once it has trained to recognize these signs (Samanta et al., 2020).

Predictive modeling for blood glucose monitoring
Ensemble models for blood glucose monitoring

In order to increase the precision and dependability of forecasts, ensemble models have been frequently employed in blood glucose forecasting. Combining the forecasts of several different individual models results in ensemble predictions, which are frequently more reliable and accurate than the predictions of any one individual model alone. A few popular ensemble techniques for predicting blood sugar levels are as:

1. **Bagging (bootstrap aggregating):** Bagging is an ensemble technique that is widely used that involves training multiple models on various subsets of the training data. Independently trained models are blended based on average predictions or popular votes. Bagging aids in reducing individual model overfitting and variation (Saiti et al., 2020).

2. **Random forest:** Bagging's extension known as random forest makes use of decision trees as its fundamental foundation models. Several decision trees are trained on various subsets of the data in this ensemble technique, and the predictions are then averaged or combined through voting. The robustness, interpretability, and capacity for handling high-dimensional data of random forests are well known (Li et al., 2019).

3. **Boosting:** Boosting is a well-liked ensemble strategy that combines a number of weak models to produce a powerful predictive model. It functions by iteratively training models on the training data, with each model providing more weight to the examples that the prior models incorrectly identified (Saiti et al., 2020). The forecasts of all weak models are combined to get the final prediction. One of the popular boosting algorithms is called Ada-boost (adaptive boosting).

4. **Stacking:** Stacking is a more sophisticated ensemble strategy that utilizes a meta-learner or blender model to integrate the predictions of numerous models. The base models are trained on the training data, and the meta-learner uses the base models' predictions as input features before producing the final prediction. The ensemble can capture many patterns and interactions among the predictions from the basis models by means of stacking.

5. **Bayesian model averaging (BMA):** BMA is an ensemble method that utilizes Bayesian inference to integrate the predictions of various models. The final forecast is a weighted average of the predictions from each model, with each model given a weight that represents its credibility. BMA is a potent ensemble strategy since it considers both model and parameter uncertainty (Wasserman, 2000).

Continuous glucose monitoring and machine learning

With the help of CGM, people with diabetes can track their blood sugar levels in real time all throughout the day. It entails wearing a tiny sensor that detects the presence of glucose in interstitial fluid and wirelessly sends the information to a receiver or a smartphone. In comparison to conventional intermittent finger stick tests, CGM offers frequent glucose readings, often every few minutes, providing a more thorough view of glucose changes (Ajjan et al., 2019).

CGM data can be used to extract insightful information, enhance glucose management, and facilitate decision-making by utilizing ML approaches. An explanation of how ML can be applied to CGM is provided below.

1. **Predictive modeling:** ML algorithms that forecast future glucose levels can be taught using historical CGM data. These models can make long-term forecasts, such as the likelihood of hypoglycemia or hyperglycemic episodes, as well as short-term predictions, such as forecasting glucose levels for the upcoming few hours. These models can assist people in anticipating glucose changes and taking preventive steps by taking into account elements like time of day, recent glucose patterns, insulin dosages, and other pertinent data (Ramazi et al., 2019).

2. **Pattern recognition:** By analyzing CGM data, ML systems can spot patterns and trends in glucose levels. These algorithms can recognize individual-specific patterns, find hidden patterns that are challenging to find manually, and detect regularity or irregularity in glucose swings. This knowledge can be useful for determining the elements that affect glycemic control, comprehending the effects of lifestyle decisions, and developing treatment strategies (Kamble et al., 2021).

3. **Anomaly detection:** CGM data and ML can be used to find anomalies or outliers in glucose patterns. Algorithms can figure out a person's normal glucose range and warn them when the readings dramatically depart from that range. Early diagnosis of hypoglycemic or hyperglycemic episodes can be aided by this, allowing for prompt management to avoid consequences (Nassif et al., 2021).
4. **Decision support system:** ML can be incorporated into decision support systems to help people make well-informed decisions regarding their diets, lifestyle changes, and insulin administration. These systems can offer tailored recommendations for insulin changes or best practices to maintain stable glucose levels by taking into account CGM data together with other pertinent information like meal intake, exercise, and medication.
5. **Personalized medicine:** Personalized models that take a person's preferences and feature into account can be created using ML algorithms in the field of medicine. ML can assist individualized treatment plans and treatments that optimize glucose management by learning from CGM data in conjunction with additional patient-specific parameters including demographics, medical history, and genetic data (McCarthy, 2017).

The use of ML models in CGM applications necessitates the use of reliable and correct training data. Usually, this entails preparing the data to get rid of artifacts, normalize it, deal with missing numbers, and align it with pertinent clinical information. To maintain their performance and adjust to shifting individual demands, the models should also be regularly reviewed and updated as new data becomes available.

Continuous glucose monitoring, which makes use of ML techniques, can offer people with diabetes improved insights, real-time feedback, and personalized guidance for effectively managing their glucose levels, ultimately leading to an improvement in their overall diabetes management and quality of life.

Several CGM systems in use today that make use of ML methods are as:

1. **Dexcom G6:** The commonly used CGM device Dexcom G6 uses ML algorithms to forecast blood glucose levels and identify anomalies. It makes use of the "Dexcom G6 Prophylactic Low Glucose Suspend" (PLGS) predictive algorithm to foresee upcoming hypoglycaemia situations and to halt insulin delivery in an effort to prevent or lessen them. In order to generate personalized forecasts and guide decisions regarding insulin delivery, the system uses ML to examine past CGM data, including glucose trends, insulin dose, and other variables (Isitt et al., 2022).

The Dexcom G6 measures the amount of glucose in the interstitial fluid using a tiny, wearable sensor that is implanted under the skin. Every 5 min, the sensor wirelessly sends glucose readings to a receiver or a compatible smartphone. Real-time glucose readings, trend arrows showing the rate and direction of glucose change, and programmable low- and high-glucose warnings are also provided.

2. **Medtronic guardian connect:** The Medtronic Guardian Connect CGM system uses an ML algorithm dubbed "Smart Guard" to send out anticipatory alerts for hypoglycemia. It makes use of the information from CGM sensors to identify patterns and trends in glucose levels and forecast when the levels may drop below a predetermined threshold. In order to deliver individualized alarms and recommendations, the system uses ML to continuously learn from and adapt to the user's glucose trends (Cohen et al., 2018).

To monitor the amount of glucose in the interstitial fluid, a tiny, wearable sensor is implanted beneath the skin. Users can examine their glucose readings, trends, and alarms in real time by using a compatible smartphone and the sensor, which wirelessly sends glucose data. Additionally, it has an app that sends out forecasted notifications for low and high blood sugar readings.

3. **Senseonics Eversense:** Senseonics Eversense is an implantable CGM system that uses ML algorithms to predict blood glucose levels and identify anomalies. The system analyses CGM data and forecasts future glucose trends using a neural network-based algorithm. It continuously picks up information from the user's glucose trends and offers alerts and suggestions to help control diabetes (Irace et al., 2021).

It makes use of an implanted sensor that is positioned in the upper arm beneath the skin. A wearable smart transmitter receives the data wirelessly from the sensor, which measures blood glucose levels. A compatible smartphone app can then be contacted by the transmitter to get real-time glucose levels, trend data, and configurable alarms.

Ethical considerations in machine learning for diabetes
Privacy and security in diabetes data handling

When handling diabetic data, especially in the context of CGM systems, privacy and security are critical factors to take into account. Following are some pertinent details and sources on privacy and security in the processing of diabetes data.

1. **Data encryption:** To safeguard the transmission and storage of sensitive data, such as glucose measurements and individual health information, CGM systems should make use of robust encryption technologies. Data security and privacy are enhanced via encryption, which makes data inaccessible to unauthorized parties.
2. **User authentication:** Strong user authentication techniques, such as secure login credentials and multifactor authentication, are implemented to make sure that only persons with the proper authorization can access diabetes data (American Diabetes Association, 2021). People ought to be in charge of their diabetes data, including the power to provide knowingly permission for data collection and sharing. Data handling policies should be disclosed by CGM system providers, and users should have choices regarding how their data are shared and accessed by outside parties.
3. **Secure data storage:** Diabetes data should be stored securely to prevent data breaches and unauthorized access. Examples of secure data storage include encrypted databases and secure cloud storage (American Diabetes Association, 2021). Companies handling diabetes data should put strong security measures in place to safeguard stored data. This entails utilizing secure servers, putting access limits in place, and frequently updating software to fix any security flaws (Security Techniques, 2019).
4. **Privacy policies:** Providers of CGM systems should have explicit and lucid privacy policies that specify how they handle and safeguard user data. These rules should outline the procedures for collecting, storing, and exchanging data and give users ways to manage their data (American Diabetes Association, 2021).

5. **Data anonymization and de-identification:** In order to respect individual privacy, diabetes data containing personal health information should be suitably anonymized or deidentified. By removing personally identifiable information from the data, this procedure reduces the possibility of reidentification (Portability, Insurance, and Accountability Act, 2012).
6. **Consent for data sharing:** Users of CGM systems should be in charge of deciding whether or not to share their diabetes data with others and offer their consent after being fully informed. There should be choices for withdrawing consent and clear mechanisms for obtaining it (American Diabetes Association, 2021).

Addressing bias and fairness in machine learning models

Responsible AI development must prioritize addressing bias and fairness in ML algorithms. In order to combat bias and encourage fairness in ML models, keep in mind the following important factors.

1. **Preprocessing and data collection:** These processes have the potential to unintentionally introduce biases. Making sure that the training data used to construct the model is diverse, representative, and bias-free is crucial. Techniques for data preprocessing that can reduce bias in the dataset include data augmentation, data balance, and cautious feature selection (Buolamwini and Gebru, 2018).
2. **Bias evaluation and mitigation:** Careful analysis of bias in models is crucial. Methods like fairness metrics, audits for bias, and disparity analyses can be used to measure and track bias in model projections for various demographic groups. Several mitigating strategies, including algorithmic changes, data resampling, or the employment of fairness-aware learning algorithms, can be used once bias has been identified (Zemel et al., 2013).
3. **Models that are transparent and explainable:** Creating models that are transparent and explanatory might help you recognize and comprehend biases. Techniques for interpretability, such as feature importance analysis, rule extraction, or attention processes, can reveal how the model makes decisions and point out any biases (Ribeiro et al., 2016).
4. **Diversity and inclusion in model development:** Promoting interdisciplinary cooperation and a diversity of viewpoints can potentially reduce bias while developing models. Engaging specialists from various fields, such as ethicists, social scientists, and subject matter experts, can offer insightful information and aid in the creation of fair and impartial models (Mitchell et al., 2019).
5. **Regular monitoring and iterative improvement:** Models should be regularly checked to spot and correct any new biases or difficulties with fairness. Regular monitoring and iterative improvements should be made to models. By regularly assessing the model's performance across a range of demographic groups and seeking user feedback, biases can be found and corrected, resulting in continuous gains in fairness (Holstein et al., 2019).

It is important to note that there is current research being done on how to address bias and fairness in ML, and efforts are being made to provide more thorough frameworks and norms. When tackling bias and fairness in ML models, it is essential to adhere to ethical standards, involves a variety of stakeholders, and take legal and regulatory requirements into account.

Explainability and interpretability of ML-based diabetes solutions

When creating and implementing ML-based solutions for diabetes control, explainability and interpretability are key factors to take into account. These ideas center on the capacity to comprehend and offer insightful justifications for the judgments or forecasts produced by ML models. Explainability and interpretability are crucial in the context of diabetic solutions for a number of reasons.

1. **Trust and acceptance:** In order to effectively control diabetes, ML models must be trusted by both healthcare professionals and patients. Explainability and interpretability offer transparency, enabling stakeholders to comprehend the rationale behind a specific choice or forecast. The ML-based solution gains more acceptability and trust as a result of this transparency (Ribeiro et al., 2016).
2. **Clinical decision-making:** In the management of diabetes, clinicians' judgments are frequently dependent on the forecasts or suggestions provided by ML models. Explainability and interpretability allow medical personnel to evaluate the assumptions that underlie the model's predictions, assisting them in making wise choices and offering important information about the patient's condition (Adlung et al., 2021).
3. **Identification of biases and errors:** Inadvertent biases or errors that affect the fairness and accuracy of the predictions may be incorporated into ML models. By highlighting the characteristics or variables that significantly influence the model's decisions, explainability, and interpretability approaches can aid in the detection of these biases. Developers can correct biases and enhance the model's overall performance because of this insight (Vokinger et al., 2021).
4. **Regulatory compliance:** Adhering to regulations is essential in the healthcare industry. Diabetes solutions that use ML must abide by rules and restrictions. Explainability and interpretability enable stakeholders to evaluate whether the model complies with regulatory standards by offering insights into the model's decision-making process. This helps to assure compliance (Petersen et al., 2022).

Several strategies can be used to achieve explainability and interpretability in ML-based diabetic treatments, including.

- **Importance of features:** Determining which features have the greatest influence on the model's predictions might provide light on the variables influencing the decision-making process.
- **Rule extraction:** Human-readable rules can be extracted from intricate ML models to enable open decision-making.
- **Local explanations:** Providing local explanations at the instance level makes it easier to comprehend how particular input features affect the model's predictions.
- **Model visualization:** ML models' internal workings and decision-making processes can be better understood when they are represented visually.
- **Sensitivity analysis:** Examining how modifications to the input variables affect the model's results might show the model's behavior and boundaries for making decisions.

By using these methods, ML-based diabetic solutions can be made more transparent, interpretable, and accountable, allowing healthcare professionals and patients to better understand and utilize the models. A brief schematic representation of use of ML in medical science is given in Fig. 7.3. It is vital to keep in mind that while there are continuing research efforts and breakthroughs being made in order to achieve explainability and interpretability in ML models, the subject is still very much in flux.

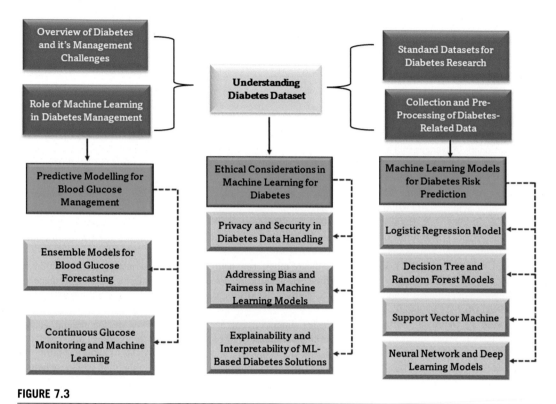

FIGURE 7.3

A schematic outline of use of machine learning in medical science.

Conclusion

The integration of ML into diabetes management holds significant potential for improving patient health outcomes and enhancing the overall quality of care. By leveraging data-driven insights, healthcare providers can make informed decisions, tailor treatment plans, and empower individuals with diabetes to actively manage their condition. It highlights the importance of interdisciplinary collaborations between data scientists, healthcare professionals, and patients to develop robust and user-friendly machine-learning solutions that can contribute to improved diabetes care. However, careful attention must be paid to ethical considerations to ensure the responsible implementation of ML in diabetes management.

The ML models have had a transformative impact on diabetes patients and overall diabetes management. These models enable early detection, personalized treatment plans, continuous monitoring, predictive analytics, decision support for healthcare providers, lifestyle modification, patient education, and advancements in research and drug development. By leveraging the power of data and AI, these models have the potential to significantly improve patient's health, quality of life, and the overall management of diabetes.

References

Adlung, L., Cohen, Y., Mor, U., Elinav, E., 2021. Machine learning in clinical decision making. Medicine 2 (6), 642−665. https://doi.org/10.1016/j.medj.2021.04.006.

Ahamed, F., Farid, F., 2019. Applying internet of things and machine-learning for personalized healthcare: issues and challenges. Proceedings − International Conference on Machine Learning and Data Engineering, iCMLDE 2018 22−29. https://doi.org/10.1109/iCMLDE.2018.00014.

Ajjan, R., Slattery, D., Wright, E., 2019. Continuous glucose monitoring: a brief review for primary care practitioners. Advances in Therapy 36 (3), 579−596. https://doi.org/10.1007/s12325-019-0870-x.

American Diabetes Association, 2021. 2. Classification and diagnosis of diabetes: standards of medical care in diabetes-2021. Diabetes Care 44, S15−S33. https://doi.org/10.2337/dc21-S002.

Arneth, B., Arneth, R., Shams, M., 2019. Metabolomics of type 1 and type 2 diabetes. International Journal of Molecular Sciences 20 (10). https://doi.org/10.3390/ijms20102467.

Barakat, N., Bradley, A.P., Barakat, M.N.H., 2010. Intelligible support vector machines for diagnosis of diabetes mellitus. IEEE Transactions on Information Technology in Biomedicine 14 (4), 1114−1120. https://doi.org/10.1109/titb.2009.2039485.

Bhatt, V., Chakraborty, S., 2021. Real-time healthcare monitoring using smart systems: a step towards healthcare service orchestration Smart systems for futuristic healthcare. Proceedings − International Conference on Artificial Intelligence and Smart Systems, ICAIS 2021 772−777. https://doi.org/10.1109/ICAIS50930.2021.9396029.

Buolamwini, J., Gebru, T., 2018. Gender shades: intersectional accuracy disparities in commercial gender classification. Proceedings of Machine Learning Research 81, 77−91. In: https://proceedings.mlr.press/.

Celis-Morales, C.A., Franzén, S., Eeg-Olofsson, K., Nauclér, E., Svensson, A.-M., Gudbjornsdottir, S., Eliasson, B., Sattar, N., 2022. Type 2 diabetes, glycemic control, and their association with dementia and its major subtypes: findings from the Swedish national diabetes register. Diabetes Care 45 (3), 634−641. https://doi.org/10.2337/dc21-0601.

Cohen, O., Abraham, S.B., Mcmahon, C.M., Agrawal, P., Vigersky, R., 2018. Real-world avoidance of glucose excursions with the guardian connect CGM system's predictive alerts. Diabetes 67 (Suppl. 1). https://doi.org/10.2337/db18-953-p.

Cowie, M.R., Blomster, J.I., Curtis, L.H., Duclaux, S., Ford, I., Fritz, F., Goldman, S., Janmohamed, S., Kreuzer, J., Leenay, M., Michel, A., Ong, S., Pell, J.P., Southworth, M.R., Stough, W.G., Thoenes, M., Zannad, F., Zalewski, A., 2017. Electronic health records to facilitate clinical research. Clinical Research in Cardiology 106 (1), 1−9. https://doi.org/10.1007/s00392-016-1025-6.

Esteva, A., Robicquet, A., Ramsundar, B., Kuleshov, V., DePristo, M., Chou, K., Cui, C., Corrado, G., Thrun, S., Dean, J., 2019. A guide to deep learning in healthcare. Nature Medicine 25 (1), 24−29. https://doi.org/10.1038/s41591-018-0316-z.

Ethem, A., 2020. Introduction to Machine Learning. Google Books.

Foster, N.C., Beck, R.W., Miller, K.M., Clements, M.A., Rickels, M.R., Dimeglio, L.A., Maahs, D.M., Tamborlane, W.V., Bergenstal, R., Smith, E., Olson, B.A., Garg, S.K., 2019. State of type 1 diabetes management and outcomes from the T1D exchange in 2016−2018. Diabetes Technology and Therapeutics 21 (2), 66−72. https://doi.org/10.1089/dia.2018.0384.

Giordano, C., Brennan, M., Mohamed, B., Rashidi, P., Modave, F., Tighe, P., 2021. Accessing artificial intelligence for clinical decision-making. Frontiers in Digital Health 3. https://doi.org/10.3389/fdgth.2021.645232.

Hallgren Elfgren, I.-M., Grodzinsky, E., Törnvall, E., 2016. The Swedish National Diabetes Register in clinical practice and evaluation in primary health care. Primary Health Care Research and Development 17 (06), 549−558. https://doi.org/10.1017/S1463423616000098.

Hassan, M.M., Mollick, S., Yasmin, F., 2022. An unsupervised cluster-based feature grouping model for early diabetes detection. Healthcare Analytics 2, 100112.

Holstein, K., Vaughan, J.W., Daumé, H., Dudík, M., Wallach, H., 2019. Improving fairness in machine learning systems: what do industry practitioners need?. In: Conference on Human Factors in Computing Systems — Proceedings. Association for Computing Machinery, United States. https://doi.org/10.1145/3290605.3300830.

Hussain Bukhari, S.N., Masoodi, F., Dar, M.A., Wani, N.I., Sajad, A., Hussain, G., 2023. Prediction of erythemato-squamous diseases using machine learning. Applications of Machine Learning and Deep Learning on Biological Data 87—96. https://doi.org/10.1201/9781003328780-6.

International Diabetes Federation, 2019. International Diabetes Federation. IDF Diabetes Atlas.

International Diabetes Federation, 2021. IDF Diabetes Atlas, tenth ed. Brussels, Belgium. Available at: https://www.diabetesatlas.org.

Inzucchi, S.E., Bergenstal, R.M., Buse, J.B., Diamant, M., Ferrannini, E., Nauck, M., Peters, A.L., Tsapas, A., Wender, R., Matthews, D.R., 2015. Management of hyperglycemia in type 2 diabetes, 2015: a patient-centered approach: update to a position statement of the American Diabetes Association and the European Association for the Study of Diabetes. Diabetes Care 38 (1), 140—149. https://doi.org/10.2337/dc14-2441.

Irace, C., Cutruzzolà, A., Tweden, K., Kaufman, F.R., 2021. Device profile of the eversense continuous glucose monitoring system for glycemic control in type-1 diabetes: overview of its safety and efficacy. Expert Review of Medical Devices 18 (10), 909—914. https://doi.org/10.1080/17434440.2021.1982380.

Isitt, J.J., Roze, S., Tilden, D., Arora, N., Palmer, A.J., Jones, T., Rentoul, D., Lynch, P., 2022. Long-term cost-effectiveness of Dexcom G6 real-time continuous glucose monitoring system in people with type 1 diabetes in Australia. Diabetic Medicine 39 (7). https://doi.org/10.1111/dme.14831.

Kamble, A., Hannan, S.A., Jain, A., Manza, R., 2021. Prediction of prediabetes, no diabetes and diabetes mellitus-2 using pattern recognition. Advances in Intelligent Systems and Computing 1187, 749—755. https://doi.org/10.1007/978-981-15-6014-9_90.

Kingsford, C., Salzberg, S.L., 2008. What are decision trees? Nature Biotechnology 26 (9), 1011—1013. https://doi.org/10.1038/nbt0908-1011.

Kleinberger, J.W., Pollin, T.I., 2015. Personalized medicine in diabetes mellitus: current opportunities and future prospects. Annals of the New York Academy of Sciences 1346 (1), 45—56. https://doi.org/10.1111/nyas.12757.

Kovatchev, N., Denev, D., Ivanov, S., 2020. A review of deep learning methods for natural language processing. Frontiers in Artificial Intelligence 3.

Li, Z., Ye, Q., Guo, Y., Tian, Z., Ling, B.W.K., Lam, R.W.K., 2019. Wearable non-invasive blood glucose estimation via empirical mode decomposition based hierarchical multiresolution analysis and random forest. In: International Conference on Digital Signal Processing, DSP. 2018. Institute of Electrical and Electronics Engineers Inc., China https://doi.org/10.1109/ICDSP.2018.8631545.

Luo, J.J., Bian, W.P., Liu, Y., Huang, H.Y., Yin, Q., Yang, X.J., Pei, D.S., 2018. CRISPR/Cas9-based genome engineering of zebrafish using a seamless integration strategy. Federation of American Societies for Experimental Biology Journal 32 (9), 5132—5142. https://doi.org/10.1096/fj.201800077RR.

Masoodi, F., Quasim, M., Hussain Bukhari, S.N., Dixit, S., Alam, S., 2023. Applications of machine learning and deep learning on biological data. Applications of Machine Learning and Deep Learning on Biological Data 1—200. https://doi.org/10.1201/9781003328780.

McCarthy, M.I., 2017. Painting a new picture of personalised medicine for diabetes. Diabetologia 60 (5), 793—799. https://doi.org/10.1007/s00125-017-4210-x.

Mitchell, M., Wu, S., Zaldivar, A., Barnes, P., Vasserman, L., Hutchinson, B., Spitzer, E., Raji, I.D., Gebru, T., 2019. Model cards for model reporting. In: FAT* 2019 — Proceedings of the 2019 Conference on Fairness, Accountability, and Transparency. Association for Computing Machinery, Inc, pp. 220—229. https://doi.org/10.1145/3287560.3287596.

Nassif, A.B., Talib, M.A., Nasir, Q., Dakalbab, F.M., 2021. Machine learning for anomaly detection: a systematic review. IEEE Access 9, 78658−78700. https://doi.org/10.1109/ACCESS.2021.3083060.

Paul, D., Sanap, G., Shenoy, S., Kalyane, D., Kalia, K., Tekade, R.K., 2021. Artificial intelligence in drug discovery and development. Drug Discovery Today 26 (1), 80−93. https://doi.org/10.1016/j.drudis.2020.10.010.

Petersen, E., Potdevin, Y., Mohammadi, E., Zidowitz, S., Breyer, S., Nowotka, D., Henn, S., Pechmann, L., Leucker, M., Rostalski, P., Herzog, C., 2022. Responsible and regulatory conform machine learning for medicine: a survey of challenges and solutions. IEEE Access 10, 58375−58418. https://doi.org/10.1109/access.2022.3178382.

Portability, Insurance, and Accountability Act, 2012. Guidance Regarding Methods for De-identification of Protected Health Information in Accordance with the Health Insurance Portability and Accountability Act (HIPAA) Privacy Rule.

Pradhan, N., Rani, G., Dhaka, V.S., Poonia, R.C., 2020. Diabetes prediction using artificial neural network. Deep Learning Techniques for Biomedical and Health Informatics 327−339. https://doi.org/10.1016/B978-0-12-819061-6.00014-8.

Rajkomar, A., Dean, J., Kohane, I., 2019. Machine learning in medicine. New England Journal of Medicine 380 (14), 1347−1358. https://doi.org/10.1056/NEJMra1814259.

Ramazi, R., Perndorfer, C., Soriano, E., Laurenceau, J.P., Beheshti, R., 2019. Multi-modal predictive models of diabetes progression. ACM-BCB 2019 − Proceedings of the 10th ACM International Conference on Bioinformatics, Computational Biology and Health Informatics 253−258. https://doi.org/10.1145/3307339.3342177.

Ribeiro, M.T., Singh, S., Guestrin, C., 2016. "Why should I trust you?" Explaining the predictions of any classifier. In: Proceedings of the ACM SIGKDD International Conference on Knowledge Discovery and Data Mining. Association for Computing Machinery, United States, pp. 1135−1144. https://doi.org/10.1145/2939672.2939778.

Russel, S., Norvig, P., 2012. Artificial Intelligence-A Modern Approach, third ed.

Saiti, K., Macaš, M., Lhotská, L., Štechová, K., Pithová, P., 2020. Ensemble methods in combination with compartment models for blood glucose level prediction in type 1 diabetes mellitus. Computer Methods and Programs in Biomedicine 196, 105628. https://doi.org/10.1016/j.cmpb.2020.105628.

Samanta, A., Saha, A., Satapathy, S.C., Fernandes, S.L., Zhang, Y.-D., 2020. Automated detection of diabetic retinopathy using convolutional neural networks on a small dataset. Pattern Recognition Letters 135, 293−298. https://doi.org/10.1016/j.patrec.2020.04.026.

Security Techniques, 2019. Security Techniques − Extension to ISO/IEC 27001 and ISO/IEC 27002 for Privacy Information Management − Requirements and Guidelines 27701.

Sharma, T., Shah, M., 2021. A comprehensive review of machine learning techniques on diabetes detection. Visual Computing for Industry, Biomedicine, and Art 4 (1). https://doi.org/10.1186/s42492-021-00097-7.

Smith, J.W., Everhart, J.E., Dickson, W.C., Knowler, W.C., Johannes, R.S., 1988. Using the ADAP learning algorithm to forecast the onset of diabetes mellitus. Proceedings − The Annual Symposium on Computer Applications in Medical Care 261−265.

Srikanthan, K., Feyh, A., Visweshwar, H., Shapiro, J.I., Sodhi, K., 2016. Systematic review of metabolic syndrome biomarkers: a panel for early detection, management, and risk stratification in the west virginian population. International Journal of Medical Sciences 13 (1), 25−38. https://doi.org/10.7150/ijms.13800.

Strack, B., DeShazo, J.P., Gennings, C., Olmo, J.L., Ventura, S., Cios, K.J., Clore, J.N., 2014. Impact of HbA1c measurement on hospital readmission rates: analysis of 70,000 clinical database patient records. BioMed Research International 2014, 1−11. https://doi.org/10.1155/2014/781670.

Swapna, G., Vinayakumar, R., Soman, K.P., 2018. Diabetes detection using deep learning algorithms. ICT Express 4 (4), 243−246. https://doi.org/10.1016/j.icte.2018.10.005.

Teli, T.A., Masoodi, F.S., Masoodi, Z., 2023. Applications of Machine Learning and Deep Learning on Biological Data. Auerbach Publications.

Thompson, J.R., Gustafsson, H.C., DeCapo, M., Takahashi, D.L., Bagley, J.L., Dean, T.A., Kievit, P., Fair, D.A., Sullivan, E.L., 2018. Maternal diet, metabolic state, and inflammatory response exert unique and long-lasting influences on offspring behavior in non-human primates. Frontiers in Endocrinology 9. https://doi.org/10.3389/fendo.2018.00161.

Tu, Y., 2019. Machine Learning in EEG Signal Processing and Feature Extraction.

UK Prospective Diabetes Study (UKPDS) Group, 1998. Effect of intensive blood-glucose control with metformin on complications in overweight patients with type 2 diabetes (UKPDS 34). The Lancet 352 (9131), 854−865. https://doi.org/10.1016/s0140-6736(98)07037-8.

Vokinger, K.N., Feuerriegel, S., Kesselheim, A.S., 2021. Mitigating bias in machine learning for medicine. Communications Medicine 1 (1). https://doi.org/10.1038/s43856-021-00028-w.

Waring, J., Lindvall, C., Umeton, R., 2020. Automated machine learning: review of the state-of-the-art and opportunities for healthcare. Artificial Intelligence in Medicine 104, 101822. https://doi.org/10.1016/j.artmed.2020.101822.

Wasserman, L., 2000. Bayesian model selection and model averaging. Journal of Mathematical Psychology 44 (1), 92−107. https://doi.org/10.1006/jmps.1999.1278.

Zemel, R., Wu, Y., Swersky, K., Pitassi, T., Dwork, C., 2013. Learning fair representations. In: 30th International Conference on Machine Learning, ICML 2013. International Machine Learning Society (IMLS), Canada, pp. 1362−1370.

Smiles 2.0: The AI dentistry frontier

Shazeena Qaiser[1] and Ambreen Hamadani[2]
[1]*Government Dental College, Srinagar, Jammu and Kashmir, India;* [2]*National Institute of Technology, Srinagar, Jammu and Kashmir, India*

Introduction

Artificial intelligence (AI) is a specialized area of computer science concerned with developing systems to analyze and solve basic and heavy tasks as a human does. It can be defined as "the theory and development of computer systems able to perform tasks normally requiring human intelligence, such as visual perception, speech recognition, decision making, and translation between languages."

Among the well-known specialties of AI, machine learning (ML) and deep learning (DL) are the most established. ML algorithms are utilized to enable machines to study on their own from data, identify patterns, and complete a specific task. In DL, multiple processing layer computer models can learn data representations at several levels of abstraction (Leite et al., 2020). AI covers a vast array of novel innovations that tend to have an impact on daily life like automobiles, robotics, financial analysis, etc. AI has also recently received accolades for its enormous potential in the medical and healthcare disciplines, being applied specifically in diagnostics, decision assistance, precision and digital medicine, hospital monitoring, and robotic assistants.

The development of AI technologies (ML and DL) has led to significant improvements in the dental domain as well. The use of AI in dentistry can be categorized as diagnosis, decision-making, treatment planning, and treatment outcome prediction. The most common application of AI in dentistry is diagnosis. Dentists can lessen their burden through AI's improved diagnostic capabilities. Dentists, on the one hand, are making more and more judgments using computer programs (Schleyer et al., 2006; Chae et al., 2011). However, dental computer applications are developing ever-smarter, more precise, and more dependable software. In dentistry, research on AI has encompassed all specialties, the most popular being endodontics and prosthodontics.

This chapter intends to highlight the research on the use of AI for diagnosis, clinical decision-making, and predicting successful treatment in all types of dental specialties, alongside identifying any current limitations in the usage of AI.

Applications of AI in dentistry

Modern dentistry makes extensive use of AI-based technology to streamline and automate time-tested procedures. By providing a range of services that enhance diagnosis accuracy, anticipate future dental

A Biologist's Guide to Artificial Intelligence. https://doi.org/10.1016/B978-0-443-24001-0.00008-7

illnesses, and suggest treatments, these systems help simplify the dentist's tasks. From identifying cavities to determining a person's gender in forensic dentistry, the dental industry utilizes AI technology in numerous fields (Fig. 8.1).

Operative dentistry

A precise evaluation and diagnosis of deep carious lesions and pulpitis is quite tough clinically as well as radiographically. The carious depth of penetration has been observed to be related to the level of bacterial penetration and the pulp's inflammatory state (Demant et al., 2021). On a high-quality image, it might be challenging for even experienced professionals to differentiate between healthy and affected areas. Visual assessment of the radiographic penetration depth can be impacted by the clinician's expertise (Shah, 2009). However, AI may be able to identify little aberrations in radiographic images that are typically difficult to see with the unaided eye and do not depend on the clinical expertise and experience of the dentist.

In a two-dimensional (2D) radiograph, each grayscale pixel represents an intensity that depicts the object density (Ding et al., 2023a). By taking into consideration the abovementioned features, an AI algorithm can analyze the behavior and make predictions to segment the tooth, detect cavities, etc. Out of the existing AI now in use, convolutional neural networks (CNNs) have shown tremendous success in image recognition (LeCun et al., 2015). They now surpass even human professionals in tasks requiring the recognition and interpretation of images, outperforming them in several instances (LeCun et al., 2015; Srinidhi et al., 2021). It has been noticed that CNNs are effective and accurate at spotting early caries lesions on bitewings (Schwendicke et al., 2020; Casalegno et al., 2019; Cantu et al., 2020). A CNN algorithm was created by Lee et al. (2018a) and Kühnisch et al. (2022) to identify dental caries on radiographs. In comparison to dentists' diagnosis, Schwendicke et al. (2021) investigated the cost-effectiveness of AI for proximal caries detection. Another study compared three CNNs for determining the penetration depth on intraoral periapical radiographs (IOPAR) in deep carious lesions and pulpitis and concluded that ResNet18 performed better than VGG19, Inception V3, as well as the comparator dentists ($P = .001$) (Zheng et al., 2021).

Vinayahalingam et al. (2021) worked on research whereby he introduced an automated caries detection on panoramic radiographs. A pretrained CNN-based DL model was used on 400 panoramic radiographs, and 87% accuracy was achieved on the test dataset. Ghaznavi Bidgoli et al. (2021) however incorporated the images of jaws and teeth. Each tooth was identified and cropped through region-CNN, and the diagnosis network was then fed into these cropped images to determine if each tooth was healthy, decayed, or restored. With accuracy scores of 92%, 88%, and 82%, respectively, this proposed network outperformed AlexNet and VGGNet16. Grad-CAM is another method frequently used for feature visualization on a radiograph as a decision-making tool (Selvaraju et al., 2020).

As demonstrated by the several studies stated above, AI shows favorable outcomes in early lesion diagnosis with precision even better than dentists. Collaboration across disciplines is necessary for this accomplishment between clinicians and computer scientists. While computer scientists create the dataset and ML algorithm, the dentists manually identify radiographs with the site of caries. Finally, the precision and accuracy of the training results are collaboratively checked and verified by clinicians and computer scientists (Chen et al., 2020). AI models may offer a potent means to detect caries and crown/root fracture, the tooth finishing line, and the prediction of any type of restorative failure. Additionally, CAD-/CAM-based techniques are employed in dentistry to produce highly precise completed dental restorations.

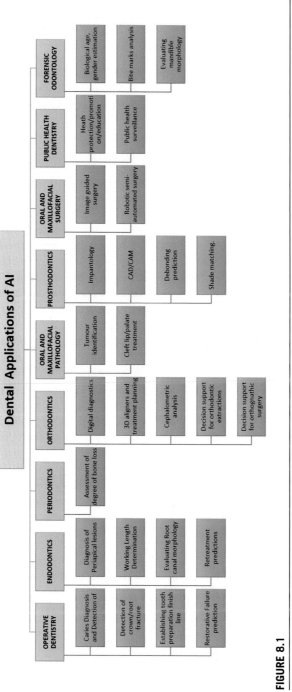

FIGURE 8.1

Applications of artificial intelligence (AI) in dentistry.

Endodontics

AI is playing a bigger role in endodontics, which incorporates analyzing the anatomical and morphological structures of the root canal system (RCS), identifying the periapical pathologies, working length determination, crown/root fracture detection, and prognostic evaluation of endodontic retreatment (Aminoshariae et al., 2021). On radiographs, artificial neural networks (ANNs) were utilized as a new approach for locating the minor apical foramen (Saghiri et al., 2012a).

The various applications in endodontics are explained in detail as follows.

Endodontic diagnosis

(i) Periapical lesion detection

To create AI models for periapical disease and periodontitis detection, the characteristics of alveolar bone resorption and periapical radiolucency can be beneficial (Hung et al., 2020). Two models were presented by Lin et al. for alveolar bone loss diagnosis (Lin et al., 2015) and assessment of the degree of bone loss (Lin et al., 2017), respectively. In relation to the extent of alveolar bone loss, (Lee et al., 2022) created a model using a DL to classify periodontally compromised teeth. Similar studies were done by Mol and van der Stelt (1992) and Carmody et al. (2001) to detect the extent of periapical lesions. A DL model can correspond to the mean diagnostic performance of 24 OMF surgeons in the process of detecting radiolucent changes on panoramic radiographs, according to Endres et al. (2020) According to Orhan et al.'s findings (Orhan et al., 2020), the AI technology could accurately identify 142 out of 153 periapical lesions. The identification of cystic lesions has been done using ANNs (Naik et al., 2016). Furthermore, Okada et al. (2015) developed a technique to distinguish periapical granulomas from cysts employing CBCT imaging; this technique is highly significant in routine dental clinical practice since it enables the granulomas to be more easily identified.

(ii) Root fractures detection

A CNN may be an effective method for locating vertical root fracture (VRF) on an OPG, based on an analysis by Fukuda et al. (2020). Moreover models were made to distinguish VRFs in teeth that were intact as well as endodontically treated using periapical radiographs and CBCT images (Johari et al., 2017) whereby detection of VRFs on CBCT images was found to be superior in terms of specificity, accuracy, and sensitivity. Shah et al. (2018) analysis of second molar fractures made use of wavelets and synthetic data. Steerable wavelets were utilized to successfully detect fractures in CBCT images despite the small sample size.

Treatment planning

(i) Evaluation of morphology of root and root canal system (RCS)

The endodontic success depends critically on a comprehensive understanding of the RCS. Typically, IOPAR, bitewings, and CBCT have been employed for this role, of which the latter has been proven to be the most precise. However, owing to radiation exposure effects, it is not advocated in daily dental practice. Additionally, evaluation of the number of canals and root canal shape could also be done using AI. Hiraiwa et al. (2019) found an 87% accuracy in the capacity of AI to identify single or multiple distal roots on lower first molars on panoramic radiographs. In their study of three-dimensional tooth segmentation, Lahoud et al. (2021) discovered that AI was both correspondingly precise and quicker than humans at identifying the root canal anatomy.

Future research will focus on increasing AI datasets to incorporate additional variances of typical dentoalveolar anatomy (Asiri and Altuwalah, 2022). In light of this, AI has the potential to help clinicians select the best endodontic files for cleaning the RCS and set the technical parameters on their endomotor to the ideal levels for completing the endodontic therapy with the fewest possible procedural errors.

(ii) Working length measurement

During root canal therapy, identifying the apical termination of the endodontic procedure is a crucial step. A comprehensive chemomechanical cleaning of *RCS* is feasible with an accurate working length (WL) assessment. When treating diseased root canal systems, it was noted that a millimeter decrease in WL can lower the success rate by 12%–14% (Chugal et al., 2003; Ng et al., 2008). Additionally, obturating the canals past the radiographic apex can have a negative impact on the treatment's result (Ng et al., 2008). Electronic apex locators even with a greater extent of accuracy are associated with faulty readings in wet canals, existing metallic restoration, or damaged cables (Martins et al., 2014). Algorithms are now being created to help clinicians locate the apical endpoint on radiographs. The precision of working length measurement can be enhanced, according to Saghiri et al. (2012a), by employing ANNs as a newer evaluation to trace the minor apical foramen. A human cadaver model was used to simulate a clinical setting in a second study by Saghiri et al. (2012b), which evaluated the accuracy of WL evaluation by ANN. When they contrasted the former with the actual measurement after extraction, they found that the root length measurements remained unchanged. They also found that while using periapical radiographs to detect anatomic constriction (96%) than an endodontist (76%) the ANN performed significantly better. As a result, an ANN can be regarded as a reliable technique for determining WL.

(iii) Retreatment predictions

Based on the study by Campo et al. (2016), the algorithm was created for the prognostic evaluation of endodontic retreatment cases, to essentially and precisely suggest the need for retreatment. The limitation was that the system's precision could only match the data's information (Aminoshariae et al., 2021).

Regenerative endodontics

The neuro-fuzzy inference method was utilized in research by Bindal et al. (2017) to evaluate the stem cells taken from dental pulp used in numerous restorative therapies. This method was able to forecast the outcome by monitoring the stem cells' survival after being treated with bacterial lipopolysaccharides in a representative clinical setting.

Periodontics

Periodontitis is unquestionably one of the *most common diseases* of mankind which necessitates its early detection and treatment to prevent mobility and tooth loss (Tonetti et al., 2017). Chang et al. (2020) created an automated approach for detecting and classifying the degree of bone loss. First, the created CNN finds the border, the teeth, and the implants. Next, by examining the percentage rate of the RBL, it automatically categorizes each tooth. This method outperformed a newer research (Jiang et al., 2022) that integrated numerous models, including automatic tooth detection and segmentation using a UNet network, object detection using the YOLO-v4 model, estimation of the extent of bone loss, and periodontitis staging. Future study is necessary to enhance the application of these suggested techniques, nevertheless. Furthermore, Krois et al. (2019) used CNN to identify periodontal bone loss (PBL) on panoramic radiographs. A suggested CNN method to automatically identify teeth with damaged periodontal tissues was assessed by Lee et al. (2018b) for its potential use and precision. According to Yauney et al. (2019), a CNN algorithm employing systemic health-related data could analyze periodontal diseases.

Orthodontics

The most popular innovation is personalized orthodontic care driven by AI (Deshmukh, 2018). For diagnostic and treatment planning, radiographs and images captured by intraoral scanners and cameras can be analyzed. This eliminates the need for several laboratory tests and the making of patient impressions, and the outcomes are frequently more accurate when compared to human perception. By simulating the differences between pre- and post-treatment facial pictures, AI also aids in the prediction of treatment outcomes. The aligners can easily be 3D printed by a specific treatment plan using precise 3D scans and virtual models. As the massive amounts of data are analyzed, an algorithm is created that smartly identifies the pressure spots particularly to each tooth or set of teeth, as well as how and under what pressure teeth must be shifted.

Thanathornwong (2018) created a Bayesian-based decision support system to use data on orthodontics as input to determine whether orthodontic treatment is necessary. To determine if extractions from lateral cephalometric radiographs are necessary, Xie et al. (2010) suggested an ANN model. A DL method for accurately and automatically recognizing cephalometric landmarks on radiographs was demonstrated by Park et al. (2019).

Oral and maxillofacial pathology

Artificial intelligence has mostly been studied to use radiographic, microscopic, and ultrasonographic pictures to detect tumors and cancer. The suitability of CNN algorithms as a method for automatically identifying tumors has been established (Aubreville et al., 2017). It is important to note that AI could play a role in the treatment of cleft lip and palate in terms of risk assessment, diagnosis, speech evaluation, and surgery. To identify oral squamous cell carcinoma and other possibly malignant conditions in intraoral optical images, Warin et al. (2022) used a CNN technique. Optical coherence tomography (OCT) has been utilized to detect benign and malignant lesions in the oral mucosa besides intraoral imaging. Malignant and dysplastic oral lesions were distinguished by James et al. (2021) using ANN and SVM models.

Prosthodontics

The primary area of AI's use in prosthodontics is restorative design. In commercially available products like those from CEREC, Sirona, 3Shape, etc., CAD/CAM has digitalized the design work. Hwang et al. (2018) and Tian et al. (2022) developed newer techniques based on 2D-GAN models to produce a crown by acquiring from technicians' designs. From 3D tooth models, 2D depth maps were used as the training data. The morphology of the generated crowns was comparable to that of normal teeth, according to Ding's research (Ding et al., 2023b), which used a 3DDCGAN network that directly used 3D data in the crown production process. A more ideal process can be achieved with great efficiency by combining AI with CAD/CAM or 3D/4D printing. Additionally, AI has been utilized for debonding prediction and shade matching (Wei et al., 2018).

Oral and maxillofacial surgery

The major use of AI in oral surgery is the introduction of robotic surgery, whereby human body movements and intellect are recreated. The temporomandibular joint (TMJ) surgery, removal of tumors and foreign objects, and dental implants are among image-guided surgery techniques that have proven effective in clinical settings.

AI has revolutionized surgery, and there are currently several robotic surgeons that do increasingly effective semi-automated operative procedures under the guidance of a qualified surgeon.

Public health dentistry

AI can improve the effectiveness and efficiency of procedures all through a larger public health system, which has been explained in Fig. 8.2.

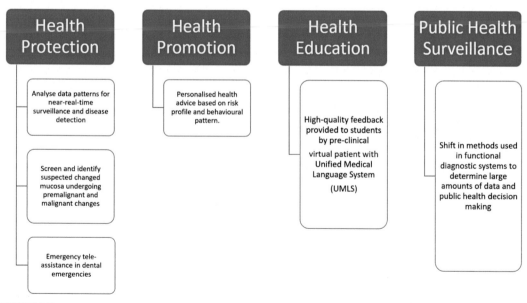

FIGURE 8.2

Applications of Artificial Intelligence (AI) in Public Health Dentistry.

Forensic odontology

The use of AI in forensic medicine/dentistry is extensive. It has proven to be successful in distinguishing between healthy and sick based on their biological age and gender. Additionally, it is used to predict the mandibular shape and analyze bite marks (Khanagar et al., 2021). The newest invention is a voice-command dental chair that eliminates the need for the dentist to manually perform any actions. All operations include voice commands. Shortly, dental chairs can monitor patients' vital signs, level of anxiety, weight, and the length of the procedure while also offering patient reassurance, notifying and warning the operating specialists to look for any differences. Ultimately among the most innovative use of AI is in "bioprinting." In the future, this technology might be utilized to restore lost oral hard and soft tissues that were lost because of pathological reasons (Khanna and Dhaimade, 2017).

Patient management

AI-powered virtual dental assistants can perform a variety of functions in the dental office with more precision and limited errors while using less staff as a result. It can assist with a variety of duties, including disease diagnosis, treatment planning, appointment scheduling, and paperwork management. In case of dental emergencies, the patient has the alternative of receiving teleassistance, especially if the specialist is not immediately available.

Ethical considerations

While AI has become more prevalent in dentistry and is potentially very beneficial, it also presents a number of ethical and societal challenges. When creating, deploying, or receiving AI applications based on a complete framework to solve the associated ethical dilemma, the industry should take the following into consideration in terms of utilization and users. Prior to implementation, AI systems must receive regulatory board approval from duly elected bodies. Dentists ought to receive training and education in the use of AI, and they ought to supervise and continually check on how well AI-based systems are working with their patient population. AI use should facilitate transparency, patient protection, and strict data management control in order to guarantee the security of patients and their data. If there is any uncertainty that the AI system is utilized for purposes other than enhancing patient health, such as when a dentist or healthcare institution has a conflict of interest, patients need to be fully informed (Roganović and Radenković, 2023).

Future scope

The offered AI models have produced promising results, but it is nonetheless mandatory to validate their generalizability and reliability using pertinent outside data gathered from recently drafted patients or gathered from other dental facilities. The pros and cons of AI have been listed in Table 8.1.

Future research intends to identify early carious lesions that are imperceptible to the human eye as well as improve the execution of AI models to professional levels. An intriguing zone that has been receiving great attention is the application of AI-enhanced algorithms to establish lifelike 3D reconstructions of tooth anatomy, root canal space, and common orofacial disorders. Cinematic rendering (CR) is the name given to this innovative reconstruction technique whereby photorealistic

Table 8.1 Pros and cons of AI.

Pros	Cons
1. Execute tasks instantly.	1. Mechanism/system complexity.
2. Logical and feasible decisions which results in anaccurate diagnosis.	2. Expensive setup.
3. Standardization of procedures is possible.	3. Appropriate training is necessary.
	4. The outcomes of AI in dentistry are not readily applicable.
	5. Data is frequently used both for training and testing, leading to "data snooping bias."
	6. The transparency of AI algorithms and data is a significant issue.

3D images can be created by CR using CBCT data sets and wide dynamic range rendering light maps to mimic natural illumination. This may upgrade the precision of the diagnosis by better presenting the anatomical details. Another potential future use is the designing of AI-guided robots to assist in actually treating patients and in the placement of dental implants. It is anticipated that a comparable technology can be created to help with normal root canal therapy or possibly endodontic microsurgery.

Conclusion

Dental professionals can use AI as a supplement to their current tools to lighten their burden and increase diagnostic accuracy, assist in decision-making, treatment planning, and estimating the prognosis. Clinical workflows can be accelerated by automated technologies, which also increases physician productivity. The practice of these technologies for supplementary views can increase diagnostic accuracy. Before neural networks may be used to assist dentists in daily practice, further research must be done on their usage in dentistry. There is some disagreement over whether using AI in dentistry is better than other approaches; it in no way takes the place of a dentist. AI must be implemented in dentistry in a safe and controlled way to guarantee that humans can still choose their course of action and direct educated decisions.

References

Aminoshariae, A., Kulild, J., Nagendrababu, V., 2021. Artificial intelligence in endodontics: current applications and future directions. Journal of Endodontics 47 (9), 1352−1357. https://doi.org/10.1016/j.joen.2021.06.003.

Asiri, A.F., Altuwalah, A.S., 2022. The role of neural artificial intelligence for diagnosis and treatment planning in endodontics: a qualitative review. Saudi Dental Journal 34 (4), 270−281. https://doi.org/10.1016/j.sdentj.2022.04.004.

Aubreville, M., Knipfer, C., Oetter, N., Jaremenko, C., Rodner, E., Denzler, J., Bohr, C., Neumann, H., Stelzle, F., Maier, A., 2017. Automatic classification of cancerous tissue in laserendomicroscopy images of the oral cavity using deep learning. Scientific Reports 7 (1). https://doi.org/10.1038/s41598-017-12320-8.

Bindal, P., Bindal, U., Lin, C.W., Kasim, N.H.A., Ramasamy, T.S.A.P., Dabbagh, A., Salwana, E., Shamshirband, S., 2017. Neuro-fuzzy method for predicting the viability of stem cells treated at different time-concentration conditions. Technology and Health Care 25 (6), 1041−1051. https://doi.org/10.3233/THC-170922.

Campo, L., Aliaga, I.J., De Paz, J.F., Garcia, A.E., Bajo, J., Villarubia, G., Corchado, J.M., 2016. Retreatment predictions in odontology by means of CBR systems. Computational Intelligence and Neuroscience 2016. https://doi.org/10.1155/2016/7485250.

Cantu, A.G., Gehrung, S., Krois, J., Chaurasia, A., Rossi, J.G., Gaudin, R., Elhennawy, K., Schwendicke, F., 2020. Detecting caries lesions of different radiographic extension on bitewings using deep learning. Journal of Dentistry 100. https://doi.org/10.1016/j.jdent.2020.103425.

Carmody, D.P., McGrath, S.P., Dunn, S.M., Van Der Stelt, P.F., Schouten, E., 2001. Machine classification of dental images with visual search. Academic Radiology 8 (12), 1239−1246. https://doi.org/10.1016/S1076-6332(03)80706-7.

Casalegno, F., Newton, T., Daher, R., Abdelaziz, M., Lodi-Rizzini, A., Schürmann, F., Krejci, I., Markram, H., 2019. Caries detection with near-infrared transillumination using deep learning. Journal of Dental Research 98 (11), 1227−1233. https://doi.org/10.1177/0022034519871884.

Chae, Y.M., Yoo, K.B., Kim, E.S., Chae, H., 2011. The adoption of electronic medical records and decision support systems in Korea. Healthcare Informatics Research 17 (3), 172−177. https://doi.org/10.4258/hir.2011.17.3.172.

Chang, H.J., Lee, S.J., Yong, T.H., Shin, N.Y., Jang, B.G., Kim, J.E., Huh, K.H., Lee, S.S., Heo, M.S., Choi, S.C., Kim, T.I., Yi, W.J., 2020. Deep learning hybrid method to automatically diagnose periodontal bone loss and stage periodontitis. Scientific Reports 10 (1). https://doi.org/10.1038/s41598-020-64509-z.

Chen, Y.W., Stanley, K., Att, W., 2020. Artificial intelligence in dentistry: current applications and future perspectives. Quintessence International 51 (3), 248−257. https://doi.org/10.3290/j.qi.a43952.

Chugal, N.M., Clive, J.M., Spångberg, L.S.W., 2003. Endodontic infection: some biologic and treatment factors associated with outcome. Oral Surgery, Oral Medicine, Oral Pathology, Oral Radiology and Endodontics 96 (1), 81−90. https://doi.org/10.1016/S1079-2104(02)91703-8.

Demant, S., Dabelsteen, S., Bjørndal, L., 2021. A macroscopic and histological analysis of radiographically well-defined deep and extremely deep carious lesions: carious lesion characteristics as indicators of the level of bacterial penetration and pulp response. International Endodontic Journal 54 (3), 319−330. https://doi.org/10.1111/iej.13424.

Deshmukh, S., 2018. AI in dentistry. Journal of the International Clinical Dental Research Organization 10, 47. https://doi.org/10.4103/jicdro.jicdro_17_18.

Ding, H., Wu, J., Zhao, W., Matinlinna, J.P., Burrow, M.F., Tsoi, J.K.H., 2023a. AI in dentistry—a review. Frontiers in Dental Medicine 4, 1085251.

Ding, H., Cui, Z., Maghami, E., Chen, Y., Matinlinna, J.P., Pow, E.H.N., Fok, A.S.L., Burrow, M.F., Wang, W., Tsoi, J.K.H., 2023b. Morphology and mechanical performance of dental crown designed by 3D-DCGAN. Dental Materials 39 (3), 320−332. https://doi.org/10.1016/j.dental.2023.02.001.

Endres, M.G., Hillen, F., Salloumis, M., 2020. Development of a deep learning algorithm for periapical disease detection in dental radiographs. Diagnostics 10. https://doi.org/10.3390/diagnostics10060430.

Fukuda, M., Inamoto, K., Shibata, N., Ariji, Y., Yanashita, Y., Kutsuna, S., Nakata, K., Katsumata, A., Fujita, H., Ariji, E., 2020. Evaluation of an artificial intelligence system for detecting vertical root fracture on panoramic radiography. Oral Radiology 36 (4), 337−343. https://doi.org/10.1007/s11282-019-00409-x.

Ghaznavi Bidgoli, S.A., Sharifi, A., Manthouri, M., 2021. Automatic diagnosis of dental diseases using convolutional neural network and panoramic radiographic images. Computer Methods in Biomechanics and Biomedical Engineering: Imaging and Visualization 9 (5), 447−455. https://doi.org/10.1080/21681163.2020.1847200.

Hiraiwa, T., Ariji, Y., Fukuda, M., Kise, Y., Nakata, K., Katsumata, A., Fujita, H., Ariji, E., 2019. A deep-learning artificial intelligence system for assessment of root morphology of the mandibular first molar on panoramic radiography. Dentomaxillofacial Radiology 48 (3), 20180218. https://doi.org/10.1259/dmfr.20180218.

Hung, K., Montalvo, C., Tanaka, R., Kawai, T., Bornstein, M.M., 2020. The use and performance of AI applications in dental and maxillofacial radiology: a systematic review. Dentomaxillofacial Radiology 49, 20190107. https://doi.org/10.1259/dmfr.20190107.

Hwang, J.J., Azernikov, S., Efros, A.A., Yu, S.X., 2018. Learning beyond human expertise with generative models for dental restorations. arXiv 23318422. https://arxiv.org.

James, B.L., Sunny, S.P., Heidari, A.E., Ramanjinappa, R.D., Lam, T., Tran, A.V., Kankanala, S., Sil, S., Tiwari, V., Patrick, S., Pillai, V., Shetty, V., Hedne, N., Shah, D., Shah, N., Chen, Z.P., Kandasarma, U., Raghavan, S.A., Gurudath, S., Nagaraj, P.B., Wilder-Smith, P., Suresh, A., Kuriakose, M.A., 2021. Validation of a point-of-care optical coherence tomography device with machine learning algorithm for detection of oral potentially malignant and malignant lesions. Cancers 13 (14), 3583. https://doi.org/10.3390/cancers13143583.

Jiang, L., Chen, D., Cao, Z., Wu, F., Zhu, H., Zhu, F., 2022. A two-stage deep learning architecture for radiographic staging of periodontal bone loss. BMC Oral Health 22 (1), 106. https://doi.org/10.1186/s12903-022-02119-z.

Johari, M., Esmaeili, F., Andalib, A., Garjani, S., Saberkari, H., 2017. Detection of vertical root fractures in intact and endodontically treated premolar teeth by designing a probabilistic neural network: an ex vivo study. Dentomaxillofacial Radiology 46 (2), 20160107. https://doi.org/10.1259/dmfr.20160107.

Khanagar, S.B., Vishwanathaiah, S., Naik, S., 2021. Application and performance of the technology in forensic odontology — a systematic review. Legal Medicine 101826.

Khanna, S.S., Dhaimade, P.A., 2017. Artificial intelligence: transforming dentistry today. Indian Journal of Basic and Applied Medical Research 6, 161—167.

Krois, J., Ekert, T., Meinhold, L., Golla, T., Kharbot, B., Wittemeier, A., Dörfer, C., Schwendicke, F., 2019. Deep learning for the radiographic detection of periodontal bone loss. Scientific Reports 9 (1), 8495. https://doi.org/10.1038/s41598-019-44839-3.

Kühnisch, J., Meyer, O., Hesenius, M., Hickel, R., Gruhn, V., 2022. Caries detection on intraoral images using artificial intelligence. Journal of Dental Research 101 (2), 158—165. https://doi.org/10.1177/00220345211032524.

Lahoud, P., EzEldeen, M., Beznik, T., Willems, H., Leite, A., Van Gerven, A., Jacobs, R., 2021. Artificial intelligence for fast and accurate 3-dimensional tooth segmentation on cone-beam computed tomography. Journal of Endodontics 47 (5), 827—835. https://doi.org/10.1016/j.joen.2020.12.020.

LeCun, Y., Bengio, Y., Hinton, G., 2015. Deep learning. Nature 521 (7553), 436—444. https://doi.org/10.1038/nature14539.

Lee, S.J., Chung, D., Asano, A., 2022. Diagnosis of tooth prognosis using AI. Diagnostics (Basel) 12, 1422. https://doi.org/10.3390/diagnostics12061422.

Lee, J.H., Kim, D.H., Jeong, S.N., Choi, S.H., 2018a. Detection and diagnosis of dental caries using a deep learning-based convolutional neural network algorithm. Journal of Dentistry 77, 106—111. https://doi.org/10.1016/j.jdent.2018.07.015.

Lee, J.H., Kim, D.H., Jeong, S.N., Choi, S.H., 2018b. Diagnosis and prediction of periodontally compromised teeth using a deep learning-based convolutional neural network algorithm. Journal of Periodontal and Implant Science 48 (2), 114—123. https://doi.org/10.5051/jpis.2018.48.2.114.

Leite, A.F., Vasconcelos, K.D.F., Willems, H., Jacobs, R., 2020. Radiomics and machine learning in oral healthcare. Proteomics — Clinical Applications 14 (3), 1900040. https://doi.org/10.1002/prca.201900040.

Lin, P.L., Huang, P.W., Huang, P.Y., Hsu, H.C., 2015. Alveolar bone-loss area localization in periodontitis radiographs based on threshold segmentation with a hybrid feature fused of intensity and the H-value of fractional Brownian motion model. Computer Methods and Programs in Biomedicine 121 (3), 117−126. https://doi.org/10.1016/j.cmpb.2015.05.004.

Lin, P.L., Huang, P.Y., Huang, P.W., 2017. Automatic methods for alveolar bone loss degree measurement in periodontitis periapical radiographs. Computer Methods and Programs in Biomedicine 148, 1−11. https://doi.org/10.1016/j.cmpb.2017.06.012.

Martins, J.N.R., Marques, D., Mata, A., Caramês, J., 2014. Clinical efficacy of electronic apex locators: systematic review. Journal of Endodontics 40 (6), 759−777. https://doi.org/10.1016/j.joen.2014.03.011.

Mol, A., van der Stelt, P.F., 1992. Application of computer-aided image interpretation to the diagnosis of periapical bone lesions. Dentomaxillofacial Radiology 21 (4), 190−194. https://doi.org/10.1259/dmfr.21.4.1299632.

Naik, M., de Ataide, I.D., Fernandes, M., Lambor, R., 2016. Future of endodontics. International Journal of Current Research 8, 016.

Ng, Y.L., Mann, V., Rahbaran, S., Lewsey, J., Gulabivala, K., 2008. Outcome of primary root canal treatment: systematic review of the literature − part 2. Influence of clinical factors. International Endodontic Journal 41 (1), 6−31. https://doi.org/10.1111/j.1365-2591.2007.01323.x.

Okada, K., Rysavy, S., Flores, A., Linguraru, M.G., 2015. Noninvasive differential diagnosis of dental periapical lesions in cone-beam CT scans. Medical Physics 42 (4), 1653−1665. https://doi.org/10.1118/1.4914418.

Orhan, K., Bayrakdar, I.S., Ezhov, M., Kravtsov, A., Özyürek, T., 2020. Evaluation of artificial intelligence for detecting periapical pathosis on cone-beam computed tomography scans. International Endodontic Journal 53 (5), 680−689. https://doi.org/10.1111/iej.13265.

Park, J.H., Hwang, H.W., Moon, J.H., Yu, Y., Kim, H., Her, S.B., Srinivasan, G., Aljanabi, M.N.A., Donatelli, R.E., Lee, S.J., 2019. Automated identification of cephalometric landmarks: part 1—comparisons between the latest deep-learning methods YOLOV3 and SSD. The Angle Orthodontist 89 (6), 903−909. https://doi.org/10.2319/022019-127.1.

Roganović, J., Radenković, M., 2023. Ethical Use of AI in Dentistry. IntechOpen. https://doi.org/10.5772/intechopen.1001828.

Saghiri, M.A., Asgar, K., Boukani, K.K., Lotfi, M., Aghili, H., Delvarani, A., Karamifar, K., Saghiri, A.M., Mehrvarzfar, P., Garcia-Godoy, F., 2012a. A new approach for locating the minor apical foramen using an artificial neural network. International Endodontic Journal 45 (3), 257−265. https://doi.org/10.1111/j.1365-2591.2011.01970.x.

Saghiri, M.A., Garcia-Godoy, F., Gutmann, J.L., Lotfi, M., Asgar, K., 2012b. The Reliability of artificial neural network in locating minor apical foramen: a cadaver study. Journal of Endodontics 38 (8), 1130−1134. https://doi.org/10.1016/j.joen.2012.05.004.

Schleyer, T.K.L., Thyvalikakath, T.P., Spallek, H., Torres-Urquidy, M.H., Hernandez, P., Yuhaniak, J., 2006. Clinical computing in general dentistry. Journal of the American Medical Informatics Association 13 (3), 344−352. https://doi.org/10.1197/jamia.M1990.

Schwendicke, F., Elhennawy, K., Paris, S., Friebertshäuser, P., Krois, J., 2020. Deep learning for caries lesion detection in near-infrared light transillumination images: a pilot study. Journal of Dentistry 92. https://doi.org/10.1016/j.jdent.2019.103260.

Schwendicke, F., Rossi, J.G., Göstemeyer, G., Elhennawy, K., Cantu, A.G., Gaudin, R., Chaurasia, A., Gehrung, S., Krois, J., 2021. Cost-effectiveness of artificial intelligence for proximal caries detection. Journal of Dental Research 100 (4), 369−376. https://doi.org/10.1177/0022034520972335.

Selvaraju, R.R., Cogswell, M., Das, A., Vedantam, R., Parikh, D., Batra, D., 2020. Grad-CAM: visual explanations from deep networks via gradient-based localization. International Journal of Computer Vision 128 (2), 336−359. https://doi.org/10.1007/s11263-019-01228-7.

Shah, N., 2009. Dental caries: the disease and its clinical management, 2nd edition. British Dental Journal 206 (9), 498. https://doi.org/10.1038/sj.bdj.2009.374.

Shah, H., Hernandez, P., Budin, F., Chittajallu, D., Vimort, J.B., Walters, R., Mol, A., Khan, A., Paniagua, B., 2018. Automatic quantification framework to detect cracks in teeth. Progress in Biomedical Optics and Imaging — Proceedings of SPIE 10578, 352. https://doi.org/10.1117/12.2293603.

Srinidhi, C.L., Ciga, O., Martel, A.L., 2021. Deep neural network models for computational histopathology: a survey. Medical Image Analysis 67, 101813. https://doi.org/10.1016/j.media.2020.101813.

Thanathornwong, B., 2018. Bayesian-based decision support system for assessing the needs for orthodontic treatment. Healthcare Informatics Research 24 (1), 22–28. https://doi.org/10.4258/hir.2018.24.1.22.

Tian, S., Wang, M., Dai, N., Ma, H., Li, L., Fiorenza, L., Sun, Y., Li, Y., 2022. DCPR-GAN: dental crown prosthesis restoration using two-stage generative adversarial networks. IEEE Journal of Biomedical and Health Informatics 26 (1), 151–160. https://doi.org/10.1109/jbhi.2021.3119394.

Tonetti, M.S., Jepsen, S., Jin, L., Otomo-Corgel, J., 2017. Impact of the global burden of periodontal diseases on health, nutrition and wellbeing of mankind: a call for global action. Journal of Clinical Periodontology 44 (5), 456–462. https://doi.org/10.1111/jcpe.12732.

Vinayahalingam, S., Kempers, S., Limon, L., Deibel, D., Maal, T., Hanisch, M., Bergé, S., Xi, T., 2021. Classification of caries in third molars on panoramic radiographs using deep learning. Scientific Reports 11 (1). https://doi.org/10.1038/s41598-021-92121-2.

Warin, K., Limprasert, W., Suebnukarn, S., Jinaporntham, S., Jantana, P., Vicharueang, S., Seal, A., 2022. AI-based analysis of oral lesions using novel deep convolutional neural networks for early detection of oral cancer. PLoS One 17 (8). https://doi.org/10.1371/journal.pone.0273508.

Wei, J., Peng, M., Li, Q., Wang, Y., 2018. Evaluation of a novel computer color matching system based on the improved back-propagation neural network model. Journal of Prosthodontics 27 (8), 775–783. https://doi.org/10.1111/jopr.12561.

Xie, X., Wang, L., Wang, A., 2010. Artificial neural network modeling for deciding if extractions are necessary prior to orthodontic treatment. The Angle Orthodontist 80 (2), 262–266. https://doi.org/10.2319/111608-588.1.

Yauney, G., Rana, A., Wong, L.C., Javia, P., Muftu, A., Shah, P., 2019. Automated process incorporating machine learning segmentation and correlation of oral diseases with systemic health. 2019 41st Annual International Conference of the IEEE Engineering in Medicine and Biology Society (EMBC). IEEE, pp. 3387–3393.

Zheng, L., Wang, H., Mei, L., Chen, Q., Zhang, Y., Zhang, H., 2021. Artificial intelligence in digital cariology: a new tool for the diagnosis of deep caries and pulpitis using convolutional neural networks. Annals of Translational Medicine 9 (9), 763. https://doi.org/10.21037/atm-21-119.

Applications and impact of artificial intelligence in veterinary sciences

Ambreen Hamadani[1], Nazir Ahmad Ganai[2], Henna Hamadani[2], Shabia Shabir[3] and Shazeena Qaiser[4]

[1]*National Institute of Technology, Srinagar, Jammu and Kashmir, India;* [2]*Sher-e-Kashmir University of Agricultural Sciences and Technology of Kashmir, Srinagar, Jammu and Kashmir, India;* [3]*Islamic University of Science and Technology (IUST), Awantipora, Jammu and Kashmir, India;* [4]*Government Dental College, Srinagar, Jammu and Kashmir, India*

Introduction

Artificial intelligence (AI) is reshaping life with its potential to mimic human intelligence while being faster and less error-prone. It is seeping into practically every sector and is transforming it for the better. Its medical and veterinary applications are wide-ranging as well (Hamadani et al., 2022a). There is thus a need to understand and discover its many capabilities. These will ensure better disease prediction, diagnosis, prognosis, and health care in veterinary practice. Since veterinary professionals are not just responsible for animal health and welfare, they are major stakeholders of One Health. The One Health concept is an interdisciplinary approach that recognizes the interconnectedness of human health, animal health, and the environment. It emphasizes the interdependence and interconnections between these three domains and the need for collaborative efforts to address health challenges. A veterinarian is a crucial link between all these, and his efforts could therefore be beneficial for all. The integration of AI with veterinary sciences would ensure the progress of this profession and the whole of mankind.

With increasing population, globalization, and climate change, the challenges for achieving One Heath are becoming more and more challenging difficult. This is why technologies like AI could come to the rescue. Among the many potential uses of AI in veterinary sciences, it is projected that it can save 30% or more time (Appleby and Basran, 2022), decrease harmful products like greenhouse gases by nearly 20% (Bajaj, 2021), improve patient outcomes by about 40% (Ahuja, 2019), and reduce treatment cost by nearly 50% (Hsieh, 2017; Ledley and Lusted, 1959).

Artificial intelligence is relatively new to veterinary sciences, but the idea that electronic computers could be used in diagnostics has been around since 1959 with the publication of Robert Ledley and Lee Lusted (Ledley and Lusted, 1959). The area of AI is vast and requires a deep understanding of advanced statistics, probability theory, computers and many other sub-fields for the development and training of algorithms. AI is a very broad term and encompasses many techniques and methodologies. Almost all of them find applications in veterinary science. Broadly they may be grouped as given in Fig. 9.1. However, their effective use and implementation do not require a working knowledge of these fields. This is also true for veterinary sciences (Appleby and Basran, 2022). Therefore, the

A Biologist's Guide to Artificial Intelligence. https://doi.org/10.1016/B978-0-443-24001-0.00009-9

FIGURE 9.1

Artificial intelligence (AI) in relation to other fields.

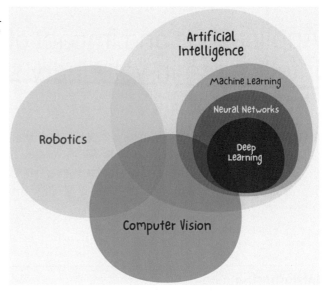

implementation of this new technology would be similar to the adoption of any new technique or tool in veterinary practice. It is important, therefore, to get a basic know-how of AI in order to know what can be adopted in practice and what its impact could be. This chapter gives a brief overview of AI techniques that are presently being used in veterinary sciences, how they are making a positive impact, and what could potentially be done in the future as well. The applications are broadly divided into subsections, which are also represented in Fig. 9.2. Technically, AI can be introduced into all aspects of veterinary practice, and this chapter is intended to introduce the reader to various practical applications like diagnostics, companion animal care, population medicine, etc.

Big data in veterinary sciences

Many veterinary procedures are now assisted by digital machines, which automatically collect and store data. These datasets are large and complex and are accumulating rapidly (VanderWaal et al., 2017). These include patient medical and vaccination records, genotypic and phenotypic data, species data, results from auto analyzers and other diagnostic tools, radiographic data, microbiome data, animal body sensors, surveillance, etc. They mostly contain all the V's of big data and are foundational in the application of AI in veterinary sciences. Researchers today are now trying to use AI so that these V's could be translated to 3 A viz. accuracy, automation, and accessibility. These include genomic data, health data, and data spanning many decades.

Additionally, global databases are now collecting and organizing data, some of them being open source. These include, AGRICOLA, which is an online database related to agriculture, PubMed: a search engine for MEDLINE database, and the Veterinary Medical Databases which presently has 7 million hospital records. These records have been compiled since 1964 and include all cases at 26

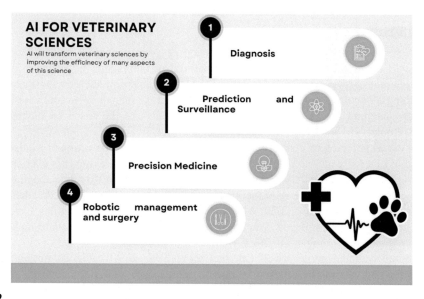

FIGURE 9.2

Applications of artificial intelligence (AI) in veterinary sciences.

universities. The World Organization for Animal Health maintains the World Animal Health Information Database, which contains records of country-specific information includes including exceptional disease reports, health status, other veterinary services, animal population, and vaccinations (University of Pretoria, 2023). Some portals which give open access to data include the European Union Open Data, RCVS Knowledge, and World Animal Health Information System. The use of big data in veterinary sciences would reduce health risks, improve management, facilitate policy decisions, and ensure animal welfare. A centralized database for consolidating the health records of sheep called Smart Sheep Breeder (Hamadani and Ganai, 2022) has also been developed by scientists.

However, for effective utilization of the data generated, it is important to strengthen the data collection process and ensure it is as accurate as possible (Hamadani et al., 2023; Franco, 2013). There is also a need to identify all additional data sources so that all associated factors can be captured. These may include pedigree data, genomic data, and even environmental data. It is also important to streamline the data and integrate it to ensure accessibility and regulation. And finally, it is important to find appropriate data analytic solutions so that inferences could be generated from them because conventional strategies alone would not suffice. Such strategies are now being implemented at organizational or national levels to facilitate the integration of AI with veterinary sciences.

Such strategies impact a myriad of areas and ensure that new technologies are utilized in a positive way. For example, the European Medicines Agency (EMA) and the Heads of Medicines Agencies (HMA) in 2022 came up with a Veterinary Big Data strategy for 2027, which defines a roadmap for promoting data-driven, progressive, digital innovations in veterinary medicine in the European Union (EU) (Parliament, 2022).

AI in diagnoses

Accurate diagnosis of animal ailments has never been easy since animals cannot speak. For this reason, using AI as an aid to diagnosis has the potential of increasing the accuracy of diagnosis manyfold. In humans, AI is being used in all fields of specialization such as anesthesiology, cardiology, radiology, oncology, infectious, and deficiency disease management and veterinary medicine is rapidly catching up. In fact, successful animal trials are the first step in many medical experiments (Franco, 2013).

A number of technologies are currently in practice, for example, the detection of Addison's disease in dogs, which is otherwise hard to diagnose in dogs. This is because its symptoms overlap with many other diseases. A major solution to this is an AI-based algorithm, which simplifies the diagnosis. This algorithm detects even the slightest abnormalities in the blood test, which aids in the identification of Addison's disease (Reagan et al., 2020; Mcevoy and Amigo, 2013). Algorithms have also been developed that can classify dog faces even in complex backgrounds, which would make it easy to identify diseased animals (Liang et al., 2023; Hamadani et al., 2013).

Many tools are available online as well. One such tool is *Sofie*, an AI-based veterinarians' assistant, which is now helping veterinarians make accurate diagnoses. *Sofie* searches a database of veterinary textbooks, journals, and conference proceedings using IBM Watson to assist with the diagnosis. Early diagnosis of economically important diseases like mastitis is critical for ensuring optimal production.

Imaging

Imaging is an important part of veterinary diagnostics and machine learning is very useful for efficiently analyzing the data. This becomes especially useful when there is a shortage of trained staff for visualization and analysis. However, the use of AI for imaging is relatively new with the first paper published only a decade ago (Mcevoy and Amigo, 2013; Abdul Ghafoor and Sitkowska, 2021). ML has the ability to analyze and interpret radiographic and other images with high precision, compare them with other cases and even prioritize high-risk patents. This is useful for streamlining the treatment process.

Today, there are more than 300 AI-enabled devices, which have been approved by FDA for radiology studies. These AI-enabled systems are useful for the diagnosis of a wide range of disorders like heart problems, embryo disorders, brain scans, and tissue and bone disorders only to name a few (Hennessey et al., 2022).

AI for disease prediction and surveillance

Artificial intelligence today is helping in the early prediction of diseases in animals. This term is not just limited to the forecasting of an animal's future with respect to whether or not it will develop a disease but also how that disease will progress. This is crucial for life-threatening diseases. For instance, today, veterinarians are able to predict which cats are most likely to develop chronic kidney disease (CKD). They are able to prevent permanent damage to kidneys by treating the cat before the onset of the disease (Bradley et al., 2019). This way the cat's health and quality of life could be improved. This is possible using an AI algorithm that was trained on 20-year data obtained from more than 100,000 cats. The data contained book work like creatinine, blood urea nitrogen, urine specific gravity, and age of the cats, which was fed to a RNN. Disease outbreaks can be devastating in animals especially poultry where sudden outbreaks of highly infectious diseases can wipe the whole flock out and cause great economic losses to the farmer. ML algorithms have been developed for many such diseases, for example, avian influenza to predict the onset of the disease based on risk factors (Yoo et al., 2022). Even for economically important diseases like mastitis, AI is playing a key role in predicting the risk based on factors like udder dimensions, udder temperature, etc (Abdul Ghafoor and Sitkowska, 2021).

Machine learning algorithms today are capable of analyzing images and learning parts of the anatomy. As a result, the algorithm can judge which part of the body has some abnormalities.

In addition to the prediction of diseases, AI is also showing promise in the assessment of mortality and morbidity risk due to particular diseases, outbreak prediction, and surveillance of contagious diseases. In addition, the multidimensional data collected are finding hitherto unknown patterns, which are useful for health policy and planning (Schwalbe and Wahl, 2020). These patterns and discoveries are not possible using conventional statistical techniques. For instance, a bat species was identified by AI, which was previously not included in the rabies reservoir list. This specie was identified using a trait profile (Worsley-Tonks et al., 2020). In addition to the detection of disease, AI can be used for pathogen detection in foods, air, water, etc.

Syndromic surveillance is done for monitoring animal health based on clinical records. The entire process of analysis is automated using data obtained from clinical and postmortem reports. Predictions can be made at the level of the individual or the clinic or at the global level (Arguello-Casteleiro et al., 2019).

An algorithm called *DeepTag* was developed, which reads veterinarians' notes to assign specific codes. Each code represents an ailment or a symptom. This algorithm is trained to predict the disease that might be relevant to the patient.

Veterinary precision medicine

Precision medicine is also called personalized medicine or molecularly guided medicine. This area is widely popular in both human medicine and veterinary sciences. The ability to provide patient-specific treatment is no longer limited to human health care. With the advancement in technology, it is becoming possible to custom-tailor therapeutics so that recovery chances are improved greatly. This is especially true for some conditions, wherein genomic and molecular analysis helps in identifying the actual drivers of the condition alongside the correlated factors responsible for it (Lloyd et al., 2016). This kind of genomic analysis also opens the window for providing patient-specific care.

The availability of affordable next-generation sequencing and molecular technologies, electronic records, digital data, use of sensors, devices for tracking, and the availability of sophisticated bioinformatics tools for interpreting large datasets are making precision medicine possible.

These data are then supplemented by data about other factors likely to have been involved in disease development and progression, for example, environmental exposures, metabolic pathways, and all other relevant details. The analysis of large genomes for extracting precise and targeted information is now possible using AI. Veterinarians occupy a unique position and have a broad perspective. They are in a position to address the gaps in understanding of this science due to their ability to change both human and animal lives for the better. They can also bridge the gap by better understanding the mechanisms of animals and their subsequent applicability to humans. This includes primary screening, confirmatory diagnosis and subsequent treatment regimens. The outcomes of such leadership could be replicated in humans with minor modifications.

Precision medicine can potentially revolutionize the veterinary sciences. A homologous genomic structure of animals that are domesticated is an important starting point in this study. This is because the molecular variants of purebreds can be used to study specific diseases as they are linked.

Precision medicine is also seeing breakthroughs in drug development. It focuses on the discovery of small molecules that can target specific mutations. One example of this is imatinib, a selective tyrosine kinase inhibitor, which targets the BCR-ABL fusion gene that has been associated with chronic myelogenous leukemia (Druker et al., 2001). Other drugs include epidermal growth factor receptor inhibitors for the treatment of breast and colorectal cancers, sunitinib, targeting tyrosine kinase receptors, etc.

Tumor screening for genetic mutations for improved differential diagnosis is also possible today. Targeted drug delivery can drastically reduce side effects (Chamundeeswari et al., 2019).

There are many studies that prove the advantages of precision medicine. Many prospective clinical trials are now assessing the hypothesis that precision medicine can improve treatment outcomes in patients with cancers (Von Hoff et al., 2010).

For instance, in the treatment of feline blood cancers, AI is being used to find the right chemotherapy for every animal based on individual drug response (Bohannan et al., 2021). Precision medicine is most popular in oncology for improving patient outcomes. Precision medicine enables the capture of patient-to-heterogeneous and complex medical data for complex diseases like cancer and since every case is different, its administration improves the chances of a positive prognosis.

The AI-based approach uses fine-needle aspiration of cancer cells and then with the help of AI and molecular, cellular, and clinical data, predicts that anticancer drugs that would be the best for the treatment of the cancer.

Blood cancer in dogs is life-threatening, and the whole process is difficult for the animal and the patient. AI can reduce the risk involved and provide the most suitable treatment options to the patients as per their specific needs. By calculating the precise amount of dosage, AI can also reduce the treatment cost and the time wasted in determining the right course of action. Precision medicine is also important for the treatment of common diseases like diabetes, obesity, and infectious diseases.

Robots in veterinary sciences

AI in veterinary hospitals can be useful for monitoring animal health as well as their behavior. There is the availability of software that can raise alarm as soon as any warning signs are demonstrated by the animal.

AI is an umbrella, which encompasses robotics and machine learning. All are improving surgeries and their outcomes. Robots like RIBA, a nursing assistant, provide assistance in lifting and handling animals, which is otherwise very hectic in a veterinary hospital (Mukai et al., 2010). Some robots (Akimana, 2016) can carry and transport supplies and equipment up to 450 kg while driving autonomously. Cleaning robots that disinfect hospitals and clinics are also available today. Some robots developed for surgeries which can also be used for veterinary surgeries are given in Table 9.1.

The Da Vinci surgery system, which was one of the first robotic systems for surgery, is gaining popularity in animal surgeries as well (Schlake et al., 2020). It is much less invasive than the

Table 9.1 Some robots developed for performing surgeries.

No.	Robot name	Use	References
1	ROBODOC	Assists in orthopedic surgeries, including joint replacements	Yu et al. (2018)
2	Da Vinci surgical system	Performs minimally invasive procedures in various specialties	Wei and Cerfolio (2019)
3	ROSA	Aids in neurosurgical procedures	Lefranc and Peltier (2016)
4	MAKOplasty	Facilitates precise joint resurfacing and replacements	Werner et al. (2014)
5	AESOP	Positions endoscopes or laparoscopes	Kraft et al. (2004)
6	TITAN	Enables remote surgery	Liu et al. (2011)
7	ZEUS surgical system	Provides robotic assistance in various procedures	Marescaux and Rubino (2003)
8	HUGO	Performs minimally invasive procedures with enhanced precision	Gueli Alletti et al. (2022)
9	RARP	Assists in various veterinary surgical procedures	Schlake et al. (2020)
10	SurgiBot	A robotic system for minimally invasive surgeries	Peters et al. (2018)
11	Robotized laser delivery system	Combines robotics with laser technology for precise surgeries	Ding et al. (2016)
12	VetSim	A virtual reality-based robotic simulator for training	VetSim Open (2023)

conventional surgeries and is more precise. It improves surgical outcomes and also reduces hospital stay times greatly. Robotic assistants shall also become a norm very soon. One such assistant is *Poli*, which is an autonomous robot for manipulation. It can pick up, pass, and deliver various hospital supplies.

Microbots also find huge applications in animals as they act as microrobots. Microscale or nanoscale robots may either be controlled by force or they are autonomous. The force involved for the first kind of robot is photonic, magnetic, or even acoustic. On the other hand, more autonomous robots can be either passive or active. The operator generally has little control over the passive robot. There are many different types of robots in literature. These include magnetic swimmers, enzyme catalysis-driven, inchworms, liposome-based systems, optical-driven microrobots, etc (Gotovtsev, 2023).

These tiny robots are able to through bodily fluids and deliver target-specific drugs with extreme precision. They are also important for manipulation as well analysis of cells or tissues without the use of any invasive technique. One such robot is an optoelectronic microrobot, which response to light patterns that are projected so as to control electric fields. This can thus be programmed to pick up cells or particles and isolate them for micromanipulation. Microbots also can be used extensively to study the rumen microflora. Scientists are now designing flexible and soft robots that are motorless (Zhang et al., 2019). These can swim and move like bacteria wherever an electric field is applied. This would be useful both for bettering our understanding of the patterns of bacteria and also in treatment.

AI and the future of veterinary medicine

Artificial intelligence and automation are all set to revamp veterinary sciences (Gotovtsev, 2023) and transform it. This shall improve veterinary health care a great deal. Animal welfare, owner satisfaction, and advanced research are all becoming possible using AI today. Patient care is important in veterinary sciences because of the inability of the actual patient to speak and AI is increasing the standard of patient care a great deal. Most importantly, it is helping in saving lives both directly and indirectly by reducing zoonosis. Food security is the natural additional benefit.

The emergence of new technologies is also changing the way we look at veterinary sciences (Hamadani et al., 2022). It is calling for the incorporation of new skills and is also impacting the age-old techniques of examining, training, diagnosing, and treating patients. The overall management is also changing and its impact is profound. This is true for both animal and human medicine.

An important question that arises is whether AI can ever completely replace veterinarians. Nothing can ever replace an able veterinarian as his/her judgment, expertise, and empathy for the diagnosis and treatment are central. However, the skillset of vets today needs additional skills so that they can keep pace with the current trends and provide the best possible health care. AI after all is only a tool in the hands of humans and unleashing its power will only make the veterinarians' services better. It will increase their efficiency and speed. It will also improve creativity and facilitate new ideas by freeing their minds from the more mundane tasks to focus on what is really important for human and animal welfare (Hamadani et al., 2022b).

Robots as pets

Robots as pets are an emerging concept that has gained some attention and interest in recent years. While traditional pets like cats and dogs offer companionship and emotional support, robots as pets aim to provide similar benefits using advanced technology. Companion robots are designed to simulate the presence of a pet, providing companionship and emotional support to their owners. These robots often have interactive features, such as responding to touch, voice commands, and displaying emotions through facial expressions or sounds. Robots as pets can offer certain advantages over traditional pets. They do not require feeding, grooming, or exercise like living animals. They can also be programmed to perform specific tasks or provide assistance to people with disabilities or special needs.

While robots can simulate aspects of pet companionship, they may lack the genuine emotional connection and unpredictability that comes with real pets. Robots lack the ability to genuinely reciprocate emotions or display the same level of loyalty and affection. Additionally, robots cannot experience physical sensations like touch or warmth, which are important elements of human—animal interactions. Owning a robot as a pet also raises ethical questions about the treatment of nonliving beings. Some argue that using robots for companionship purposes may devalue the importance of genuine animal welfare and the bond between humans and living animals.

While robotics technology has advanced significantly, the current capabilities of robots as pets are still limited compared to living animals. Most companion robots on the market are designed for simple interactions and entertainment rather than providing deep emotional connections.

Conclusion

In conclusion, the integration of AI in veterinary sciences holds tremendous potential for advancing healthcare for animals and improving veterinary practice. AI technologies such as machine learning, computer vision, and natural language processing can contribute to faster and more accurate diagnoses, personalized treatment plans, and enhanced monitoring of animal health. By analyzing vast amounts of data, AI algorithms can detect patterns and identify subtle indicators of disease or health conditions that may go unnoticed by human observers. This can lead to early detection and intervention, potentially saving animal lives and improving treatment outcomes. AI-powered diagnostic tools can assist veterinarians in interpreting medical images, analyzing laboratory results, and providing real-time recommendations based on evidence-based guidelines. Additionally, AI can enable remote monitoring of animals, allowing veterinarians to track vital signs, behavior, and other health parameters without the need for constant physical presence. This can be particularly beneficial in rural or underserved areas where access to veterinary care may be limited. Moreover, AI-powered decision support systems can aid veterinarians in making informed treatment decisions by considering various factors, including medical history, genetic data, and treatment response data. While AI in veterinary sciences offers great promise, it is important to acknowledge the need for human expertise and ethical considerations. Veterinary professionals will continue to play a crucial role in interpreting AI-generated results, making clinical judgments, and ensuring the well-being of animals. Privacy and data security measures should also be prioritized to protect sensitive animal health information. In summary, AI has the potential to revolutionize veterinary sciences by augmenting the capabilities of veterinarians, enhancing diagnostic accuracy, enabling remote monitoring, and improving treatment outcomes. By harnessing the power of AI, veterinary medicine can advance toward more personalized and efficient care for animals, ultimately benefiting animal health and welfare.

Abbreviations

ABL1   Abelson
AGRICOLA   AGRICultural Online Access
AI   Artificial intelligence
BCR   Breakpoint cluster region
CKD   Chronic kidney disease
EMA   The European Medicines Agency
FDA   Food and Drug Administration
HMA   Heads of Medicines Agencies
IBM   International Business Machines
MEDLINE   Medical Literature Analysis and Retrieval System Online
MEDLINE   MEDLARS online
ML   Machine learning
RCVS   Royal College of Veterinary Surgeons
RIBA   Robot for interactive body assistance
RNN   Recurrent neural network

References

Abdul Ghafoor, N., Sitkowska, B., 2021. MasPA: a machine learning application to predict risk of mastitis in cattle from AMS sensor data. AgriEngineering 3 (3), 575–584. https://doi.org/10.3390/agriengineering3030037.

Ahuja, A.S., 2019. The impact of artificial intelligence in medicine on the future role of the physician. PeerJ 7 (10), e7702. https://doi.org/10.7717/peerj.7702.

Akimana, B.-T., 2016. A Survey of Human-Robot Interaction in the Internet of Things.

Appleby, R.B., Basran, P.S., 2022. Artificial intelligence in veterinary medicine. Journal of the American Veterinary Medical Association 260 (8), 819–824. https://doi.org/10.2460/javma.22.03.0093.

Arguello-Casteleiro, M., Jones, P.H., Robertson, S., Irvine, R.M., Twomey, F., Nenadic, G., 2019. Exploring the automatisation of animal health surveillance through natural language processing. Lecture Notes in Computer Science 11927, 213–226. https://doi.org/10.1007/978-3-030-34885-4_17.

Bajaj, S., 2021. India Needs Widespread Adoption of Artificial Intelligence to Improve Crop Productivity. https://agriculturepost.com/opinion/india-needs-widespread-adoption-of-artificial-intelligence-to-improve-crop-productivity/. (Accessed 10 June 2023).

Bohannan, Z., Pudupakam, R.S., Koo, J., Horwitz, H., Tsang, J., Polley, A., Han, E.J., Fernandez, E., Park, S., Swartzfager, D., Qi, N.S.X., Tu, C., Rankin, W.V., Thamm, D.H., Lee, H.R., Lim, S., 2021. Predicting likelihood of in vivo chemotherapy response in canine lymphoma using ex vivo drug sensitivity and immunophenotyping data in a machine learning model. Veterinary and Comparative Oncology 19 (1), 160–171. https://doi.org/10.1111/vco.12656.

Bradley, R., Tagkopoulos, I., Kim, M., Kokkinos, Y., Panagiotakos, T., Kennedy, J., De Meyer, G., Watson, P., Elliott, J., 2019. Predicting early risk of chronic kidney disease in cats using routine clinical laboratory tests and machine learning. Journal of Veterinary Internal Medicine 33 (6), 2644–2656. https://doi.org/10.1111/jvim.15623.

Chamundeeswari, M., Jeslin, J., Verma, M.L., 2019. Nanocarriers for drug delivery applications. Environmental Chemistry Letters 17 (2), 849–865. https://doi.org/10.1007/s10311-018-00841-1.

Ding, Y., Warton, J., Kovacevic, R., 2016. Development of sensing and control system for robotized laser-based direct metal addition system. Additive Manufacturing 10, 24–35. https://doi.org/10.1016/j.addma.2016.01.002.

Druker, B.J., Talpaz, M., Resta, D.J., Peng, B., Buchdunger, E., Ford, J.M., Lydon, N.B., Kantarjian, H., Capdeville, R., Ohno-Jones, S., Sawyers, C.L., 2001. Efficacy and safety of a specific inhibitor of the BCR-ABL tyrosine kinase in chronic myeloid leukemia. New England Journal of Medicine 344 (14), 1031–1037. https://doi.org/10.1056/NEJM200104053441401.

Franco, N., 2013. Animal experiments in biomedical research: a historical perspective. Animals 3 (1), 238–273. https://doi.org/10.3390/ani3010238.

Gotovtsev, P., 2023. Microbial cells as a microrobots: from drug delivery to advanced biosensors. Biomimetics 8 (1), 109. https://doi.org/10.3390/biomimetics8010109.

Gueli Alletti, S., Chiantera, V., Arcuri, G., Gioè, A., Oliva, R., Monterossi, G., Fanfani, F., Fagotti, A., Scambia, G., 2022. Introducing the new surgical robot HUGO™ RAS: system description and docking settings for gynecological surgery. Frontiers in Oncology 12. https://doi.org/10.3389/fonc.2022.898060.

Hamadani, A., Ganai, N.A., 2022. Development of a multi-use decision support system for scientific management and breeding of sheep. Scientific Reports 12 (1). https://doi.org/10.1038/s41598-022-24091-y.

Hamadani, H., Khan, A., Banday, M., Ashraf, I., Handoo, N., Shah, A., Hamadani, A., 2013. Bovine mastitis - a disease of serious concern for dairy farmers. International Journal of Livestock Research 3 (1), 42. https://doi.org/10.5455/ijlr.20130213091143.

Hamadani, A., Ganai, N.A., Mudasir, S., Shanaz, S., Alam, S., Hussain, I., 2022a. Comparison of artificial intelligence algorithms and their ranking for the prediction of genetic merit in sheep. Scientific Reports 12 (1). https://doi.org/10.1038/s41598-022-23499-w.

Hamadani, A., Ganai, N.A., Raja, T., Alam, S., Andrabi, S.M., Hussain, I., Ahmad, H.A., 2022b. Outlier removal in sheep farm datasets using winsorization. Bhartiya Krishi Anusandhan Patrika. https://doi.org/10.18805/bkap397.

Hamadani, A., Ganai, N.A., Bashir, J., 2023. Artificial neural networks for data mining in animal sciences. Bulletin of the National Research Centre 47 (1). https://doi.org/10.1186/s42269-023-01042-9.

Hennessey, E., DiFazio, M., Hennessey, R., Cassel, N., 2022. Artificial intelligence in veterinary diagnostic imaging: a literature review. Veterinary Radiology & Ultrasound 63 (S1), 851–870. https://doi.org/10.1111/vru.13163.

Hsieh, P., 2017. AI in Medicine: Rise of the Machines.

Kraft, B.M., Jäger, C., Kraft, K., Leibl, B.J., Bittner, R., 2004. The AESOP robot system in laparoscopic surgery: increased risk or advantage for surgeon and patient? Surgical Endoscopy 18 (8), 1216–1223. https://doi.org/10.1007/s00464-003-9200-z.

Ledley, R.S., Lusted, L.B., 1959. Reasoning foundations of medical diagnosis. Science 130 (3366), 9–21. https://doi.org/10.1126/science.130.3366.9.

Lefranc, M., Peltier, J., 2016. Evaluation of the ROSA™ Spine robot for minimally invasive surgical procedures. Expert Review of Medical Devices 13 (10), 899–906. https://doi.org/10.1080/17434440.2016.1236680.

Liang, J., Cai, W., Xu, Z., Zhou, G., Li, J., Xiang, Z., 2023. A fine-grained image classification approach for dog feces using MC-SCMNet under complex backgrounds. Animals 13 (10), 1660. https://doi.org/10.3390/ani13101660.

Liu, H., Yue, C., Zhang, W., Zhu, X., Yang, G., Jia, Z., 2011. Association of the KAP 8.1 gene polymorphisms with fibre traits in inner Mongolian cashmere goats. Asian-Australasian Journal of Animal Sciences 24 (10), 1341–1347. https://doi.org/10.5713/ajas.2011.11120.

Lloyd, K.C.K., Khanna, C., Hendricks, W., Trent, J., Kotlikoff, M., 2016. Precision medicine: an opportunity for a paradigm shift in veterinary medicine. Journal of the American Veterinary Medical Association 248 (1), 45–48. https://doi.org/10.2460/javma.248.1.45.

Marescaux, J., Rubino, F., 2003. The ZEUS robotic system: experimental and clinical applications. Surgical Clinics of North America 83 (6), 1305–1315. https://doi.org/10.1016/S0039-6109(03)00169-5.

Mcevoy, F.J., Amigo, J.M., 2013. Using machine learning to classify image features from canine pelvic radiographs: evaluation of partial least squares discriminant analysis and artificial neural network models. Veterinary Radiology & Ultrasound 54 (2), 122–126. https://doi.org/10.1111/vru.12003.

Mukai, T., Hirano, S., Nakashima, H., Kato, Y., Sakaida, Y., Guo, S., Hosoe, S., 2010. Development of a nursing-care assistant robot RIBA that can lift a human in its arms. In: IEEE/RSJ 2010 International Conference on Intelligent Robots and Systems, IROS 2010 - Conference Proceedings, pp. 5996–6001. https://doi.org/10.1109/IROS.2010.5651735.

Parliament, European, 2022. Regulation (EU) 2019/6 of the European Parliament and of the Council. European Union.

Peters, B.S., Armijo, P.R., Krause, C., Choudhury, S.A., Oleynikov, D., 2018. Review of emerging surgical robotic technology. Surgical Endoscopy 32 (4), 1636–1655. https://doi.org/10.1007/s00464-018-6079-2.

Reagan, K.L., Reagan, B.A., Gilor, C., 2020. Machine learning algorithm as a diagnostic tool for hypoadrenocorticism in dogs. Domestic Animal Endocrinology 72, 106396. https://doi.org/10.1016/j.domaniend.2019.106396.

Schlake, A., Dell'Oglio, P., Devriendt, N., Stammeleer, L., Binetti, A., Bauwens, K., Terriere, N., Saunders, J., Mottrie, A., de Rooster, H., 2020. First robot-assisted radical prostatectomy in a client-owned Bernese mountain dog with prostatic adenocarcinoma. Veterinary Surgery 49 (7), 1458–1466. https://doi.org/10.1111/vsu.13448.

Schwalbe, N., Wahl, B., 2020. Artificial intelligence and the future of global health. The Lancet 395 (10236), 1579—1586. https://doi.org/10.1016/S0140-6736(20)30226-9.

University of Pretoria, 2023. Veterinary Science Open Access Resources: Datasets, Databases and Repositories. https://library.up.ac.za/c.php?g=1085098&p=7950806. (Accessed 10 June 2023).

VanderWaal, K., Morrison, R.B., Neuhauser, C., Vilalta, C., Perez, A.M., 2017. Translating big data into smart data for veterinary epidemiology. Frontiers in Veterinary Science 4. https://doi.org/10.3389/fvets.2017.00110.

VetSim Open, 2023. VetSim Open. https://www.vetsim.org/open-vetsim.

Von Hoff, D.D., Stephenson, J.J., Rosen, P., Loesch, D.M., Borad, M.J., Anthony, S., Jameson, G., Brown, S., Cantafio, N., Richards, D.A., Fitch, T.R., Wasserman, E., Fernandez, C., Green, S., Sutherland, W., Bittner, M., Alarcon, A., Mallery, D., Penny, R., 2010. Pilot study using molecular profiling of patients' tumors to find potential targets and select treatments for their refractory cancers. Journal of Clinical Oncology 28 (33), 4877—4883. https://doi.org/10.1200/JCO.2009.26.5983.

Wei, B., Cerfolio, R.J., 2019. Surgical approaches to remove the esophagus: robotic. Shackelford's Surgery of the Alimentary Tract: 2 Volume Set 424—430. https://doi.org/10.1016/B978-0-323-40232-3.00186-2.

Werner, S.D., Stonestreet, M., Jacofsky, D.J., 2014. Makoplasty and the accuracy and efficacy of robotic-assisted arthroplasty. Surgical Technology International 24, 302—306.

Worsley-Tonks, K.E.L., Escobar, L.E., Biek, R., Castaneda-Guzman, M., Craft, M.E., Streicker, D.G., White, L.A., Fountain-Jones, N.M., 2020. Using host traits to predict reservoir host species of rabies virus. PLoS Neglected Tropical Diseases 14 (12). https://doi.org/10.1371/journal.pntd.0008940.

Yoo, D.s., Song, Y.h., Choi, D.w., Lim, J.S., Lee, K., Kang, T., 2022. Machine learning-driven dynamic risk prediction for highly pathogenic avian influenza at poultry farms in Republic of Korea: daily risk estimation for individual premises. Transboundary and Emerging Diseases 69 (5), 2667—2681. https://doi.org/10.1111/tbed.14419.

Yu, F., Li, L., Teng, H., Shi, D., Jiang, Q., 2018. Robots in orthopedic surgery. Annals of Joint 3, 15. https://doi.org/10.21037/aoj.2018.02.01.

Zhang, S., Scott, E.Y., Singh, J., Chen, Y., Zhang, Y., Elsayed, M., Chamberlain, M.D., Shakiba, N., Adams, K., Yu, S., Morshead, C.M., Zandstra, P.W., Wheeler, A.R., 2019. The optoelectronic microrobot: a versatile toolbox for micromanipulation. Proceedings of the National Academy of Sciences 116 (30), 14823—14828. https://doi.org/10.1073/pnas.1903406116.

Advancing precision agriculture through artificial intelligence: Exploring the future of cultivation

10

Rohitashw Kumar, Muneeza Farooq and Mahrukh Qureshi

College of Agricultural Engineering and Technology, Sher-e-Kashmir University of Agricultural Sciences and Technology of Kashmir, Srinagar, Jammu and Kashmir, India

Introduction

The global population has reached 7.8 billion and is projected to reach 9.9 billion by 2050, putting immense pressure on the agricultural industry to produce more food than ever before (Bhat and Huang, 2021). Insufficient access to food affects around 821 million people worldwide, and according to the estimates of food and agriculture organizations, agriculture production would need to expand by 70% by 2050 in order to feed the world's rising population (Bhat and Huang, 2021). This escalating demand, coupled with the need for sustainable and efficient farming practices, has spurred the development of precision agriculture, a transformative approach to farming. Precision agriculture, also known as smart farming, is an innovative approach that leverages advanced technologies to optimize agricultural practices. Precision agriculture utilizes advanced technologies, including artificial intelligence (AI), to optimize farming operations, maximize yields, reduce resource wastage, and enhance sustainability. In recent years, the integration of AI has revolutionized the field, enabling farmers to make data-driven decisions, enhance productivity, and minimize resource wastage (Torky and Hassanein, 2020). The emergence of precision agriculture, combined with advancements in AI, has transformed the way farmers approach crop production.

John McCarthy first introduced the term "AI" during the Dartmouth Conference in 1956. He described it as the scientific and engineering field dedicated to developing intelligent machines or, more precisely, intelligent computer programs. AI falls within the domain of computer science and incorporates various techniques such as machine learning, deep learning (DL) algorithms, Internet of Things (IoT), image processing, artificial neural networks (ANNs), wireless sensor networks (WSN), and robotics (Fig. 10.1).

The primary goal of AI is to learn from data and extrapolate information in an effort to mimic human intelligence (Javaid et al., 2023). By integrating AI technologies, farmers can optimize resource allocation, improve crop yields, and minimize environmental impact. AI, with its ability to process vast amounts of data, analyze patterns, and make accurate predictions, has emerged as a game-changer in precision agriculture. By leveraging AI algorithms, farmers can gain valuable insights into crop health, soil conditions, weather patterns, and various other factors influencing farm productivity. These

A Biologist's Guide to Artificial Intelligence. https://doi.org/10.1016/B978-0-443-24001-0.00010-5

Copyright © 2024 Elsevier Inc. All rights are reserved, including those for text and data mining, AI training, and similar technologies.

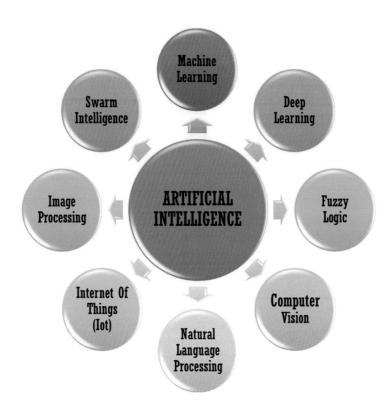

FIGURE 10.1

Different techniques used in artificial intelligence.

insights enable farmers to make data-driven decisions and implement targeted interventions, leading to improved efficiency, increased yields, and minimized environmental impact.

One of the key pillars of AI-driven precision agriculture is remote sensing (RS). RS technologies, such as satellite imagery, drones, and ground-based sensors, capture data on crop growth, moisture levels, nutrient content, and pest infestations (Brown, 2015; Ampatzidis and Partel, 2019). Landsat, Sentinel GeoEye, and WorldView satellite systems are a few examples of satellites providing agriculture-related RS data (Brown, 2015). These data are then processed by AI algorithms that identify patterns, detect anomalies, and generate actionable recommendations. For instance, AI-powered image analysis can identify specific crop diseases or nutrient deficiencies, allowing farmers to intervene precisely and apply targeted treatments, thus reducing the overall use of pesticides and fertilizers (Dharmaraj and Vijayanand, 2018). Furthermore, AI enables predictive analytics in precision agriculture. By analyzing historical data and real-time information, AI models can forecast crop yields, predict pest outbreaks, estimate water requirements, and optimize harvesting schedules. These predictive capabilities empower farmers to anticipate challenges and take proactive measures, such as adjusting irrigation schedules, implementing early pest management strategies, or planning optimal resource allocation, ultimately leading to higher yields and improved profitability.

AI also plays a significant role in autonomous farming systems, enabling the development of self-driving tractors, robotic harvesters, and automated irrigation systems (Javaid et al., 2023). These autonomous machines utilize AI algorithms to navigate fields, monitor crop conditions, and perform tasks such as planting, spraying, and harvesting with remarkable precision. Weeds and plant diseases can be detected by utilizing autonomous and semi-autonomous equipment such as unmanned aerial vehicles (UAV) and robots (Boursianis et al., 2022). By automating repetitive and labor-intensive operations, farmers can save time, reduce costs, and allocate their resources more efficiently.

In addition to on-field applications, AI-powered platforms and mobile applications provide farmers with user-friendly interfaces to monitor and manage their farms remotely. These platforms integrate data from various sources, including weather forecasts, soil sensors, and equipment telemetry, providing farmers with real-time insights into their farms' performance. Farmers can access personalized recommendations, monitor crop health, track resource usage, and even remotely control farm machinery, all from the convenience of their smartphones or computers. Moreover, AI-driven analytics contribute to sustainable farming practices. By optimizing resource allocation, such as water, fertilizers, and pesticides, precision agriculture minimizes waste and reduces environmental impact. AI algorithms consider multiple variables, including soil type, weather patterns, and plant characteristics, to determine the precise amount of inputs required, thereby minimizing overuse and leaching into the ecosystem. This chapter explores the applications, benefits, and challenges of precision agriculture powered by AI.

Understanding precision agriculture

Precision agriculture, also known as smart farming, involves the use of advanced technologies to collect and analyze data, enabling farmers to make informed decisions and implement precise actions for crop management. This approach aims to optimize productivity, minimize input wastage, and enhance sustainability. Precision agriculture also refers to the practice of using advanced technologies and data-driven techniques to optimize agricultural operations with precision and efficiency. It involves the integration of various technologies, such as GPS, RS, drones, and data analytics, to collect and analyze information about soil conditions, crop growth, weather patterns, and other relevant factors. By gaining a deeper understanding of the variability within a field, farmers can make more informed decisions about irrigation, fertilization, pest management, and harvesting, among other processes. This approach enables farmers to apply the right inputs, in the right quantities, at the right time, and in the right locations, thus maximizing yields, minimizing waste, and reducing environmental impact. The key principle behind precision agriculture is to treat each field or even each part of a field as unique, recognizing that different areas may have different requirements. This targeted and site-specific approach allows farmers to optimize their resources, improve productivity, and ultimately contribute to sustainable and responsible agricultural practices. Some of the benefits of precision agriculture are shown in Fig. 10.2.

Need for AI in precision agriculture

Since farmers are under a lot of pressure to meet the rising demand, they need a strategy to enhance output. Due to the restricted availability of fertile soil, farming will also require innovation. To help farmers reduce their risks or, at the very least, manage them, solutions must be developed. Climate

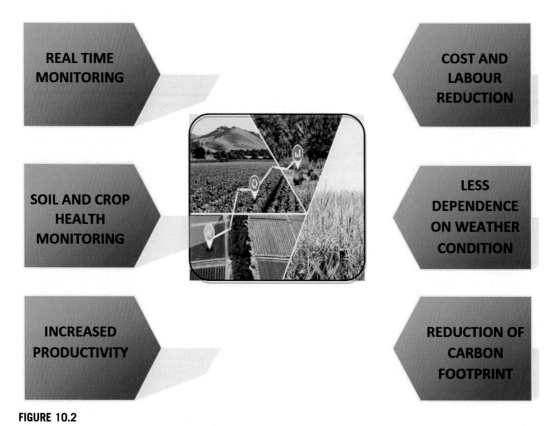

REAL TIME
MONITORING

COST AND
LABOUR
REDUCTION

SOIL AND CROP
HEALTH
MONITORING

LESS
DEPENDENCE
ON WEATHER
CONDITION

INCREASED
PRODUCTIVITY

REDUCTION OF
CARBON
FOOTPRINT

FIGURE 10.2

Benefits of adopting precision agriculture.

change, extensive use of chemicals, and pest and weed infestation are a threat to crop productivity and hence need effective resolution. One of the most interesting possibilities is the widespread use of AI in agriculture. Among the uses of AI in agriculture are predicting the spread of diseases, simulating crop development, assessments of nutrient and agricultural chemicals loss, etc.

Artificial intelligence plays a crucial role in precision agriculture by processing and interpreting vast amounts of data collected from various sources, including RS, sensors, and farm machinery. Data collection accuracy and the effective use of AI technologies are markers of the status of the agriculture sector's development and its potential for future expansion (Vazquez et al., 2021). With AI, it is simple to comprehend how rain, solar radiation, and wind and will affect agricultural growing cycles. Forecasts of the weather will be useful to farmers since they can plan and evaluate when to plant seeds. AI solutions may help farmers increase the quality of their goods, decrease waste, and ensure speedier access to markets. Self-driving bots are being developed by several businesses to manage labor-intensive agricultural procedures. These agricultural robots are a complement to human labor and may provide work with improved quality, cheaper prices, and increased productivity. The adoption of AI in agriculture is depicted in Fig. 10.3.

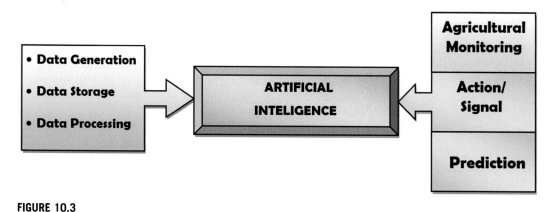

FIGURE 10.3

Adoption of artificial intelligence in agriculture.

Application of AI in precision agriculture

Artificial intelligence can be applied in agriculture to improve the management of crops and productivity by promptly recognizing plant diseases and effectively delivering agrochemicals. Yield predictions, weather forecasting, monitoring of soil and crops, pest infestation, etc. are some of the ways in which AI can aid the farmer. Some of the applications (Table 10.1) are discussed below.

Data collection and analysis

Artificial intelligence technologies facilitate efficient data collection and analysis, providing farmers with valuable insights to enhance decision-making. Sensor data are also increasing at a great pace (Hamadani et al., 2022a). Drones equipped with multispectral or hyperspectral cameras can capture high-resolution images of fields, allowing AI algorithms to analyze plant health, detect diseases, and assess nutrient deficiencies. Satellite imagery further assists in monitoring large-scale agricultural areas, identifying spatial variations in crop growth, and predicting yield potential. AI is being used for making many kinds of predictions on animal farms (Hamadani et al., 2022b,c, 2023). Soil sensors and weather stations provide real-time data on soil moisture, temperature, and weather conditions, aiding in irrigation scheduling and pest management. In order to determine the dosage that would have the least negative impact on the environment while optimizing productivity, AI makes use of datasets to analyze the environmental effects of using different doses and types of fertilizers (Javaid et al., 2023). Forecasting algorithms may be used to estimate parameters like precipitation and evapotranspiration. Machine learning models may be trained to offer significant information regarding soil water content, temperature, and overall state when paired with soil samples and other data. Data may be used by farmers to irrigate their crops more effectively. These automated technologies can keep track of farming requirements, making agriculture less labor-intensive while becoming more productive (Jose et al., 2021).

Table 10.1 Some applications of artificial intelligence in precision agriculture.

S.NO.	Application	Description	References
1	Crop yield prediction	• Identification and forecasting of crop yield • Integrates weather forecasts with AI, to help farmers decide what type of crop can be grown and when to plant seeds • Helps farmers decide which crops will be the most profitable • Agrochemicals reduce the likelihood of crop failure	Sharma et al. (2021), Orn et al. (2020), and Zhang and Wang (2020)
2	Crop and soil monitoring	• Detects pests, problems in the soil, and nutrient deficiencies in the soil • Satellite images, the IoT, and drones used to gather data and processed by AI-based apps • May aid farmers in crop data assessment, operation tracking, and weather monitoring	Kodama and Hata (2019), Sishodia et al. (2020), and Tang et al. (2018)
3	Prediction of weather	• AI-based modern weather forecasting methods make it simpler for farmers to make the right choices when planning their crops • Helps to organize farming operations and crop varieties effectively	Sehgal et al. (2017) and Sinwar et al. (2019)
4	Weed and pest control	• Using picture recognition technology, different insect and pests may be recognized and handled • Conduct remedial measures such as mechanical weed removal or pesticide application • AI may quickly detect and spray specific plants	McAllister et al. (2019), Partel et al. (2019), and Chen et al. (2020)
5	Intelligence spraying	• Detect infestations and accurately spray agro-chemicals in the appropriate area, decreasing consumption • AI assists farmers in identifying sustainable resource utilization patterns in order to avoid water scarcity and land damage • Drone spraying of chemicals increases effectiveness while decreasing human labor and the workload	Abbas et al. (2020), Chen et al. (2021), and Kim et al., 2020
6	Self-driving tractors and robotics	• More precise technology • Less prone to errors, and may work for longer periods of time • Laborious tasks such as gathering fruits and vegetables etc. can be easily done	Barrile et al. (2022) and Husti (2019)
7	Water management and smart irrigation	• Helps to utilize water more effectively, avoiding both under and overirrigation	Kamienski et al. (2019), Krishnan et al. (2022), and Lowe et al. (2022)

Table 10.1 Some applications of artificial intelligence in precision agriculture.—cont'd

S.NO.	Application	Description	References
8	Cultivation and harvesting of crops	• Aid in the design of a smart water management system for sustainable water usage by AI, DL, and IoT • Applications include water use optimization, wastewater treatment, water pollution control, automated water quality and level monitoring, aquaponics, and hydroponics • Farmers are supported by AI-powered robots and drones for tasks such as cultivation and harvesting • The most prevalent applications of AI in agriculture include predictive analytics, robotics, UAVs, and autonomous farm vehicles • AI technology aids in the early detection of issues, allowing farmers to prevent significant crop loss or damage	Ennouri et al. (2021) and Sarkar et al. (2022)

Crop monitoring and management

Artificial intelligence–powered algorithms enable farmers to monitor and manage crops with precision. Through machine learning techniques, AI systems analyze data collected from sensors, drones, and satellites to monitor plant growth, detect anomalies, and predict yield outcomes. By combining this information with historical data and weather forecasts, AI algorithms generate recommendations for optimal resource allocation, such as irrigation, fertilization, and pest control. Farmers can benefit from a data-driven approach that enables them to optimize crop health, minimize expenses, and maximize yields. By considering various soil and atmospheric factors like soil type, pH level, nutrient content (nitrogen, phosphate, potassium), organic carbon, calcium, magnesium, sulfur, manganese, copper, iron, depth, temperature, rainfall, and humidity, crop prediction methods are employed to forecast the most suitable crop for cultivation (Dahikar and Rode, 2014).

Artificial intelligence has emerged as a transformative technology in the realm of precision agriculture, particularly in the domain of crop monitoring and management. McKinion and Lemmon (1985) first proposed using AI techniques in crop management. Through the integration of advanced algorithms and machine learning techniques, AI has revolutionized the way farmers analyze and optimize their agricultural practices. AI-powered sensors, drones, and satellite imagery enable real-time monitoring of crop health, growth patterns, and environmental conditions. These data are then processed by sophisticated AI models that can identify disease outbreaks, detect nutrient deficiencies, predict yield potentials, and recommend precise interventions. By harnessing AI, farmers can make informed decisions regarding irrigation, fertilization, pest control, and harvesting, leading to enhanced productivity, reduced costs, and minimized environmental impact. For instance, Prakash et al. created a fuzzy logic-based soybean crop management system "Prithvi" that gave guidance on crop selection, fertilizer use, and pest-related concerns. The application of AI in crop monitoring and management has the potential to revolutionize agriculture, enabling farmers to achieve higher yields, optimize resource allocation, and contribute to sustainable food production in an increasingly dynamic and challenging world.

Weed and pest control

Artificial intelligence contributes to effective weed and pest management in precision agriculture. Weed infestations decrease the farmer's profitability considerably by reducing the yields of crops by 50% or more (Neil Harker, 2001; Fahad et al., 2015). Weed management in many crops like wheat, maize, barley, etc., can be accomplished by employing computer-based decision-making and GPS-controlled patch spraying by utilizing UAV or drone data for digital image analysis and online weed identification (John et al., 2020). Machine learning algorithms can be trained to recognize weed species and distinguish them from crops, allowing for targeted herbicide application. Similarly, AI-based pest detection systems can identify and monitor insect populations, enabling early intervention and minimizing the use of pesticides. By precisely targeting weeds and pests, farmers can minimize chemical usage, reduce environmental impact, and enhance crop health.

The application of AI in weed and pest control in precision agriculture has revolutionized the way farmers manage these challenges. AI algorithms and technologies offer precise and targeted solutions, minimizing the use of chemicals and reducing environmental impact. Through image analysis, AI can accurately identify and distinguish between crops and weeds, enabling farmers to implement specific control measures. Machine learning models trained on extensive datasets of weed images can recognize different weed species and growth stages, allowing farmers to select the most effective control methods. AI-powered robotic systems equipped with computer vision technology can autonomously detect and remove weeds, reducing the need for manual labor and herbicide usage. In pest control, AI utilizes sensor networks and real-time data analysis to monitor pest populations and predict outbreaks. By integrating data from various sources, AI algorithms provide early warnings to enable protective pest control measures. Decision support systems based on AI offers farmers recommendation for pest control, considering factors like pest thresholds, alternative management techniques, and sustainability by leveraging AI in weed and pest control, precision agriculture optimizes interventions, reduces reliance on broad-spectrum pesticides, and enhances overall crop health and productivity. DSSs are also used in animals for similar reasons (Hamadani and Ganai, 2022).

Irrigation and soil management

Artificial intelligence systems that have undergone training and possess knowledge about historical weather patterns, suitable crops, and soil conditions can automate irrigation processes and enhance overall crop yield. In one study, an ANN model was developed by integrating satellite data and features extracted from current soil maps to predict soil texture, specifically sand, clay, and silt concentrations (Zhao et al., 2009). DL-based tools and sensors utilizing satellite imagery are available to monitor soil conditions and improve its value per acre (Sharma, 2021). RS techniques can also be employed to map water resources, providing information on soil moisture levels and guiding farmers in selecting appropriate crops.

Automated irrigation systems play a crucial role in irrigation management, monitoring, and conservation of resources with minimal human intervention. Smart irrigation helps reduce water wastage by regulating the amount of water supplied to plants based on their needs. Various irrigation systems, such as drip and sprinklers, can be automated and controlled through sensors, timers, and other devices. Intelligent irrigation controllers utilize data on weather conditions, soil characteristics, evaporation rates, and plant water requirements to automatically adjust the watering schedule to suit specific site conditions (Sivarethinamohan et al., 2021).

Autonomous farming

Artificial intelligence–driven automation revolutionizes farming operations by replacing manual labor with autonomous machinery. Robotic systems equipped with AI technologies, computer vision, and machine learning algorithms can perform tasks such as seeding, planting, weeding, and harvesting with precision and efficiency. These machines can navigate fields, identify and interact with plants, and execute tasks according to predefined algorithms. The robots utilize AI models to recognize and differentiate between crops, weeds, and soil conditions, enabling precise and targeted actions. AI, self-driving tractors, and IoT might help tackle one of farming's most pressing problems of labor shortage. A driverless tractor is made up of many machine-tractor components that perform plowing, cultivating, and cropping (Senkevich et al., 2019). This technology is less expensive and more accurate. The software is connected to sensors, radar, and GPS systems in a driverless tractor, which enables them to sow seeds, apply pesticides, and harvest crops.

Additionally, AI algorithms analyze data from sensors and satellite imagery to monitor crop health, growth rates, and environmental factors. This information is used to make real-time decisions regarding irrigation, fertilization, and pest control, optimizing resource utilization and maximizing crop yields. The application of AI in autonomous farming has transformed precision agriculture by enabling the development of sophisticated and efficient farming systems. Autonomous farming powered by AI not only increases efficiency and productivity but also reduces labor requirements and minimizes the use of chemical inputs, promoting sustainable and environmentally friendly agriculture practices. It has also minimized human errors and physical strain. Through the integration of AI algorithms, robotics, and sensor technologies, autonomous farming has become a reality.

Benefits of precision agriculture using AI
Increased productivity and yield

Precision agriculture powered by AI enables farmers to optimize crop productivity and maximize yields. By leveraging AI algorithms for data analysis and decision-making, farmers can make accurate and timely interventions in crop management. Precise irrigation scheduling, targeted nutrient application, and early pest detection contribute to healthier crops and improved yield.

The application of AI in precision agriculture has significantly contributed to increased productivity and higher crop yields. AI algorithms, combined with advanced sensors and data analytics, have enabled farmers to gather and process vast amounts of information about their crops, soil conditions, weather patterns, and other relevant factors. By analyzing these data, AI models can provide valuable insights and recommendations for optimizing farming practices. For example, AI can accurately determine the optimal timing for planting, irrigation, and harvesting based on real-time environmental conditions and crop growth stages. AI can also identify and predict the presence of pests, diseases, or nutrient deficiencies, allowing farmers to take proactive measures to mitigate risks. Furthermore, AI-powered machinery and robotics can perform tasks with precision and efficiency, such as applying fertilizers or pesticides in targeted areas, leading to optimized resource allocation and reduced waste. Ultimately, the application of AI in precision agriculture empowers farmers to make data-driven decisions, streamline operations, and achieve higher levels of productivity and yield, thereby enhancing food production and meeting the growing global demand.

Resource efficiency

Artificial intelligence—based precision agriculture promotes resource-efficient farming practices. By precisely allocating resources such as water, fertilizers, and pesticides based on crop requirements, farmers can minimize wastage and reduce environmental impact. AI-driven irrigation systems monitor soil moisture levels and weather forecasts to determine optimal irrigation schedules, preventing both under and overirrigation. Similarly, AI algorithms analyze soil nutrient levels and crop needs to provide recommendations for targeted fertilization, reducing excess nutrient application.

The application of AI in precision agriculture has brought about significant improvements in resource efficiency. By leveraging AI algorithms and advanced technologies, farmers can optimize the use of key resources such as water, fertilizers, and energy. AI-based sensors and data analysis systems provide real-time monitoring of soil moisture levels, enabling precise irrigation strategies that prevent overwatering and minimize water waste. AI algorithms can also analyze soil composition and nutrient levels to determine precise fertilizer requirements, reducing unnecessary applications and minimizing environmental impact. Additionally, AI-driven models can analyze weather patterns and historical data to optimize planting schedules and maximize resource utilization. Furthermore, the use of AI-powered machinery and robotics helps streamline operations, reducing human error and ensuring efficient use of energy and labor resources. By enhancing resource efficiency through AI, precision agriculture not only reduces costs and minimizes environmental impact but also contributes to sustainable farming practices and the conservation of natural resources.

Cost reduction and economic viability

Precision agriculture using AI can lead to cost reductions and improved economic viability for farmers. By optimizing resource usage, minimizing chemical inputs, and reducing labor requirements through automation, farmers can lower production costs. AI-powered decision support systems provide insights into the market trends, enabling farmers to make informed decisions about crop selection, timing and pricing, ultimately increasing profitability.

The application of AI in precision agriculture has proven to be instrumental in reducing costs and enhancing the economic viability of farming operations. AI-powered technologies enable farmers to make data-driven decisions that optimize resource allocation and minimize waste, ultimately leading to cost savings. For instance, AI algorithms analyze a wealth of data, including weather patterns, soil conditions, and crop health, to determine the optimal timing and dosage for inputs such as fertilizers and pesticides. By precisely applying these inputs only when and where needed, farmers can minimize their expenditure on agrochemicals while ensuring effective crop protection. Additionally, AI-driven machinery and robotics automate various tasks, reducing labor costs and increasing operational efficiency. Autonomous robots equipped with AI capabilities can perform activities such as planting, weeding, and harvesting with high precision, reducing the need for manual labor and associated expenses.

Furthermore, AI aids in predicting and mitigating potential risks enabling farmers to proactively manage challenges and reduce losses. AI models can analyze historical data and current conditions to forecast disease outbreaks, pest infestations, or adverse weather events. This allows farmers to take preventive measures, implement targeted interventions, and minimize crop damage. By mitigating risks and optimizing operations, AI helps improve the overall economic viability of precision agriculture, making it a sustainable and profitable approach for farmers.

Environmental sustainability

AI-driven precision agriculture promotes environmentally sustainable farming practices. By minimizing the use of water, fertilizers, and pesticides through targeted application, farmers can mitigate the negative impacts on ecosystems and reduce water pollution. Additionally, precise resource management helps preserve soil health, prevent erosion, and conserve biodiversity.

The application of AI in precision agriculture plays a crucial role in promoting environmental sustainability. AI-powered technologies enable farmers to adopt practices that reduce the environmental impact of farming operations. By leveraging AI algorithms and sensor technologies, farmers can monitor and optimize resource usage, such as water and fertilizers, to minimize waste and pollution. AI models analyze data from sensors, satellite imagery, and historical records to provide precise recommendations on irrigation schedules, fertilizer application rates, and crop rotation strategies. This precision and efficiency in resource management help conserve water resources, reduce nutrient runoff, and decrease the use of agrochemicals, thereby minimizing environmental pollution and promoting sustainable farming practices.

Furthermore, AI aids in the early detection and management of pests, diseases, and weed infestations. AI models can analyze vast amounts of data to identify patterns and indicators of crop health issues, enabling farmers to take proactive measures before the problem escalates. This reduces the reliance on broad-spectrum pesticides and allows for targeted interventions, minimizing the impact on beneficial organisms and promoting biodiversity in agricultural ecosystems. Additionally, AI-powered drones and robots can precisely apply pesticides or perform targeted weed control, reducing the overall quantity of chemicals used. By integrating AI in precision agriculture, farmers can achieve a balance between productivity and environmental sustainability, contributing to the conservation of ecosystems, biodiversity, and long-term agricultural viability.

Challenges and considerations
Data quality and integration

The success of AI in precision agriculture relies on the availability of high-quality, diverse data from multiple sources. However, integrating data from various platforms and ensuring its accuracy, consistency, and compatibility can be challenging. Standardized data formats, data sharing agreements, and interoperability among different systems are necessary to overcome these challenges.

Infrastructure and connectivity

Effective implementation of AI in precision agriculture requires reliable infrastructure and connectivity, which can be a limitation in remote or underdeveloped areas. Access to high-speed Internet, sensor networks, and technological support is essential for seamless data collection, transmission, and analysis.

Technical expertise and training

The successful adoption of AI technologies in precision agriculture requires farmers to have technical expertise or access to trained professionals. Providing training programs and support to farmers, especially in rural areas, is crucial for them to understand and utilize AI-based systems effectively.

Ethical and regulatory considerations

The ethical and regulatory aspects of AI in precision agriculture require attention. Issues such as data privacy, ownership, transparency, and potential bias in AI algorithms need to be addressed. Ensuring transparent and accountable AI systems, along with the development of appropriate regulatory frameworks, can help address these concerns.

Conclusion

Precision agriculture utilizing AI holds immense potential to revolutionize the agriculture sector. Through the integration of AI technologies in data collection, analysis, and decision-making, farmers can optimize productivity, conserve resources, and promote sustainability. Overcoming challenges related to data integration, infrastructure, technical expertise, and ethical considerations will be crucial for the widespread adoption and responsible implementation of AI in precision agriculture. As AI technologies continue to evolve and improve, precision agriculture is poised to make significant contributions to global food security and sustainable farming practices. By harnessing the capabilities of AI, farmers are able to analyze vast amounts of data, identify patterns, and make data-driven decisions with unprecedented accuracy and efficiency. AI-powered technologies such as machine learning and computer vision enable farmers to automate tasks, predict crop health and yield, detect diseases or pests early on, and optimize resource allocation. This not only leads to higher productivity and profitability but also promotes environmental sustainability by reducing the use of water, fertilizers, and pesticides.

The integration of AI in precision agriculture has the potential to significantly transform the agricultural industry, allowing farmers to overcome challenges associated with population growth, climate change, and resource scarcity. With AI, farmers can gain deeper insights into their fields, adapt their practices to changing conditions, and make informed decisions in real time. Furthermore, AI-driven systems can facilitate knowledge sharing and collaboration among farmers, enabling them to learn from each other's experiences and collectively improve agricultural practices. As we continue to advance AI technologies and refine their applications in precision agriculture, we can expect even greater advancements in productivity, sustainability, and food security, ensuring a brighter future for the global farming community.

Abbreviations

AI Artificial intelligence
ANN Artificial neural networks
DEM Digital elevation model
DL Deep learning
GPS Global Positioning System
RS Remote sensing
UAV Unmanned aerial vehicle
WSN Wireless sensor networks

References

Abbas, I., Liu, J., Faheem, M., Noor, R.S., Shaikh, S.A., Solangi, K.A., Raza, S.M., 2020. Different sensor based intelligent spraying systems in agriculture. Sensors and Actuators, A: Physical 316. https://doi.org/10.1016/j.sna.2020.112265.

Ampatzidis, Y., Partel, V., 2019. UAV-based high throughput phenotyping in citrus utilizing multispectral imaging and artificial intelligence. Remote Sensing 11 (4). https://doi.org/10.3390/rs11040410.

Barrile, V., Simonetti, S., Citroni, R., Fotia, A., Bilotta, G., 2022. Experimenting agriculture 4.0 with sensors: a data fusion approach between remote sensing, UAVs and self-driving tractors. Sensors 22 (20). https://doi.org/10.3390/s22207910.

Bhat, S.A., Huang, N.F., 2021. Big data and AI revolution in precision agriculture: survey and challenges. IEEE Access 9, 110209−110222. https://doi.org/10.1109/ACCESS.2021.3102227.

Boursianis, A.D., Papadopoulou, M.S., Diamantoulakis, P., Liopa-Tsakalidi, A., Barouchas, P., Salahas, G., Karagiannidis, G., Wan, S., Goudos, S.K., 2022. Internet of things (IoT) and agricultural unmanned aerial vehicles (UAVs) in smart farming: a comprehensive review. Internet of Things 18. https://doi.org/10.1016/j.iot.2020.100187.

Brown, M.E., 2015. Satellite remote sensing in agriculture and food security assessment. Procedia Environmental Sciences 29. https://doi.org/10.1016/j.proenv.2015.07.278.

Chen, C.J., Huang, Y.Y., Li, Y.S., Chang, C.Y., Huang, Y.M., 2020. An AIoT based smart agricultural system for pests detection. IEEE Access 8, 180750−180761. https://doi.org/10.1109/ACCESS.2020.3024891.

Chen, C.J., Huang, Y.Y., Li, Y.S., Chen, Y.C., Chang, C.Y., Huang, Y.M., 2021. Identification of fruit tree pests with deep learning on embedded drone to achieve accurate pesticide spraying. IEEE Access 9, 21986−21997. https://doi.org/10.1109/ACCESS.2021.3056082.

Dahikar, S.S., Rode, S.V., 2014. Agricultural crop yield prediction using artificial neural network approach. International Journal of Innovative Research in Electrical, Electronics, Instrumentation and Control Engineering 2 (1), 683−686.

Dharmaraj, V., Vijayanand, C., 2018. Artificial intelligence (AI) in agriculture. International Journal of Current Microbiology and Applied Sciences 7 (12), 2122−2128. https://doi.org/10.20546/ijcmas.2018.712.241.

Ennouri, K., Smaoui, S., Gharbi, Y., Cheffi, M., Ben Braiek, O., Ennouri, M., Triki, M.A., 2021. Usage of artificial intelligence and remote sensing as efficient devices to increase agricultural system yields. Journal of Food Quality 2021. https://doi.org/10.1155/2021/6242288.

Fahad, S., Hussain, S., Chauhan, B.S., Saud, S., Wu, C., Hassan, S., Tanveer, M., Jan, A., Huang, J., 2015. Weed growth and crop yield loss in wheat as influenced by row spacing and weed emergence times. Crop Protection 71, 101−108. https://doi.org/10.1016/j.cropro.2015.02.005.

Hamadani, A., Ganai, N.A., 2022. Development of a multi-use decision support system for scientific management and breeding of sheep. Scientific Reports 12 (1). https://doi.org/10.1038/s41598-022-24091-y.

Hamadani, A., Ganai, N.A., Farooq, S.F., Bhat, B.A., 2022a. Big data management: from hard drives to DNA drives. Indian Journal of Animal Sciences 90 (2), 134−140. https://doi.org/10.56093/ijans.v90i2.98761.

Hamadani, A., Ganai, N.A., Mudasir, S., Shanaz, S., Alam, S., Hussain, I., 2022b. Comparison of artificial intelligence algorithms and their ranking for the prediction of genetic merit in sheep. Scientific Reports 12 (1). https://doi.org/10.1038/s41598-022-23499-w.

Hamadani, A., Ganai, N.A., Alam, S., Mudasir, S., Raja, T.A., Hussain, I., Ahmad, H.A., 2022c. Artificial intelligence techniques for the prediction of body weights in sheep. Indian Journal of Animal Research. https://doi.org/10.18805/IJAR.B-4831.

Hamadani, A., Ganai, N.A., Bashir, J., 2023. Artificial neural networks for data mining in animal sciences. Bulletin of the National Research Centre 47 (1). https://doi.org/10.1186/s42269-023-01042-9.

Husti, I., 2019. Possibilities of using robots in agriculture. Hungarian Agricultural Engineering 35, 59−67. https://doi.org/10.17676/hae.2019.35.59.

Javaid, M., Haleem, A., Khan, I.H., Suman, R., 2023. Understanding the potential applications of artificial intelligence in agriculture sector. Advanced Agrochem 2 (1), 15−30. https://doi.org/10.1016/j.aac.2022.10.001.

John, K.N., Valentin, V., Abdullah, B., Bayat, M., Kargar, M.H., Zargar, M., 2020. Weed mapping technologies in discerning and managing weed infestation levels of farming systems. Research on Crops 21 (1), 93−98. https://doi.org/10.31830/2348-7542.2020.015.

Jose, A., Nandagopalan, S., Akana, C.M.V.S., 2021. Artificial Intelligence techniques for agriculture revolution: a survey. Annals of the Romanian Society for Cell Biology 2580−2597.

Kamienski, C., Soininen, J.-P., Taumberger, M., Dantas, R., Toscano, A., Salmon Cinotti, T., Filev Maia, R., Torre Neto, A., 2019. Smart water management platform: IoT-based precision irrigation for agriculture. Sensors 19 (2). https://doi.org/10.3390/s19020276.

Kim, J., Seol, J., Lee, S., Hong, S.W., Son, H.I., 2020. An intelligent spraying system with deep learning-based semantic segmentation of fruit trees in orchards. In: Proceedings - IEEE International Conference on Robotics and Automation. Institute of Electrical and Electronics Engineers Inc., South Korea, pp. 3923−3929. https://doi.org/10.1109/ICRA40945.2020.9197556.

Kodama, T., Hata, Y., 2019. Development of classification system of rice disease using artificial intelligence. In: Proceedings - 2018 IEEE International Conference on Systems, Man, and Cybernetics, SMC 2018. Institute of Electrical and Electronics Engineers Inc., Japan, pp. 3699−3702. https://doi.org/10.1109/SMC.2018.00626. http://ieeexplore.ieee.org/xpl/mostRecentIssue.jsp?punumber=8615119.

Krishnan, S.R., Nallakaruppan, M.K., Chengoden, R., Koppu, S., Iyapparaja, M., Sadhasivam, J., Sethuraman, S., 2022. Smart water resource management using artificial intelligence—a review. Sustainability 14 (20). https://doi.org/10.3390/su142013384.

Lowe, M., Qin, R., Mao, X., 2022. A review on machine learning, artificial intelligence, and smart technology in water treatment and monitoring. Water 14 (9), 1384. https://doi.org/10.3390/w14091384.

McAllister, W., Osipychev, D., Davis, A., Chowdhary, G., 2019. Agbots: weeding a field with a team of autonomous robots. Computers and Electronics in Agriculture 163. https://doi.org/10.1016/j.compag.2019.05.036.

McKinion, J.M., Lemmon, H.E., 1985. Expert systems for agriculture. Computers and Electronics in Agriculture 1 (1), 31−40. https://doi.org/10.1016/0168-1699(85)90004-3.

Neil Harker, K., 2001. Survey of yield losses due to weeds in central Alberta. Canadian Journal of Plant Science 81 (2), 339−342. https://doi.org/10.4141/p00-102.

Orn, D., Duan, L., Liang, Y., Siy, H., Subramaniam, M., 2020. Agro-AI education: artificial intelligence for future farmers. In: SIGITE 2020 - Proceedings of the 21st Annual Conference on Information Technology Education. Association for Computing Machinery, Inc, United States, pp. 54−57. https://doi.org/10.1145/3368308. 3415457. http://dl.acm.org/citation.cfm?id=3368308.

Partel, V., Charan Kakarla, S., Ampatzidis, Y., 2019. Development and evaluation of a low-cost and smart technology for precision weed management utilizing artificial intelligence. Computers and Electronics in Agriculture 157, 339−350. https://doi.org/10.1016/j.compag.2018.12.048.

Sarkar, M.R., Masud, S.R., Hossen, M.I., Goh, M., 2022. A comprehensive study on the emerging effect of artificial intelligence in agriculture automation. In: 2022 IEEE 18th International Colloquium on Signal Processing and Applications, CSPA 2022 - Proceeding. Institute of Electrical and Electronics Engineers Inc., Bangladesh, pp. 419−424. https://doi.org/10.1109/CSPA55076.2022.9781883. http://ieeexplore.ieee.org/xpl/mostRecentIssue.jsp?punumber=9781838.

Sehgal, G., Gupta, B., Paneri, K., Singh, K., Sharma, G., Shroff, G., 2017. Crop planning using stochastic visual optimization. arXiv. https://arxiv.org.

Senkevich, S., Kravchenko, V., Duriagina, V., Senkevich, A., Vasilev, E., 2019. Optimization of the parameters of the elastic damping mechanism in class 1,4 tractor transmission for work in the main agricultural operations. Advances in Intelligent Systems and Computing 866, 168−177. https://doi.org/10.1007/978-3-030-00979-3_17.

Sharma, R., 2021. Artificial intelligence in agriculture: a review. In: Proceedings - 5th International Conference on Intelligent Computing and Control Systems, ICICCS 2021. Institute of Electrical and Electronics Engineers Inc., India, pp. 937−942. https://doi.org/10.1109/ICICCS51141.2021.9432187. http://ieeexplore.ieee.org/xpl/mostRecentIssue.jsp?punumber=9432068.

Sharma, S., Gahlawat, V.K., Rahul, K., Mor, R.S., Malik, M., 2021. Sustainable innovations in the food industry through artificial intelligence and big data analytics. Logistics 5 (4). https://doi.org/10.3390/logistics5040066.

Sinwar, D., Dhaka, V.S., Sharma, M.K., Rani, G., 2019. AI-based Yield Prediction and Smart Irrigation. Springer Science and Business Media LLC, pp. 155−180. https://doi.org/10.1007/978-981-15-0663-5_8.

Sishodia, R.P., Ray, R.L., Singh, S.K., 2020. Applications of remote sensing in precision agriculture: a review. Remote Sensing 12 (19), 1−31. https://doi.org/10.3390/rs12193136.

Sivarethinamohan, R., Yuvaraj, D., Shanmuga Priya, S., Sujatha, S., 2021. Captivating Profitable Applications of Artificial Intelligence in Agriculture Management. Springer Science and Business Media LLC, pp. 848−861. https://doi.org/10.1007/978-3-030-68154-8_73.

Tang, D., Feng, Y., Gong, D., Hao, W., Cui, N., 2018. Evaluation of artificial intelligence models for actual crop evapotranspiration modeling in mulched and non-mulched maize croplands. Computers and Electronics in Agriculture 152, 375−384. https://doi.org/10.1016/j.compag.2018.07.029.

Torky, M., Hassanein, A.E., 2020. Integrating blockchain and the internet of things in precision agriculture: analysis, opportunities, and challenges. Computers and Electronics in Agriculture 178. https://doi.org/10.1016/j.compag.2020.105476.

Vazquez, J.P.G., Torres, R.S., Perez, D.B.P., 2021. Scientometric analysis of the application of artificial intelligence in agriculture. Journal of Scientometric Research 10 (1), 55−62. https://doi.org/10.5530/JSCIRES.10.1.7.

Zhang, L., Wang, S., 2020. Input-output analysis of agricultural economic benefits based on big data and artificial intelligence. Journal of Physics: Conference Series 1574 (1). https://doi.org/10.1088/1742-6596/1574/1/012121.

Zhao, Z., Chow, T.L., Rees, H.W., Yang, Q., Xing, Z., Meng, F.R., 2009. Predict soil texture distributions using an artificial neural network model. Computers and Electronics in Agriculture 65 (1), 36−48. https://doi.org/10.1016/j.compag.2008.07.008.

Artificial intelligence in animal farms for management and breeding

Henna Hamadani[1], Ambreen Hamadani[2] and Shabia Shabir[3]

[1]*Sher-e-Kashmir University of Agricultural Sciences and Technology of Kashmir, Srinagar, Jammu and Kashmir, India;*
[2]*National Institute of Technology, Srinagar, Jammu and Kashmir, India;* [3]*Islamic University of Science and Technology (IUST), Awantipora, Jammu and Kashmir, India*

Introduction

Livestock farming plays a significant role in meeting the global demand for food, supporting livelihoods, and contributing to economic development (Banda and Tanganyika, 2021; Hamadani et al., 2022a). Livestock provides a valuable source of protein, including meat, poultry, fish, eggs, and dairy products, which are crucial components of a balanced diet. Livestock farming helps promote food security and provides nutritional diversity to meet the dietary needs of a growing global population (Reynolds et al., 2015; Varijakshapanicker et al., 2019). It also contributes to the cycling of nutrients in agricultural systems. Livestock waste, such as manure, can be used as organic fertilizer (vermicompost), replenishing the soil with essential nutrients and improving soil fertility. This nutrient cycling helps maintain soil health, increases agricultural productivity, reduces the need for synthetic fertilizers, and contributes to the cycling of nutrients in agricultural systems. Additionally, vermicompost production also adds to the farmer's income (Hamadani et al., 2020d).

Livestock can utilize land resources that are not suitable for crop cultivation. Grazing animals on pasturelands and utilizing marginal lands for forage production can efficiently convert nonarable land into valuable food resources. They are capable of converting low-quality protein into high-quality protein (Broderick, 2018). Integrating animal farming with crop production through mixed farming systems can promote sustainable land use and optimize resource utilization (Hamadani et al., 2021b). Various by-products from the agriculture and food industries are utilized in livestock farming, which helps in reducing waste, promoting sustainability, and socioeconomic upliftment (Hamadani et al., 2020c; Hamadani, 2023). For example, animal feed can be produced from agricultural residues, food processing by-products (Bhandari and Bahadur, 2019), and other biomass that would otherwise go to waste. This minimizes environmental impact and improves resource efficiency.

Livestock farming contributes to national and international trade, creating economic opportunities and generating foreign exchange earnings (Baltenweck et al., 2020). Countries with efficient livestock sectors can export meat, dairy products, and other animal-derived commodities, contributing to their overall economic growth and trade balance. Conventional and alternate livestock and poultry resources contribute to this. The cultural and social significance of animal farming can be also adjudged by the fact that it is rooted in the cultural and social fabric of many societies. Livestock and poultry rearing is often intertwined with traditional practices, festivals, and cultural celebrations by playing a role in preserving cultural heritage, supporting rural communities, and maintaining social cohesion (Alders et al., 2021; Hamadani et al., 2020e; Hamadani and Khan, 2013).

Despite all this, livestock farming today faces challenges related to sustainability, animal welfare, and environmental impact (Godde et al., 2020), which were irrelevant a few decades back. These include population explosion, climate change, antibiotic resistance, new and emerging animal diseases, food insecurity, loss of genetic variance, depleting and degrading resources, etc (Eeswaran et al., 2022). It is important to address these challenges through responsible and sustainable practices, such as implementing efficient resource management, adopting animal welfare standards, and embracing innovative technologies that minimize environmental footprint. Additionally, some major constraints faced in livestock management and breeding is that it is labor intensive and time-consuming and involves drudgery of repetitive tasks leading to inefficiency and low productivity.

Artificial intelligence (AI) has the potential to revolutionize livestock management and breeding with its vast applicability. The algorithms of machine learning (ML) are helping to transform the conventional inefficient system of farming into smart farming. Livestock is monitored and managed smartly with AI-based systems to enhance animal welfare, improve health, control diseases, increase production, and provide useful insights to farm managers.

AI and big data in livestock farms

Artificial intelligence has the potential to significantly improve all aspects of animal farming practices by providing innovative solutions for monitoring, management, and decision-making (Hamadani et al., 2023). However, before the major applications of AI in animal management and breeding are discussed, it is important to understand the relevance of big data in animal management (Hamadani et al., 2020b). Big data in animal farms empowers farmers and offers valuable insights and tools to improve animal health, productivity, and overall farm management. By leveraging the power of data analytics, farmers can make informed decisions, enhance efficiency, and contribute to sustainable and profitable animal farming practices. AI is the obvious choice for the analysis of data generated because it deals very efficiently with data that has volume, variety, and velocity (Wong et al., 2019). Under the conditions of large volumes of data, AI allows the delegation of difficult pattern recognition, learning, and other tasks to computer-based approaches. In addition, AI can handle the great velocity of data, by enabling rapid computer-based decisions each leading to other subsequent decisions. Variety is mitigated by capturing, structuring, and understanding unstructured data using AI and other analytics.

Artificial intelligence consists of all computing systems that can perform tasks that require human intelligence, such as decision-making, object detection, solving complex problems, and so on. The realm of AI also prompts the inception of ML and deep learning, as well as predictive analytics (Wong et al., 2019). All this is collectively referred to as data science. This is a data-intensive science, and the data gathered do not only benefit the farm but also researchers, companies, and other stakeholders use it to make inferences that can help farms around the world to make informed decisions on individual animal performance.

Artificial intelligence includes various techniques viz. ML, deep learning, natural language processing, fuzzy logic, robotics, and expert systems. ML is a subcategory of AI, which is mainly divided into supervised learning, reinforcement learning, and unsupervised learning (Rashidi et al., 2019). Supervised learning is a very popular area of ML in animal sciences. Artificial neural networks also come under the umbrella of ML, which is used both for supervised and unsupervised learning. These derive inspiration from the natural neural network of the human nervous system. Deep learning is the process of the implementation of neural networks on high-dimensional data to gain insights and form solutions. Data from livestock are currently being used for satellite imagery, data tracking, and analysis applications (Hamadani and Ganai, 2022). It is proving to be useful for the improvement of animal health and food safety, identification, traceability, and labor reduction. The most prominent areas where AI is being used to bring about improvement are discussed below.

Identification of animals

Animal identification involves identifying animals individually, which allows monitoring and tracking each animal separately. It has a variety of applications, which include record keeping, efficient farm management, ownership verification, biosecurity control, registration, insurance, and security of animals from theft. Identification of livestock is also needed by different regulatory bodies, which ensure safety and traceability and help in improving the product quality (Fuentes et al., 2022; Hossain et al., 2022). In many livestock farms, the path of each animal is monitored from birth onward by using an identification system that helps in keeping track of different parameters and maintains the information base of each animal on the farm.

In earlier times, small-scale farmers would remember few important details about their animals without any proper record system. However, in some organized farms, manual record system would be maintained to keep track of the animals and which would also help in decision-making. Now with the modernization of farming practices and increased use of automated technology, the use of electronic identification systems is becoming necessary, especially in large commercial livestock farms. Many management operations are automatized in precision and smart farming leading to increased efficiency and reduced labor.

Automation on livestock farms tailors activities to the needs of individual animals, which need the subsystems to recognize the animals individually as they interact with these systems (Hamadani and Khan, 2015). Hence, electronic identification of animals becomes a prerequisite for all these systems to generate data on each animal and record that information for further processing to draw useful inferences. Integration of AI into these systems helps track animals intelligently with increased ease and efficiency. Identification of animals can be automatized by engaging computer vision. All the required information about individual animals can be accessed easily. Even the smaller species like poultry can also be identified individually using AI. AI-driven software is now being developed, which has the potential to noninvasively and unobtrusively track individual animals' location as well as estimate their position in an area (Congdon et al., 2022).

In addition to identification, AI has also been used in counting as well as in describing animals using camera-trap images (Norouzzadeh et al., 2018). Applications of AI on the data obtained by a thermal camera for locating animals during nocturnal hours have also been reported (Munian et al., 2022). This can be helpful in locating predators on livestock farms and preventing the loss of livestock due to predation. Detection of cattle using images captured from drones has also been attempted (Rivas et al., 2018). Acoustics has been used to identify gender (Hamadani et al., 2017), breed, and species of animals and birds, and the same can be automatized by employing the techniques of AI (Yeo et al., 2011; Pabico et al., 2015).

Animal monitoring

Monitoring animal behavior and physiology is important for their welfare and management. Any abnormality detected in the behavior or physiology indicates either disease or any other abnormal situation which needs attention. Animals when managed in a way that their welfare is taken into consideration perform better. However, it is not easy to continuously monitor animal behavior and physiology manually. AI can be employed to continuously monitor animal behavior, physiological parameters, and environmental conditions (Congdon et al., 2022). Computer vision and AI have been employed for automatic behavior analysis in animals (Bernardes et al., 2021). ML algorithms can analyze large datasets generated by sensors, wearable devices, and video surveillance to detect early signs of illness, stress, or abnormal behavior. This allows for timely intervention and preventive measures to ensure optimal animal health (Ezanno et al., 2021). AI algorithms can analyze this data in real time to detect abnormalities, stress, or signs of illness. Farmers can receive timely alerts and take appropriate actions to ensure the health and well-being of their animals. This would be useful for AI can leverage historical and real-time data to develop predictive models for livestock health and performance. By analyzing factors such as genetics, environmental conditions, feed quality, and management practices, AI can generate insights and forecasts regarding animal growth, production outcomes, and disease susceptibility. This enables farmers to make informed decisions on breeding, nutrition, and overall management strategies. Various sensors embedded in wearable devices can collect data on animal behavior, health, and activity. Various sensors used in animal monitoring are given in Fig. 11.1. For example, sensors like image analysis sensors, walk-over-weigh sensors, GPS, acetometers, and thermometers are used for live weight estimation; calculation of reproductive efficiency, growth rates and maternal parentage; monitoring of calving, activity, health, rumination, lameness, body temperature, intake, heat detection, resting, feeding, calving, etc (Aquilani et al., 2022). While it remains relatively easy to implement in barn and pasture settings, free-range farming presents additional challenges; herd supervision can become more difficult as cattle are not contained. Supervision of animals in the barn or shed is relatively easy, while it is difficult in the pasture setup. To address this issue, virtual fence collars are used to remotely monitor and manage livestock in grazing areas (Versluijs et al., 2023). A sensor suite deployed with long-range radio (LoRa) can be used for tracking the activity and location of livestock in pasture lands (Dos Reis et al., 2021).

Disease detection and prevention

Artificial intelligence algorithms can analyze complex data patterns, including environmental factors, animal health records, and disease outbreaks, to predict and detect disease risks. By utilizing big data analytics and ML, AI can provide early warnings of potential disease outbreaks, enabling farmers to take proactive measures such as quarantine, vaccination, or treatment, thus reducing the spread of diseases and minimizing economic losses. These can help in the following:

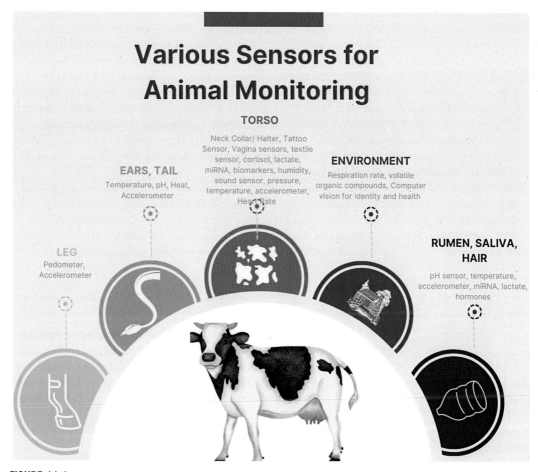

Various Sensors for Animal Monitoring

TORSO
Neck Collar/ Halter, Tattoo Sensor, Vagina sensors, textile sensor, cortisol, lactate, miRNA, biomarkers, humidity, sound sensor, pressure, temperature, accelerometer, Heart Rate

EARS, TAIL
Temperature, pH, Heat, Accelerometer

ENVIRONMENT
Respiration rate, volatile organic compounds, Computer vision for identity and health

LEG
Pedometer, Accelerometer

RUMEN, SALIVA, HAIR
pH sensor, temperature, accelerometer, miRNA, lactate, hormones

FIGURE 11.1

Various sensors for animal monitoring.

- **Early disease detection:** AI algorithms can analyze various data sources, including animal health records, environmental factors, and sensor data, to detect early signs of disease. By identifying patterns, anomalies, and correlations in the data, AI can provide early warnings of potential disease outbreaks. This allows farmers to take prompt action, such as isolating affected animals, implementing biosecurity measures, or initiating targeted treatment protocols.
- **Real-time monitoring:** AI-powered systems can continuously monitor animal health using sensors, cameras, and wearable devices. These technologies can capture data on vital signs, behavior patterns, feed intake, and other health-related indicators. AI algorithms can process these data in real time, detecting deviations from normal behavior or physiological ranges. Real-time monitoring enables immediate response to any health issues, minimizing the spread of diseases and reducing the impact on overall herd or flock health.

- **Disease risk assessment:** AI can assist in assessing disease risks by analyzing a combination of historical data, environmental factors, and disease prevalence information. By considering variables such as weather conditions, geographical factors, animal movements, and biosecurity practices, AI algorithms can identify high-risk areas or periods for specific diseases. This information helps farmers prioritize prevention measures and allocate resources accordingly.
- **Disease prediction and modeling:** AI can utilize ML techniques to develop predictive models for disease outbreaks. By analyzing historical data, such as disease incidence, environmental conditions, and animal population dynamics, AI algorithms can forecast the likelihood and severity of future disease outbreaks. Such models enable farmers to take proactive measures, such as adjusting management practices or implementing preventive measures, to minimize the impact of diseases on animal health and productivity.
- **Surveillance and early warning systems:** AI can contribute to surveillance systems for animal diseases by analyzing diverse data sources, including animal health databases, laboratory reports, and social media data. By employing natural language processing and ML algorithms, AI can detect signals or outbreaks of emerging or contagious diseases. Early warning systems powered by AI can alert authorities and stakeholders, facilitating rapid response and containment of diseases.

Precision nutrition and feed management

AI can optimize animal nutrition by analyzing data on animal characteristics, feed ingredients, and environmental factors. By considering factors such as animal genetics, growth stage, and performance goals, AI algorithms can recommend personalized nutrition plans and feed formulations. This helps maximize feed efficiency, improve animal growth rates, and reduce waste, resulting in more sustainable, cost-effective feeding practices for personalized animal nutrition (Saxena and Parasher, 2019).

Artificial intelligence can significantly enhance precision nutrition and feed management in animal farming. Broadly, the application of AI is stated as under:

- **Nutrient analysis and formulation:** AI algorithms can analyze vast amounts of data, including animal characteristics, dietary requirements, ingredient composition, and nutritional databases, to optimize feed formulations. By considering factors such as animal age, weight, growth stage, and performance goals, AI can recommend personalized and precise nutrient compositions for individual animals or groups. This ensures that animals receive the necessary nutrients while minimizing waste and optimizing feed efficiency.
- **Feed quality assessment:** AI can help assess feed quality and safety by analyzing data from various sources, such as lab reports, ingredient specifications, and historical data. ML algorithms can identify patterns and anomalies that indicate variations in feed quality, such as contamination or nutrient deficiencies. AI-powered systems can provide real-time feedback on feed quality, enabling farmers to make informed decisions about feed purchases and adjust feed rations accordingly.
- **Real-time feed management:** AI can assist in real-time feed management by integrating data from sensors, feeding systems, and environmental conditions. By monitoring feed intake, animal behavior, and environmental factors, AI algorithms can optimize feed delivery schedules and

adjust feed quantities based on individual animal needs. This helps prevent overfeeding or underfeeding, reduces feed waste, and improves feed conversion efficiency.
- **Predictive modeling for feed optimization:** AI can develop predictive models that consider multiple variables, such as animal performance data, weather forecasts, feed prices, and ingredient availability. By analyzing historical and real-time data, AI algorithms can forecast animal growth, feed requirements, and production outcomes. This enables farmers to optimize feed management strategies, adjust feed rations, and make informed decisions on feed purchasing and allocation (Liebe and White, 2019).
- **Impact assessment:** AI can evaluate the environmental impact of feed management practices by analyzing data on feed ingredients, energy usage, water consumption, and greenhouse gas emissions. By assessing the lifecycle impacts of different feed formulations and management strategies, AI can help identify opportunities to minimize the environmental footprint of livestock production. This includes optimizing ingredient sourcing, reducing feed waste, and promoting sustainable feed production practices.

Automation for precision farming

Artificial intelligence—powered systems can automate various tasks in animal farming, such as feeding, milking, and waste management. Robotic devices equipped with AI capabilities can carry out these tasks efficiently and accurately, freeing up human labor (Carolan, 2020) and ensuring consistent and precise operations. AI can also integrate data from multiple sources, such as sensors and cameras, to provide real-time insights and automated decision-making. It can precision farming practices by optimizing resource allocation and utilization (Hamadani and Khan, 2015). By analyzing data on animal behavior, environmental conditions, and resource availability, AI algorithms can provide insights for efficient resource management. This includes optimizing grazing patterns, water usage, energy consumption, and waste management. Precision farming techniques can reduce environmental impact and promote sustainable livestock production.

AI can revolutionize automated monitoring and management systems in animal farming by enabling real-time data collection, analysis, and decision-making. Here is how AI can be applied in this context.

1. **Sensor-based data collection:** AI can leverage sensors and IoT devices to collect real-time data on various parameters, such as animal behavior, health indicators, environmental conditions, and feed consumption. Sensors can be integrated into wearable devices, feeding systems, environmental monitoring systems, and more. AI algorithms can process and analyze these data, providing valuable insights into animal well-being, productivity, and farm conditions.
2. **Behavior analysis:** AI algorithms can analyze sensor data to monitor animal behavior patterns and detect deviations from normal behavior. By identifying changes in activity levels, feeding patterns, or social interactions, AI can alert farmers to potential health issues, stress, or abnormal behavior that may indicate a problem. Timely intervention based on AI-driven behavior analysis can help prevent disease outbreaks, improve animal welfare, and optimize production outcomes.
3. **Automated systems:** AI can optimize and automate feeding systems in animal farming. By integrating data from sensors, animal profiles, and nutritional requirements, AI algorithms can adjust feed quantities, feeding schedules, and formulations based on individual animal needs and

performance goals. This ensures optimal feed utilization, reduces feed wastage, and promotes healthy growth. AI-powered feeding systems can also monitor feed availability and track consumption rates and robotic milking systems: Robotic milking systems are becoming more common in dairy farming (Hamadani and Khan, 2015). These systems use sensors and robotic arms to clean and milk cows without human intervention. They can monitor the cow's health and milk production and even track individual cow data.

4. Automated feeding systems are used in both dairy and livestock farming. These systems can automatically distribute feed to animals based on preprogrammed schedules or individual animal needs. They help optimize feed efficiency and reduce labor requirements. In poultry farming, automated systems are used for egg sorting and grading. These systems use computer vision and sensors to detect defects, measure size, and sort eggs based on various parameters such as weight and quality. AI can optimize environmental conditions in animal farming by monitoring and controlling factors such as temperature, humidity, ventilation, and lighting. By analyzing real-time sensor data and historical patterns, AI algorithms can adjust environmental parameters to provide optimal conditions for animal comfort, health, and productivity. Automated systems controlled by AI can regulate environmental parameters, detect abnormalities, and notify farmers in case of deviations that may impact animal well-being.

5. **Animal farm robots:** Robots are being increasingly used on animal farms to streamline operations, enhance productivity, and improve animal welfare. A few robots developed for animal farming are listed in Table 11.1.

Genetic improvement and breeding

Animal genetics and breeding is a data-intensive science and techniques like the least squares, and best linear unbiased predictions have been used over many years for selection and breeding (Hamadani et al., 2019, 2020a, 2021a; Rather et al., 2019; Baba et al., 2020a,b). AI can accelerate genetic improvement in animal breeding programs by analyzing large genomic datasets. ML algorithms can identify genetic markers associated with desired traits, predict breeding values, and optimize mating strategies. This enables breeders to make informed decisions to enhance desirable traits, reduce genetic disorders, and improve overall breeding efficiency (Hamadani et al., 2022b,c). Implementing AI in genetic improvement and breeding programs requires reliable genomic data, advanced AI algorithms, and collaborations between breeders, researchers, and technology providers. It is crucial to ensure data privacy, maintain genetic diversity, and consider ethical and welfare aspects associated with breeding practices. By leveraging AI in genetic improvement and breeding, animal farmers can accelerate genetic progress, enhance production traits, and achieve sustainable and resilient livestock populations.

1. **Genomic selection:** AI algorithms can analyze large-scale genomic data to predict the genetic potential of animals. By leveraging ML techniques, AI can identify genetic markers associated with desirable traits, such as milk production, meat quality, disease resistance, or reproductive performance. AI-powered genomic selection models can assist breeders in selecting animals with the highest genetic merit for breeding, accelerating the rate of genetic improvement in livestock populations.

Table 11.1 Some robots used in animal farms. Genetic improvement and breeding.		
Robot name	**Utility**	**References**
MilkoMatic	MilkoMatic is a state-of-the-art milking robot designed to automate the milking process on dairy farms. Equipped with advanced sensors and robotic arms, MilkoMatic can identify cows, attach to their udders, and efficiently milk them. It also monitors milk quality, ensuring optimal hygiene standards.	https://milkomatic.com/
Sheepdog robot	Sheepdog robot is a vision-based robot for detecting different predators among livestock in pastures and also distinguishes them from other animal species.	Riego del Castillo et al. (2022)
FeedMaster	FeedMaster is a feeding robot that ensures precise and timely distribution of feed on animal farms. It employs advanced programming and sensors to accurately dispense the right amount of feed to each animal. FeedMaster also monitors consumption patterns, enabling personalized nutrition plans for improved animal health.	Kabir et al. (2022)
Clean Sweep	Clean Sweep is a robotic cleaner designed to maintain cleanliness in barns and pens. This agile robot autonomously navigates the area, collecting animal waste and debris. Equipped with efficient waste disposal mechanisms, clean Sweep promotes a hygienic environment, minimizing the risk of diseases.	Shukla et al. (2022)
T-Moov	T-Moov T- Moov, created by Tibot, is a revolutionary poultry robot designed to address the specific needs of poultry farmers. It has been meticulously developed in collaboration with industry experts to excel in challenging environments characterized by factors like poor litter conditions or flat feed chains. The primary objective of T- Moov is to ensure continuous movement among your birds and minimize the occurrence of floor eggs.	https://www.tibot.fr/en/solutions/poultry-robot-tmoov/
Xo robot	Developed by Octopus poultry robotics. XO is an invaluable tool for poultry farmers, providing assistance in risk prevention and contamination treatment. It ensures continuous treatment of litter and diligent monitoring of the environment and the well-being of the animals.	https://www.octopusbiosafety.com/es/salud-animal/xo/

2. **Breeding value estimation:** AI can estimate breeding values, which reflect an animal's genetic superiority for specific traits, by integrating data from pedigree records, phenotypic data, and genomic information. AI algorithms can consider complex genetic interactions and environmental factors to calculate accurate breeding values. This enables breeders to make informed decisions regarding mating pairs, selection of breeding stock, and genetic progress tracking.
3. **Genetic diversity analysis:** AI algorithms can analyze genomic data to assess the genetic diversity within a population and identify inbreeding levels. By evaluating patterns of genetic variation and relatedness, AI can help breeders maintain optimal levels of genetic diversity to prevent inbreeding depression and promote the long-term health and performance of the population. AI-powered systems can provide recommendations on mating strategies and breeding plans to manage genetic diversity effectively.

4. **Parentage verification and pedigree reconstruction:** AI can utilize genomic data to verify parentage and reconstruct pedigrees accurately. By comparing genetic profiles, AI algorithms can determine parent–offspring relationships, identify potential errors in pedigree records, and assist in pedigree reconstruction. This ensures accurate genetic evaluations and supports the selection of superior breeding stock.

5. **Trait prediction and optimization:** AI can predict the performance of animals for various traits based on genetic and environmental data. By considering factors such as genotype, management practices, and environmental conditions, AI algorithms can provide insights into the potential performance of offspring for specific traits. This helps breeders optimize mating decisions, select animals with desired trait combinations, and design breeding programs targeted toward specific production goals.

Decision support systems

Artificial intelligence–based decision support systems can provide real-time recommendations and predictions to farmers. By integrating data on animal health, environmental conditions, and market trends, these systems can assist in making informed decisions related to breeding, nutrition, disease management, and resource allocation. This helps optimize farm operations, improve productivity, and enhance profitability (Hamadani and Ganai, 2022). AI-powered decision support systems can greatly enhance decision-making processes in animal farming. Here is how AI can be applied in decision support systems for animal farm management.

1. **Data integration:** AI can integrate and analyze data from various sources, including sensor data, farm management systems, weather forecasts, market trends, and genetic information. By aggregating and processing these data, AI algorithms can provide a comprehensive view of the farm's operations and performance. This enables farmers to make informed decisions based on a holistic understanding of multiple factors influencing their operations.

2. **Performance monitoring and analysis:** AI can monitor and analyze key performance indicators (KPIs) related to animal health, growth, production, and efficiency. By processing real-time and historical data, AI algorithms can detect trends, identify areas of improvement, and provide insights into farm performance. Farmers can track KPIs, such as feed conversion ratio, growth rates, disease incidence, and reproductive efficiency, to optimize management strategies and improve overall farm productivity.

3. **Risk assessment and management:** AI can assess and manage risks associated with animal farming. By analyzing data on disease prevalence, weather conditions, market prices, and financial indicators, AI algorithms can identify potential risks and provide risk mitigation strategies. AI-powered systems can assist in making decisions related to disease prevention, biosecurity measures, market timing, and financial planning, reducing uncertainties and improving farm resilience.

4. **Resource optimization:** AI can optimize resource allocation and utilization in animal farming. By analyzing data on feed availability, energy consumption, labor costs, and equipment usage, AI algorithms can provide recommendations for efficient resource management. Farmers can optimize feed rations, adjust staffing levels, schedule maintenance tasks, and plan production

cycles based on AI-driven insights, leading to improved cost-effectiveness and resource efficiency.

5. **Financial analysis and forecasting:** AI can assist in financial analysis and forecasting for animal farming operations. By integrating data on costs, revenues, market trends, and production outputs, AI algorithms can generate financial models and forecasts. This helps farmers evaluate profitability, assess the economic impact of management decisions, and plan for future investments or expansions.

6. **Real-time recommendations and alerts:** AI-powered decision support systems can provide real-time recommendations and alerts to farmers. By analyzing real-time sensor data, market information, and performance indicators, AI algorithms can suggest immediate actions or adjustments. For example, AI systems can provide alerts for deviations in environmental parameters, recommend adjustments to feed quantities, or suggest timely interventions for health issues. Real-time recommendations enable proactive decision-making and rapid responses to optimize farm performance.

Improving animal production using AI

Artificial intelligence can play a significant role in the production and optimization of animal products in several ways (Aharwal et al., 2021), the most important being the production of animal protein. This is done by optimizing the four most important aspects of animal farming viz feeding, breeding, weeding, and heeding. This is achieved through the following management practices:

1. **Quality control and traceability:** AI technologies can improve quality control and traceability in animal production. Computer vision systems can be used to identify and sort meat products based on quality parameters such as color, marbling, and texture. AI algorithms can also track and trace animal products throughout the supply chain, providing information on the origin, production methods, and safety standards. This helps ensure product integrity, reduce fraud, and enhance consumer confidence.

2. **Sustainability and resource optimization:** AI can contribute to the sustainable production of animal protein by optimizing resource utilization and reducing environmental impact. AI-powered systems can monitor and optimize energy usage, water consumption, and waste management in animal farming operations. By analyzing data on environmental conditions, feed efficiency, and production parameters, AI algorithms can identify opportunities to reduce resource waste, minimize greenhouse gas emissions, and improve overall sustainability (Nti et al., 2022).

3. **Processing and manufacturing:** AI technologies can automate and optimize various processes involved in the production of animal products. Computer vision systems can be used to assess the quality of raw materials, such as meat or dairy, and guide automated sorting and processing procedures. AI algorithms can analyze data from sensors and imaging technologies to monitor production parameters, such as temperature, humidity, and processing time, ensuring consistent quality and reducing product defects.

4. **Quality control and inspection:** AI can improve quality control processes by analyzing data from visual inspections, sensors, and other measurement devices. Computer vision algorithms can identify and classify defects, such as blemishes or abnormalities in animal products, allowing

for real-time quality assessments and automatic rejection of substandard items. This helps maintain consistent quality standards and reduces the risk of defective products reaching consumers.

5. **Food safety and traceability:** AI technologies can enhance food safety measures in animal products. ML algorithms can analyze data from various sources, including production records, supply chain information, and historical data on safety incidents, to identify patterns and potential risks. AI-powered systems can help detect and prevent foodborne illnesses, ensure compliance with safety regulations, and provide traceability information to track products from farm to fork.

6. **Sensory evaluation and consumer experience:** AI can contribute to sensory evaluation and product development for animal products. ML algorithms can analyze sensory data, such as taste, texture, and aroma, to identify consumer preferences and develop optimized product formulations. AI can also assist in personalized recommendations, such as suggesting suitable meat cuts or dairy products based on individual preferences and dietary requirements.

7. **Product labeling and allergen detection:** AI can aid in accurate labeling and allergen detection in animal products. Natural language processing algorithms can analyze product labels and ingredient lists, ensuring compliance with labeling regulations and identifying potential allergens. This helps prevent mislabeling and supports consumers with dietary restrictions or allergies in making informed purchasing decisions.

8. **Consumer insights and market trends:** AI technologies can analyze vast amounts of data from various sources, such as social media, online reviews, and consumer feedback, to gain insights into consumer preferences and market trends. This information can help animal product manufacturers understand consumer demands, develop targeted marketing strategies, and tailor their products to meet evolving consumer needs.

By leveraging AI in the production and management of animal products, businesses can enhance product quality, safety, and consumer satisfaction. However, it is important to ensure data privacy, comply with regulatory requirements, and address ethical considerations associated with AI implementation in the food industry.

Conclusion

The integration of AI into animal management and breeding has emerged as a transformative approach with immense potential. AI offers the capability to analyze vast amounts of data from various sources, including genomic information, sensor data, and historical records, to drive informed decision-making and improve animal productivity, health, and welfare. By leveraging advanced algorithms and ML techniques, AI can enhance precision in breeding programs by identifying desirable genetic traits, predicting breeding values, and optimizing mating strategies. AI-powered systems enable real-time monitoring of animal health, behavior, and environmental conditions, facilitating early disease detection, targeted interventions, and optimized resource allocation. Additionally, AI-driven decision support systems aid in risk assessment, financial analysis, and resource optimization, thereby promoting sustainability, efficiency, and profitability in animal farming. With its ability to handle complex datasets and generate actionable insights, AI is poised to revolutionize animal management and breeding practices, driving advancements in livestock production, genetic improvement, and sustainable farming practices.

References

Aharwal, B., Roy, B., Meshram, S., Yadav, A., 2021. Worth of artificial intelligence in the epoch of modern livestock farming: a review. Agricultural Science Digest - A Research Journal 43 (Of). https://doi.org/10.18805/ag.d-5355.

Alders, R.G., Campbell, A., Costa, R., Guèye, E.F., Ahasanul Hoque, M., Perezgrovas-Garza, R., Rota, A., Wingett, K., 2021. Livestock across the world: diverse animal species with complex roles in human societies and ecosystem services. Animal Frontiers 11 (5), 20–29. https://doi.org/10.1093/af/vfab047.

Aquilani, C., Confessore, A., Bozzi, R., Sirtori, F., Pugliese, C., 2022. Review: precision Livestock Farming technologies in pasture-based livestock systems. Animal 16 (1). https://doi.org/10.1016/j.animal.2021.100429.

Baba, M.A., Ahanger, S.A., Hamadani, A., Rather, M.A., Shah, M.M., 2020a. Factors affecting wool characteristics of sheep reared in Kashmir. Tropical Animal Health and Production 52 (4), 2129–2133. https://doi.org/10.1007/s11250-020-02238-1.

Baba, J.A., Hamadani, A., Shanaz, S., Rather, M.A., 2020b. Factors affecting wool characteristics of corriedale sheep in temperate region of Jammu and Kashmir. Indian Journal of Small Ruminants (The) 26 (2). https://doi.org/10.5958/0973-9718.2020.00035.5.

Baltenweck, I., Enahoro, D., Frija, A., Tarawali, S., 2020. Why is production of animal source foods important for economic development in Africa and Asia? Animal Frontiers 10 (4), 22–29. https://doi.org/10.1093/af/vfaa036.

Banda, L.J., Tanganyika, J., 2021. Livestock provide more than food in smallholder production systems of developing countries. Animal Frontiers 11 (2), 7–14. https://doi.org/10.1093/af/vfab001.

Bernardes, R.C., Lima, M.A.P., Guedes, R.N.C., da Silva, C.B., Martins, G.F., 2021. Ethoflow: computer vision and artificial intelligence-based software for automatic behavior analysis. Sensors 21 (9). https://doi.org/10.3390/s21093237.

Bhandari, B., Bahadur, 2019. Crop Residue as Animal Feed. https://doi.org/10.13140/RG.2.2.20372.04486.

Broderick, G.A., 2018. Review: optimizing ruminant conversion of feed protein to human food protein. Animal 12 (8), 1722–1734. https://doi.org/10.1017/s1751731117002592.

Carolan, M., 2020. Automated agrifood futures: robotics, labor and the distributive politics of digital agriculture. Journal of Peasant Studies 47 (1), 184–207. https://doi.org/10.1080/03066150.2019.1584189.

Congdon, J.V., Hosseini, M., Gading, E.F., Masousi, M., Franke, M., MacDonald, S.E., 2022. The future of artificial intelligence in monitoring animal identification, health, and behaviour. Animals 12 (13). https://doi.org/10.3390/ani12131711.

Dos Reis, B.R., Easton, Z., White, R.R., Fuka, D., 2021. A LoRa sensor network for monitoring pastured livestock location and activity. Translational Animal Science 5 (2), 1–9. https://doi.org/10.1093/tas/txab010.

Eeswaran, R., Nejadhashemi, A.P., Faye, A., Min, D., Prasad, P.V.V., Ciampitti, I.A., 2022. Current and future challenges and opportunities for livestock farming in West Africa: perspectives from the case of Senegal. Agronomy 12 (8). https://doi.org/10.3390/agronomy12081818.

Ezanno, P., Picault, S., Beaunée, G., Bailly, X., Muñoz, F., Duboz, R., Monod, H., Guégan, J.-F., 2021. Research perspectives on animal health in the era of artificial intelligence. Veterinary Research 52 (1). https://doi.org/10.1186/s13567-021-00902-4.

Fuentes, S., Gonzalez Viejo, C., Tongson, E., Dunshea, F.R., 2022. The livestock farming digital transformation: implementation of new and emerging technologies using artificial intelligence. Animal Health Research Reviews 23 (1), 59–71. https://doi.org/10.1017/S1466252321000177.

Godde, C.M., Boone, R.B., Ash, A.J., Waha, K., Sloat, L.L., Thornton, P.K., Herrero, M., 2020. Global rangeland production systems and livelihoods at threat under climate change and variability. Environmental Research Letters 15 (4). https://doi.org/10.1088/1748-9326/ab7395.

Hamadani, H., 2023. Socio-economic status of dairy farmers in the Srinagar district of Jammu and Kashmir. Asian Journal of Dairy and Food Research.

Hamadani, A., Ganai, N.A., 2022. Development of a multi-use decision support system for scientific management and breeding of sheep. Scientific Reports 12 (1). https://doi.org/10.1038/s41598-022-24091-y.

Hamadani, H., Khan, A., 2013. Domestic geese (Anser anser domesticus) as companion birds. Indian Pet Journal-Online Journal of Canine 4, 18–25.

Hamadani, H., Khan, A., 2015. Automation in livestock farming—A technological revolution. International Journal of Advanced Research 3, 1335–1344.

Hamadani, H., Khan, A.A., Hamadani, A., Rafiq, A., 2017. Practical methods of gender identification in Kashmir geese. Indian Journal of Animal Sciences 87 (5), 653–655. http://epubs.icar.org.in/ejournal/index.php/IJAnS/article/view/70274/29828.

Hamadani, A., Ganai, N.A., Khan, N.N., Shanaz, S., Ahmad, T., 2019. Estimation of genetic, heritability, and phenotypic trends for weight and wool traits in Rambouillet sheep. Small Ruminant Research 177, 133–140. https://doi.org/10.1016/j.smallrumres.2019.06.024.

Hamadani, A., Ganai, N.A., Rather, M.A., Raja, T.A., Shabir, N., Ahmad, T., Shanaz, S., Aalam, S., Shabir, M., 2020a. Estimation of genetic and phenotypic trends for wool traits in Kashmir Merino sheep. Indian Journal of Animal Sciences 90 (6), 893–897. http://epubs.icar.org.in/ejournal/index.php/IJAnS/article/view/104998.

Hamadani, A., Ganai, N.A., Farooq, S.F., Bhat, B.A., 2020b. Big data management: from hard drives to DNA drives. Indian Journal of Animal Sciences 90 (2), 134–140. http://epubs.icar.org.in/ejournal/index.php/IJAnS/article/view/98761/39230.

Hamadani, H., Khan, A.A., Wani, S.A., Khan, H.M., Banday, M.T., Wani, S.A., 2020c. Economics of milk production and profitability of different cow unit sizes in Srinagar. Indian Journal of Animal Sciences 90 (7), 1065–1069. http://epubs.icar.org.in/ejournal/index.php/IJAnS/article/view/106683/41932.

Hamadani, H., Parrah, J.D., Hassan, N., Dar, R.A., Sheikh, F.D., Shah, R.M., Reshi, P.A., Haq, S.A., 2020d. Study of the socioeconomic status of women vermicompost-producing farmers in Kashmir valley. International Journal of Current Microbiology and Applied Sciences 9 (4), 1486–1491. https://doi.org/10.20546/ijcmas.2020.904.175.

Hamadani, H., Khan, A.A., Banday, M.T., 2020e. Kashmir Anz geese breed. World's Poultry Science Journal 76 (1), 144–153. https://doi.org/10.1080/00439339.2020.1711293.

Hamadani, A., Ganai, N.A., Rather, M.A., Genetic, 2021a. Phenotypic and heritability trends for body weights in Kashmir Merino sheep. Small Ruminant Research 205.

Hamadani, H., Rashid, S.M., Parrah, J.D., Khan, A.A., Dar, K.A., Ganie, A.A., Gazal, A., Dar, R.A., Ali, A., 2021b. Traditional farming practices and its consequences. In: Microbiota and Biofertilizers. Ecofriendly Tools for Reclamation of Degraded Soil Environs, vol 2, pp. 119–128. https://doi.org/10.1007/978-3-030-61010-4_6.

Hamadani, A., Ganai, N.A., Rather, M.A., Shanaz, S., Ayaz, A., Mansoor, S., Nazir, S., 2022a. Livestock and poultry breeds of Jammu and Kashmir and Ladakh. Indian Journal of Animal Sciences 92 (4), 409–416. https://epubs.icar.org.in/index.php/IJAnS/article/view/124009.

Hamadani, A., Ganai, N.A., Mudasir, S., Shanaz, S., Alam, S., Hussain, I., 2022b. Comparison of artificial intelligence algorithms and their ranking for the prediction of genetic merit in sheep. Scientific Reports 12 (1). https://doi.org/10.1038/s41598-022-23499-w.

Hamadani, A., Ganai, N.A., Alam, S., Mudasir, S., Raja, T.A., Hussain, I., Ahmad, H.A., 2022c. Artificial intelligence techniques for the prediction of body weights in sheep. Indian Journal of Animal Research. https://doi.org/10.18805/ijar.b-4831.

Hamadani, A., Ganai, N.A., Bashir, J., 2023. Artificial neural networks for data mining in animal sciences. Bulletin of the National Research Centre 47 (1). https://doi.org/10.1186/s42269-023-01042-9.

Hossain, M.E., Kabir, M.A., Zheng, L., Swain, D.L., McGrath, S., Medway, J., 2022. A systematic review of machine learning techniques for cattle identification: datasets, methods and future directions. Artificial Intelligence in agriculture 6, 138–155. https://doi.org/10.1016/j.aiia.2022.09.002.

Kabir, M., Sultana, N., Noman, A., Hossain, S., Miraz, M., Deb, G., 2022. FeedMaster: a least-cost feed formulation app for minimizing the cost and maximizing milk yield. Journal of Advanced Veterinary and Animal Research 9 (3). https://doi.org/10.5455/javar.2022.i605.

Liebe, D.M., White, R.R., 2019. Analytics in sustainable precision animal nutrition. Animal Frontiers 9 (2), 16–24. https://doi.org/10.1093/af/vfz003.

Munian, Y., Martinez-Molina, A., Miserlis, D., Hernandez, H., Alamaniotis, M., 2022. Intelligent system utilizing HOG and CNN for thermal image-based detection of wild animals in nocturnal periods for vehicle safety. Applied Artificial Intelligence 36 (1). https://doi.org/10.1080/08839514.2022.2031825.

Norouzzadeh, M.S., Nguyen, A., Kosmala, M., Swanson, A., Palmer, M.S., Packer, C., Clune, J., 2018. Automatically identifying, counting, and describing wild animals in camera-trap images with deep learning. Proceedings of the National Academy of Sciences 115 (25). https://doi.org/10.1073/pnas.1719367115.

Nti, E.K., Cobbina, S.J., Attafuah, E.E., Opoku, E., Gyan, M.A., 2022. Environmental sustainability technologies in biodiversity, energy, transportation and water management using artificial intelligence: a systematic review. Sustainable Futures 4. https://doi.org/10.1016/j.sftr.2022.100068.

Pabico, J.P., Gonzales, A.M.V., Villanueva, M.J.S., Mendoza, A.A., 2015. Automatic Identification of Animal Breeds and Species Using Bioacoustics and Artificial Neural Networks. https://doi.org/10.48550/ARXIV.1507.05546.

Rashidi, H.H., Tran, N.K., Betts, E.V., Howell, L.P., Green, R., 2019. Artificial intelligence and machine learning in pathology: the present landscape of supervised methods. Academic Pathology 6. https://doi.org/10.1177/2374289519873088.

Rather, M.A., Shanaz, S., Ganai, N.A., Bukhari, S., Hamadani, A., Nabi Khan, N., Yousuf, S., Baba, A., Raja, T.A., Khan, H.M., 2019. Genetic evaluation of wool traits of Kashmir Merino sheep in organized farms. Small Ruminant Research 177, 14–17. https://doi.org/10.1016/j.smallrumres.2019.06.003.

Reynolds, L.P., Wulster-Radcliffe, M.C., Aaron, D.K., Davis, T.A., 2015. Importance of animals in agricultural sustainability and food security. Journal of Nutrition 145 (7), 1377–1379. https://doi.org/10.3945/jn.115.212217.

Riego del Castillo, V., Sánchez-González, L., Campazas-Vega, A., Strisciuglio, N., 2022. Vision-based module for herding with a sheepdog robot. Sensors 22 (14). https://doi.org/10.3390/s22145321.

Rivas, A., Chamoso, P., González-Briones, A., Corchado, J., 2018. Detection of cattle using drones and convolutional neural networks. Sensors 18 (7). https://doi.org/10.3390/s18072048.

Saxena, P., Parasher, Y., 2019. Application of artificial neural network (ANN) for animal diet formulation modeling. Procedia Computer Science 152, 261–266. https://doi.org/10.1016/j.procs.2019.05.018.

Shukla, A., ., K., Gaurav, P., Kumar, R., 2022. Clean sweep: the floor cleaning robot. International Journal for Research in Applied Science and Engineering Technology 10 (7), 1279–1283. https://doi.org/10.22214/ijraset.2022.45414.

Varijakshapanicker, P., Mckune, S., Miller, L., Hendrickx, S., Balehegn, M., Dahl, G.E., Adesogan, A.T., 2019. Sustainable livestock systems to improve human health, nutrition, and economic status. Animal Frontiers 9 (4), 39–50. https://doi.org/10.1093/af/vfz041.

Versluijs, E., Niccolai, L.J., Spedener, M., Zimmermann, B., Hessle, A., Tofastrud, M., Devineau, O., Evans, A.L., 2023. Classification of behaviors of free-ranging cattle using accelerometry signatures collected by virtual fence collars. Frontiers in Animal Science 4. https://doi.org/10.3389/fanim.2023.1083272.

Wong, Z.S.Y., Zhou, J., Zhang, Q., 2019. Artificial intelligence for infectious disease big data analytics. Infection, Disease & Health 24 (1), 44–48. https://doi.org/10.1016/j.idh.2018.10.002.

Yeo, C.Y., Al-Haddad, S.A.R., Ng, C.K., 2011. Animal voice recognition for identification (ID) detection system. In: Proceedings - 2011 IEEE 7th International Colloquium on Signal Processing and its Applications, CSPA 2011, pp. 198–201. https://doi.org/10.1109/CSPA.2011.5759872.

Food manufacturing, processing, storage, and marketing using artificial intelligence

12

O.H. Onyijen, S. Oyelola and O.J. Ogieriakhi

Department of Mathematical and Physical Sciences, College of Basic and Applied Sciences, Glorious Vision University, Ogwa, Edo State, Nigeria

Introduction

The food industry is essential to ensuring that food products are produced, processed, stored, and marketed. It also plays a role in ensuring that food products are safe and of good quality. Furthermore, the food industry contributes to the economic well-being of a nation by providing jobs and generating revenue. Artificial intelligence (AI) has emerged as a potent tool to transform many industries, including food manufacturing and marketing, as a result of the quickening pace of technological development. AI encompasses a range of technologies that enable machines to perform tasks that would typically require human intelligence, such as problem-solving, learning, and decision-making. In the context of food manufacture, processing, storage, and marketing, AI can be leveraged to enhance efficiency, quality, and sustainability throughout the entire value chain. This cutting-edge technology empowers food industry stakeholders to optimize processes, reduce waste, ensure food safety, and cater to consumer demands more effectively. The food industry is transforming with the integration of AI into various stages of food manufacture, processing, storage, and marketing. AI technologies are revolutionizing conventional practices, enabling businesses to optimize processes, improve efficiency, and meet consumer demands more effectively (Ribeiro et al., 2020).

The benefits of using AI in the food industry include increased productivity, efficiency, and accuracy (Rajnish et al., 2022). AI can optimize material movement by analyzing data on inventory levels, production schedules, and transportation routes. It can also be used for microbial control to prevent contamination and ensure food safety. AI can help with sustainable innovations, such as logistics, supply chain, marketing, and production patterns. Additionally, AI can be used for sentiment analysis of user reviews on food and beverage groups, food sale prediction, and customer behavior analysis. Despite the several benefits, it is pertinent to also look at the challenges of using AI in the food industry such as displacement of certain jobs and the need for retraining workers. There are also concerns about the ethical and responsible use of AI in the job market. The use of AI in food processing can also present new hazards that need to be addressed. For example, the deployment of robots, sensors, and machine learning (ML) technologies for factory cleaning tasks can pose safety risks.

A Biologist's Guide to Artificial Intelligence. https://doi.org/10.1016/B978-0-443-24001-0.00012-9

183

Several techniques have been deployed for food manufacturing, storage, processing, and marketing such as ML. It is used in the food industry for sentiment analysis of user reviews on food and beverage groups, food sale prediction, and customer behavior analysis (Irfan et al., 2022; Yanfi et al., 2022). Artificial neural networks (ANNs) are a type of ML that is modeled after the structure and function of the human brain. In the food industry, ANNs are used for sustainable innovations in marketing, production patterns, logistics, and supply chains (Sharma et al., 2021). Various algorithms are used in the food industry, including intelligent optimization algorithms, unsupervised predictors, and classifier models based on ML such as logistic regression, Naive Bayes (NB), random forest (RF), and support vector machine (SVM) (Sharma et al., 2021; Yanfi et al., 2022).

This study explores how AI is transforming the different aspects of the food industry, from manufacturing to marketing. It will delve into specific applications of AI in each stage, highlighting the benefits it brings and the challenges it addresses. By harnessing the power of AI, food businesses can gain a competitive edge in a rapidly evolving marketplace while meeting the demands for safe, high-quality, and sustainable food products.

Food manufacturing

Food manufacture is a complex process that involves various stages, including ingredient selection, recipe development, quality control, and production optimization. With the rapid advancements in AI technology, the food manufacturing industry has witnessed significant transformations. AI applications are revolutionizing the way food products are manufactured, enhancing efficiency, quality, and sustainability (Wang et al., 2019). Wang et al. discusses how AI technologies such as ML, neural networks, and expert systems can be employed in ingredient analysis, recipe development, process optimization, and quality control (Wang et al., 2019). Ma and Zhang explore how AI technologies can be used for ingredient analysis, formulation optimization, process monitoring, and quality assurance. Their study highlights the potential of AI in reducing production costs, enhancing food safety, and improving the overall efficiency of food manufacture (Ma and Zhang, 2020). AI can be used in food manufacturing to improve efficiency, reduce waste, and enhance supply chain management. AI can enable autonomous and remote monitoring of manufacturing efficiency and product end-of-life cycles, and can be utilized to effectively analyze data for further development of manufacturing processes. In the food and beverage industry, AI can help with supply chain management through logistics and predictive analytics, and can be used to improve various aspects of the manufacturing process (Agbai, 2020; Jacobs, 2023).

Application of AI in food manufacturing

Some of the specific applications of AI in food manufacturing include:

1. Sorting and grading of food products: Sorting and grading of food products is an important process in the food industry to ensure quality and consistency. AI is being increasingly used in this process to improve efficiency and accuracy. AI can be used to analyze images of food products to identify defects or inconsistencies, and to sort and grade products based on color, size, and texture (Agbai, 2020). AI can also be used to predict the quality of raw materials and optimize the use of ingredients, which can help reduce waste and improve product quality (Bendre et al., 2022).

Additionally, AI can be used to monitor and control the temperature and humidity of food storage facilities to prevent spoilage (Bendre et al., 2022).

2. Predictive maintenance and machinery inspection: AI is being increasingly used in the food industry for predictive maintenance and machinery inspection. AI can be used to predict equipment failures and detect defects in machinery, which can help prevent breakdowns and reduce downtime. Furthermore, AI can be used to monitor and control the temperature and humidity of food storage facilities to prevent spoilage (Carvalho et al., 2019; Andy, 2022).

3. Material movement: It is an important process in the food industry to ensure that raw materials and finished products are transported efficiently and safely. AI is being increasingly used in this process to improve efficiency and reduce waste. It can be used to optimize material movement by analyzing data on inventory levels, production schedules, and transportation routes. In the same vein, AI can be used to monitor and control the temperature and humidity of food storage facilities to prevent spoilage (Sonwani et al., 2022).

4. Production planning: AI is being increasingly used in the food industry for production planning to optimize production processes and improve efficiency. With the increasing demand and competition in the food industry, AI technologies are being embraced to maximize profits and explore new ways to serve consumers (Garre et al., 2020). AI can be used for efficient supply chain management, which includes inventory management, demand forecasting, and transportation optimization. AI can also be used for microbial control to prevent contamination and ensure food safety. Additionally, AI can be used for food quality monitoring using chemical and biological sensors to provide the best quality food products (Garre et al., 2020).

5. Field service: It is an important aspect of the food industry to ensure that equipment is installed, maintained, and repaired in a timely and efficient manner. AI is increasingly used in this process to improve efficiency and reduce downtime. It can be used to predict equipment failures and detect defects in machinery, which can help prevent breakdowns and reduce downtime. Additionally, AI can be used to optimize field service by analyzing data on equipment performance, maintenance schedules, and technician availability. The use of AI in field service has the potential to improve efficiency, reduce downtime, and improve product quality in the food industry. However, there are challenges to implementing AI in field service, including the need for accurate data and the need for skilled technicians to interpret the data.

6. Quality control: It is a critical process in the food industry to ensure that products meet the required standards of safety, quality, and consistency. AI is being increasingly used in this process to improve efficiency and accuracy. AI can be used to analyze images of food products to identify defects or inconsistencies, and to sort and grade products based on color, size, and texture (Hemamalini et al., 2022). AI can also be used to predict the quality of raw materials and optimize the use of ingredients, which can help reduce waste and improve product quality. Additionally, AI can be used to monitor and control the temperature and humidity of food storage facilities to prevent spoilage (Sonwani et al., 2022).

7. Reclamation: It is an important process in the food industry to reduce waste and improve sustainability. AI is being increasingly used in this process to improve efficiency and accuracy. AI can be used to optimize reclamation by analyzing data on inventory levels, production schedules, and transportation routes. Additionally, AI can be used to predict the quality of raw materials and optimize the use of ingredients, which can help reduce waste and improve product quality. AI can

also be used to monitor and control the temperature and humidity of food storage facilities to prevent spoilage.

AI can also be used to improve safety in the food manufacturing industry by using computer vision to monitor the frontline for risks such as slips, trips, and falls, missing PPE, and working at heights.

Benefits of AI in food manufacturing

Artificial intelligence has been identified to offer several benefits in the food industry, which can be enumerated as follows.

1. Currently, the employment of AI has become ubiquitous in the food processing sector, as it facilitates the enhancement of demand-supply chain management, meticulous logistics, predictive analysis, and precision within the system.
2. The process of converting demand-supply chain management systems into digital format ultimately necessitates the analysis of returns and provides a greater comprehension of the prevailing circumstances. This is where AI comes into play as it has the ability to evaluate vast quantities of data that are beyond the realm of human potential.
3. Artificial intelligence facilitates the industrial sector in reducing the duration required for introducing a product into the market and enhancing compliance with expectations.
4. Automated ordering is expected to result in a reduction of labor expenses, acceleration of the pace of manufacturing, and enhancement of product quality, thereby having a positive impact on the overall manufacturing process.

Implementation of AI in food manufacturing

There are several case studies and experimental results showcasing the implementation of AI in food manufacturing. Some of these studies are:

The study of (Bendre et al., 2022) on "Artificial Intelligence in Food Industry": In "A Current Panorama," the effective application of AI in numerous areas of the food industry is highlighted, including sorting, grading, food quality, cleaning, effective supply chain management, microbial control, and various methods of food analysis. Chemical and biological sensors are used for food quality monitoring, as well as the application of AI to provide the best quality food products.

Konur et al. (2021) in their work "Toward design and implementation of Industry 4.0 for food manufacturing" proposed the use of smart manufacturing, the internet of things, AI, ML, and big data in Industry 4.0.

Furthermore, the study of (Bandyopadhyay et al., 2021) on "Application of Artificial Intelligence in Food Industry-A Review" examines how a key role for AI is being played in the production of products that are healthy to eat. The use of AI in agriculture, food production, the food industry, pesticides, contamination, crop management, irrigation, sorting, and grading are all highlighted in the study.

Satwekar et al. on "Digital by design approach to develop a universal deep learning AI architecture for automatic chromatographic peak integration" outlines a "Digital by Design" management strategy for developing and implementing an AI (AI)-based chromatography peak integration process for the healthcare sector. In addition to reporting on the use of a convolutional neural network (CNN) model to predict analytical variability for integrating chromatography peaks, the study makes a potential GxP framework for using AI in the healthcare sector that incorporates elements on data management, model management, and human-in-the-loop processes (Satwekar et al., 2023).

Food processing

Food processing plays a crucial role in transforming raw ingredients into safe, nutritious, and convenient food products. With the advancements in AI, the food processing industry is experiencing significant changes. AI technologies offer new opportunities to optimize processes, improve quality control, enhance efficiency, and ensure food safety (Carvalho et al., 2019). AI techniques such as ML, computer vision, and robotics can be employed in various aspects of food processing, including quality monitoring, process optimization, predictive maintenance, and automation. Carvalho et al. (2019) gave a detailed study of how AI techniques such as ML algorithms, data mining, and computer vision can be utilized for quality control, process optimization, and automation in food processing operations.

Application of AI in food processing

Artificial intelligence has numerous applications in food processing, including:

Sorting and grading: AI can be used to analyze images of food products to identify defects or inconsistencies, and to sort and grade products based on color, size, and texture (Bendre et al., 2022). Deep learning-based approaches have been employed for quality inspection and grading of food products, such as fruits, vegetables, and grains. These systems use computer vision techniques to analyze visual features and detect defects, ensuring consistent quality.

Food quality: can be used for food quality monitoring using chemical and biological sensors to provide the best quality food products (Bendre et al., 2022).

Efficient supply chain management: AI can optimize material movement by analyzing data on inventory levels, production schedules, and transportation routes (Bendre et al., 2022).

Microbial control: It can be used for microbial control to prevent contamination and ensure food safety (Bendre et al., 2022).

Food Safety and Traceability: AI is used for enhancing food safety and traceability by identifying potential contaminants, pathogens, or adulterants in food products. It can also improve supply chain transparency and facilitate quick recalls in case of contamination events (Abdullah and Ali, 2021).

Cleaning: It can be used for factory cleaning tasks, including the deployment of robots, sensors, and ML technologies (Bendre et al., 2022).

Chromatographic data processing: AI can be used to predict analytical variability for integrating chromatography peaks and propose a potential GxP framework for using AI in the healthcare industry (Satwekar et al., 2023).

Predictive Analytics for Shelf-Life Prediction: AI techniques, including ML and data analytics, are used to predict the shelf life of food products. By analyzing various factors such as temperature, humidity, and storage conditions, predictive models can estimate the remaining shelf life and optimize inventory management (Esquerre et al., 2018).

Optimization of Food Processing Parameters: AI algorithms, such as genetic algorithms and neural networks, are applied to optimize food processing parameters. These techniques help in maximizing efficiency, reducing energy consumption, and improving product quality.

Flavor and Recipe Optimization: AI-powered systems are employed to develop new flavors and optimize recipes based on consumer preferences. Natural language processing (NLP) and ML techniques analyze large datasets of recipes and customer feedback to generate new recipes or enhance existing ones (Elmasri and Park, 2021).

AI potential in food processing

Artificial intelligence has the potential to improve processing speed, accuracy, and safety in the food industry. In order to design and implement an AI-based solution for the chromatography peak integration process in the healthcare industry, a study on the processing of chromatographic data has highlighted the need for a reliable technological solution and suggested a "Digital by Design" managerial approach (Satwekar et al., 2023). The study described the use of a CNN model to forecast analytical variability for integrating chromatography peaks and proposed a potential GxP framework for using AI in the healthcare sector that includes components on data management, model management, and human-in-the-loop processes (Satwekar et al., 2023). A review of the impactful use of AI in the food industry highlighted the use of AI in diverse areas of the food sector, including sorting, grading, food quality, cleaning, efficient supply chain management, and microbial control (Agbai, 2020; Mavani et al., 2022). Chemical and biological sensors are used for food quality monitoring, and AI can optimize material movement by analyzing data on inventory levels, production schedules, and transportation routes (Agbai, 2020). Despite the challenges of implementing AI in the food industry, research into optimizing production processes using AI is ongoing. Another review highlighted the use of AI in sorting and grading of food products. Machines with AI can recognize food items before processing them, reducing the time and labor required for sorting and grading. Additionally, the study emphasized the application of AI in agriculture, food production, the food industry, pesticides, contamination, crop management, irrigation, and effective supply chain management. The key AI-related influences on the food processing industry were examined through empirical research. According to the study, technology applications like automation and AI are used to improve processing and make it possible to provide customers with high-quality goods at lower costs. The use of AI in the food processing industry is quickly changing how customer inquiries are handled, enabling analysis of needs and requirements, and putting more of an emphasis on packaging, high quality, and shelf life. An approach for designing and implementing Industry 4.0 for the food manufacturing industry was suggested in a study (Konur et al., 2021). Similar food manufacturing industries and other SME industries can gain from the suggested strategy and lessons presented. The study focused on Industry 4.0's use of big data, AI, the internet of things, smart manufacturing, and ML (Rahman et al., 2023). IDTs were used in a presentation to discuss the crucial job of cleaning food factories. The talk discussed the advantages and difficulties of using robotics, sensors, and ML technologies for factory cleaning chores as well as the rising significance of efficient industrial cleaning amid a pandemic around the world.

Food storage using AI

Food storage is a critical component of the food supply chain, ensuring the preservation of food products and maintaining their quality and safety. With the advent of AI, the field of food storage has witnessed significant advancements. AI technologies offer new possibilities to monitor and manage storage conditions, detect anomalies, and optimize inventory management. AI technologies such as ML, data analytics, and internet of things (IoT) has be utilized to monitor storage conditions, optimize inventory levels, and detect anomalies in real-time. Kim gave insights on the use of technologies such as ML, data mining, and optimization algorithms to improve the management of cold storage facilities, optimize temperature control, and reduce energy consumption (Kim, 2023). The study emphasizes the potential of AI in enhancing food quality, safety, and sustainability in the storage process. AI techniques, such as ML and intelligent decision support systems, can be employed to optimize storage conditions, predict product shelf life, and prevent food spoilage (Mavani, et al., 2022). provides an overview of the applications of AI in the food supply chain, including food storage. In their research, AI technologies such as ML, IoT, and big data analytics was used to monitor storage conditions, optimize inventory levels, and improve traceability. The authors emphasize the potential of AI in enhancing food quality, safety, and efficiency in storage operations as shown in Table 12.1.

The role of AI in food storage practices
AI plays a crucial role in the improvement of food storage practices as well. The roles of AI in food storage practices may be summed as under.

1. Inventory management

Inventory management is a critical aspect of the food processing industry, and AI has been instrumental in optimizing inventory levels, reducing waste, and improving overall efficiency.

  i. AI-based Demand Forecasting and Inventory Optimization: AI techniques, such as ML and deep learning, are used to forecast demand patterns and optimize inventory levels accordingly. These models analyze historical sales data, market trends, and other factors to predict future demand, enabling more accurate inventory planning.
  ii. Reinforcement Learning for Inventory Control: Reinforcement learning algorithms, a subset of AI, are applied to optimize inventory control policies. These models learn from interactions with the environment, making sequential decisions on inventory replenishment based on rewards and penalties, ultimately achieving optimal inventory levels (Saha and Kumar, 2021).

Table 12.1 Potential artificial intelligence (AI) applications in food safety to enhance public health.

Subfields	Specific applications	AI branch
Surveillance	Real-time outbreak and sickness detection from surveillance data	Analytical tool
	Real-time detection of outbreaks and foodborne illness using data from social media	Natural language processing
Source tracking and source attribution	Foodborne illness and outbreak sources can be identified by identifying likely sources	Analytical tool
Food hazards prediction	Improving diagnostic accuracy	Analytical tool

iii. AI-enabled Supply Chain Visibility and Coordination: AI-based systems enable real-time visibility across the supply chain, allowing better coordination and synchronization of inventory management. These systems leverage advanced analytics, IoT, and AI algorithms to track inventory, monitor supplier performance, and identify potential bottlenecks for proactive inventory adjustments (Jacobs, 2023)

iv. Predictive Analytics for Inventory Optimization: Predictive analytics, including AI and statistical modeling, are utilized for inventory optimization. These techniques leverage historical data, external factors, and demand patterns to forecast future inventory requirements accurately. By identifying optimal reorder points and order quantities, inventory holding costs can be minimized while ensuring sufficient stock availability.

v. AI-enabled Just-in-Time (JIT) Inventory Management: AI algorithms play a crucial role in implementing just-in-time inventory management strategies. By analyzing real-time data from various sources, including point-of-sale systems, production schedules, and supply chain information, AI models optimize inventory levels to meet demand, reducing holding costs and waste (Liu et al., 2020).

2. Shelf life Prediction

Shelf life prediction is an important aspect of food processing and preservation, and AI techniques have proven valuable in this area.

i. Machine Learning-based Shelf Life Prediction: ML algorithms, such as SVM, ANN, and RF, have been employed to predict the shelf life of food products. These models utilize various input parameters, such as temperature, humidity, and product characteristics, to estimate shelf life (Iorliam et al., 2021; Albert-Weiss and Osman, 2022)

ii. Kinetic models, combined with AI techniques, have been used to predict the shelf life of food products. These models incorporate chemical reactions, microbial growth, and quality changes over time to estimate the remaining shelf life. AI algorithms, such as genetic algorithms and neural networks, are employed for parameter estimation and model optimization.

iii. IoT and AI for real-time shelf-life monitoring: The integration of AI with IoT devices enables real-time monitoring of environmental conditions and quality parameters during storage and transportation. AI algorithms analyze the data collected from IoT sensors to predict the remaining shelf life and ensure timely interventions for maintaining food safety and quality (Torres-Sánchez et al., 2020).

iv. Quality index and AI for shelf life prediction: Quality indices are developed based on sensory evaluation and analytical measurements to assess the quality of food products. AI algorithms, such as fuzzy logic, neural networks, and expert systems, are used to predict the shelf life by correlating the changes in quality indices with storage conditions and time.

v. Deep learning-based shelf-life prediction: Deep learning algorithms, such as CNN and long short-term memory (LSTM) networks, have shown promise in predicting the shelf life of food products. These models leverage the power of deep learning architectures to analyze complex data, such as images and time-series measurements, for accurate shelf life estimation.

3. Food safety monitoring

AI plays a crucial role in food safety monitoring, enabling real-time monitoring, early detection of contaminants, and ensuring the overall safety of food products.

i. Detection of contaminants and pathogens: AI techniques, such as ML and computer vision, are used to detect contaminants and pathogens in food products. These methods analyze visual characteristics and patterns to identify potential hazards, ensuring timely interventions.

ii. IoT-enabled food safety monitoring: The combination of AI and IoT devices enables real-time monitoring of critical parameters such as temperature, humidity, and pH during food processing and storage. AI algorithms analyze the data collected from IoT sensors to identify anomalies and potential safety risks.

iii. Predictive modeling for food safety: AI-based predictive models use historical data on food safety incidents, environmental conditions, and other relevant factors to anticipate potential risks. These models can help in proactive decision-making and implementing preventive measures.

iv. Image analysis for food quality and safety assessment: AI algorithms analyze images and videos of food products to assess their quality and safety. By detecting visual defects, spoilage, or contamination, these systems provide an objective and automated approach to ensure food safety.

v. Blockchain-based traceability and transparency: AI, along with blockchain technology, enables improved traceability and transparency in the food supply chain. AI algorithms can analyze blockchain data to track the origin, processing, and distribution of food products, facilitating rapid identification and resolution of safety issues.

Application of AI for food storage

The study of (Agbai, 2020) on "Application of artificial intelligence (AI) in food industry" reviewed that AI has been successfully deployed for applications such as sorting fresh produce, managing supply chain, food safety compliance monitoring, and effective cleaning in place systems, anticipating consumer preference and new product development with greater efficiency and savings on time and resources. Since many years ago, the use of AI in the food sector has increased for a variety of purposes, including food sorting, classification and parameter prediction, quality assurance, and food safety. Among the main methods used in the food industry are expert systems, fuzzy logic, ANNs, adaptive neuro-fuzzy inference systems (ANFIS), and ML. Prior to the application of AI, research on food has been conducted over the years to increase public understanding of food as well as to enhance the results relating to food attributes and food production (Rahman et al., 2012). A lot of benefits can be obtained by using the AI method, and its implementation in the food industry has been going on since decades ago and has been increasing till today (Rahman et al., 2012).

AgShift, an AI-based food inspection company, developed an autonomous food inspection platform that utilizes AI algorithms for food quality assessment and storage management. The system analyzes visual characteristics and quality indicators of food products, providing real-time insights to optimize storage conditions and reduce food waste.

The study of Vasantha et al. on "AI-enabled smart cold storage for fish freshness monitoring" focused on developing an AI-enabled smart cold storage system for fish freshness monitoring. The system utilized ML algorithms to analyze sensor data, including temperature, humidity, and gas composition, to determine fish freshness. The AI algorithms provided real-time monitoring and alerts to ensure optimal storage conditions and maintain fish quality. Afresh Technologies developed an AI-powered inventory management system for fresh produce in retail stores. The system uses ML algorithms to predict demand, optimize order quantities, and manage stock levels to reduce waste and ensure product freshness. It considers factors such as sales data, seasonality, and quality degradation to optimize inventory decisions. These applications and studies demonstrate the potential of AI in improving food storage practices, reducing waste, and ensuring product quality and freshness.

Food marketing

Food marketing is a dynamic and competitive field that involves promoting, distributing, and selling food products to consumers. With the advent of AI, the landscape of food marketing has experienced significant transformations. AI technologies offer new opportunities to analyze consumer behavior, personalize marketing strategies, optimize advertising campaigns, and improve customer engagement. Aggarwal and Billus (2022) provides a bibliometric analysis of the applications of AI in marketing. It discusses how AI techniques, such as ML, NLP, and recommendation systems, can be employed in food marketing to analyze consumer preferences, personalize marketing content, and optimize advertising strategies. The authors highlight the potential of AI in enhancing customer targeting, engagement, and satisfaction in food marketing. Ahmad et al. (2020) explore the applications of AI and ML in the food and beverage industry, including food marketing. They discuss how AI technologies can be utilized to analyze consumer behavior, predict market trends, personalize product recommendations, and optimize pricing and promotion strategies. Shekhawat and Tatavarthy (2020) highlights the potential benefits of AI in improving marketing effectiveness and customer satisfaction in the food industry and explores the role of AI in transforming customer engagement in marketing and advertising. It discusses how AI techniques such as NLP, sentiment analysis, and chatbots can be employed to personalize marketing communications, optimize advertising campaigns, and improve customer interaction. From their study, customer engagement and satisfaction in the food marketing industry can be enhanced. AI has emerged as a significant advantage in marketing due to its capability to analyze large volumes of data and identify patterns. This technology enables companies to segment their customers based on factors such as demographics, behavior, and preferences, thereby creating targeted marketing campaigns that are more likely to resonate with their intended audience. For instance, a coffee company can utilize AI to categorize its patrons based on their preferred coffee type, the time of day they usually purchase coffee, and their buying history. These data can then be utilized to develop personalized marketing strategies that offer discounts on the customer's preferred coffee type at the time of day they are most likely to make a purchase, similar to the Starbucks loyalty program. Fig. 12.1 shows the benefits of AI in marketing such personalized customer experience, increased productivity, higher profits, and better contents.

The use of AI in food marketing strategies

AI has been increasingly used in food marketing strategies to enhance customer targeting, personalized advertising, and predictive analytics.

   i. Personalized marketing and recommendation systems: AI-powered recommendation systems analyze customer preferences, purchase history, and online behavior to provide personalized

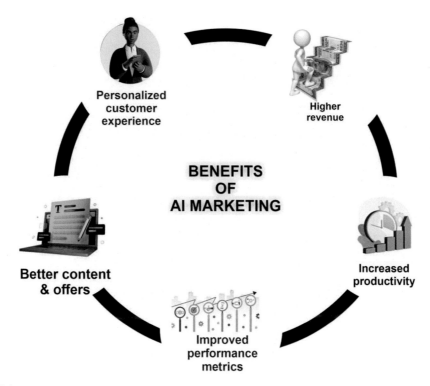

FIGURE 12.1

Benefits of artificial intelligence (AI) in food marketing such as personalized customer experience, increased productivity, higher profits, and better contents.

recommendations for food products. These systems improve customer engagement and help tailor marketing strategies to individual preferences.

ii. Customer segmentation and targeted advertising: AI techniques, such as clustering algorithms and NLP, enable advanced customer segmentation based on demographics, psychographics, and behavior patterns. This segmentation facilitates targeted advertising campaigns, ensuring that the right message reaches the right audience, increasing conversion rates, and marketing effectiveness.

iii. Social media analytics and sentiment analysis: AI algorithms are employed to analyze social media data and extract insights from customer conversations and sentiments. These insights help food marketers understand customer preferences, trends, and brand perceptions, enabling them to optimize marketing strategies and respond effectively to customer feedback (Kumar, 2020).

iv. Predictive analytics for demand forecasting: AI-based predictive analytics models leverage historical sales data, market trends, and external factors to forecast demand for food products. This information enables food marketers to plan production, manage inventory, and optimize pricing strategies, ensuring efficient supply chain management and meeting customer demand (Ren, 2021).

v. Chatbots and virtual assistants: AI-driven chatbots and virtual assistants are used in food marketing to provide personalized recommendations, answer customer queries, and facilitate interactive customer experiences. These AI-powered conversational interfaces enhance customer engagement, deliver real-time support, and help in building brand loyalty (Galitsky, 2020). Fig. 12.2 shows the various use cases of AI for marketing in the food industry.

Implementation of AI in marketing

Domino's Pizza: "Domino's AnyWare": Domino's Pizza launched the "Domino's AnyWare" campaign, which utilized AI and various technologies to enhance customer engagement and ordering convenience. The campaign included features such as voice-activated ordering through virtual assistants, ordering via social media platforms, and even ordering through smart home devices. By leveraging AI and integrating it into multiple channels, Domino's increased customer engagement and made ordering more accessible. Coca-Cola: Personalized marketing: Coca-Cola launched a personalized marketing campaign that utilized AI to create unique experiences for customers. The campaign involved personalized labels with people's names on the bottles, as well as personalized digital advertisements and social media interactions. AI algorithms analyzed consumer data and preferences to deliver customized messages, increasing consumer engagement and building brand loyalty. Starbucks:

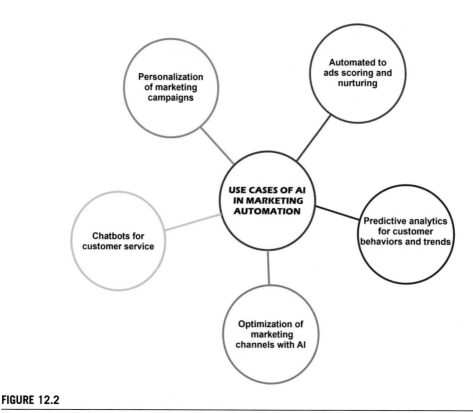

FIGURE 12.2

Uses cases of artificial intelligence (AI) for marketing in food industry.

AI-Powered personalization: Starbucks implemented an AI-powered personalization campaign that utilized customer data and AI algorithms to offer personalized recommendations and rewards. Through their mobile app, Starbucks used AI to analyze customer preferences, order history, and location to provide customized offers, suggested products, and rewards. This campaign enhanced consumer engagement by delivering tailored experiences and driving customer loyalty. McDonald's: Dynamic Digital Menu Boards: McDonald's implemented dynamic digital menu boards powered by AI to personalize menu offerings based on factors like time of day, weather conditions, and customer preferences. AI algorithms analyzed data in real time to optimize the menu board display, showcasing relevant food items and promotions. This AI-driven marketing campaign improved customer engagement by tailoring the menu offerings to individual preferences and creating a more personalized dining experience. Kellogg's utilized AI to create dynamic and personalized advertisements. AI algorithms analyzed consumer data, social media trends, and cultural cues to generate targeted advertisements. These ads were tailored to specific customer segments, making the content more relevant and engaging. The AI-driven marketing campaign helped Kellogg's increase consumer engagement and drive brand awareness (Kumar, 2020). Vacancy of store spaces in urban areas is a predicament that poses challenges for the concerned parties including landlords, city centers, and adjacent businesses. In a bid to address this issue, digital agencies such as Dept and Hello Monday have conceptualized an innovative remedy—an AI-fueled "Shoe Mirror" to convert unoccupied stores to interactive and revenue-generating advertisements. The concept entails the utilization of augmented reality to provide a personalized experience for the viewer. The Shoe Mirror operates by analyzing the attire of passersby and matching their outfit with a pair of shoes that are then displayed on their digital feet. Aberdeen, a research corporation, has indicated that businesses that leverage predictive analytics to identify customer requirements can observe an increase of 21% in their organic revenue on a yearly basis, in contrast to the average of 12% that is witnessed sans predictive analytics. A case in point is Starbucks, which employs its loyalty card and mobile application to gather and scrutinize customer data. The company announced its intention to introduce personalization in the year 2016. Since that time, Starbucks has created a highly advanced application that offers an exceptional user experience. This app is designed to document purchases, including the location and time of day they were made. By utilizing predictive analytics, Starbucks is able to personalize its marketing messages to cater to individual customers. Among its features, the application provides recommendations to users as they approach a local store and offers special promotions aimed at increasing their average order value. According to the latest State of Conversational Marketing report by Drift, chatbots are experiencing a more rapid rate of expansion than any other type of brand communication channel, displaying an increase in usage of 92% between 2019 and 2020. The beauty industry's Sephora brand was an early adopter of AI, introducing a chatbot on Kik in 2017 that provided users with beauty advice. Sephora's chatbot facilitated the narrowing of choices for consumers, beginning with a quiz about their preferences for products, which is particularly advantageous in the cosmetics industry, where the abundance of options can be overwhelming and challenging to buy without experiencing them in person. Sephora was able to gain valuable insights from its chatbot and saw enough involvement from that initial experiment that it proceeded to launch more chatbots on Messenger.

These examples demonstrate how AI-driven marketing campaigns in the food industry have successfully enhanced consumer engagement by leveraging personalization, dynamic content, and advanced analytics. By utilizing AI technologies, these brands were able to deliver more relevant and targeted marketing experiences, resulting in increased customer satisfaction and brand loyalty.

Challenges of AI in food industry

Despite the tremendous breakthrough of AI in the food industry, it is still faced with some peculiar challenges that needs to be addressed to further enhance food manufacturing, processing, storage and marketing in the industry. Some of the challenges are

i. Data quality and accessibility: AI relies on high-quality and diverse data for training and decision-making. However, obtaining accurate and comprehensive data in the food industry can be challenging due to data fragmentation, inconsistencies, and privacy concerns.

ii. Interpretability and trust: AI models often operate as black boxes, making it difficult to understand the reasoning behind their decisions. Building trust among consumers and stakeholders is crucial, requiring transparency, explainability, and ethical considerations in AI systems.

iii. Integration and adoption: Integrating AI technologies into existing food industry processes and systems can be complex and resource-intensive. Organizations need to overcome barriers related to infrastructure, workforce readiness, and cultural acceptance to fully leverage the potential of AI.

iv. Regulation and compliance: The use of AI in the food industry raises concerns related to regulatory compliance, data privacy, and ethical considerations. Developing appropriate frameworks and guidelines to ensure responsible and safe use of AI is a critical challenge.

Future directions of AI in food industry

AI can further improve personalized food experiences by considering individual preferences, dietary restrictions, and health goals. Advanced AI algorithms and data analysis techniques can enable more accurate and tailored recommendations for consumers. AI can contribute to creating more sustainable and efficient food systems by optimizing supply chain operations, reducing food waste, and promoting sustainable practices in production, distribution, and consumption. The advancements in AI-driven predictive analytics can help in forecasting consumer demand, optimizing inventory management, and ensuring efficient supply chain operations in the food industry. This can lead to improved resource allocation and cost savings. Furthermore, it can play a crucial role in ensuring food safety by detecting contaminants, monitoring product quality, and identifying potential hazards in real time. AI-powered systems can enable early detection and prevention of foodborne illnesses and improve traceability across the supply chain. The future of AI in the food industry lies in effective collaboration between humans and AI systems. Integrating human expertise with AI technologies can lead to innovative solutions, improved decision-making, and enhanced consumer experiences.

Exploring future directions in AI will require collaboration among industry stakeholders, researchers, policymakers, and consumers. By overcoming challenges and harnessing the potential of AI, the food industry can benefit from improved efficiency, sustainability, and consumer satisfaction.

Ethical considerations, data privacy concerns, and potential biases

AI systems can inherit biases from the data they are trained on, leading to unfair treatment or discrimination against certain individuals or groups. It is essential to ensure fairness in AI algorithms and mitigate bias to prevent unintended consequences and promote equitable outcomes. Also, AI systems should be accountable for their decisions and actions. It is crucial to have transparency and explainability in AI algorithms to understand how decisions are made and provide recourse in case of errors or unintended consequences. In the area of decision-making, it should adhere to ethical principles and guidelines. Organizations need to establish ethical frameworks and codes of conduct to govern the development and deployment of AI systems, considering societal impact, human rights, and ethical implications.

AI applications require access to large amounts of data, including personal and sensitive information. Collecting and using these data should adhere to privacy regulations and obtain explicit consent from individuals. AI systems should employ robust security measures to protect the confidentiality, integrity, and availability of data. It is crucial to prevent unauthorized access, data breaches, and misuse of personal information.

AI algorithms can reflect and perpetuate biases present in the training data. Biases based on race, gender, age, or other factors can result in unfair or discriminatory outcomes. Efforts should be made to identify and mitigate bias during the development and testing phases of AI systems. Biases can also arise from biased or incomplete data used to train AI models. It is essential to ensure diverse and representative datasets to avoid perpetuating or amplifying existing biases.

Addressing these concerns requires a multifaceted approach such as;

i. Ethical guidelines and regulations: Governments, industry associations, and regulatory bodies should develop clear ethical guidelines and regulations to govern the use of AI, addressing fairness, transparency, accountability, and privacy concerns.
ii. Robust data governance: Organizations should establish robust data governance practices, including data anonymization, secure storage, and compliance with privacy regulations such as GDPR or CCPA. Data should be collected and used responsibly, with clear policies on consent, purpose limitation, and data retention.
iii. Bias detection and mitigation: Developers should implement techniques to detect and mitigate biases in AI algorithms. This includes auditing the training data, evaluating model outputs for bias, and incorporating fairness metrics during the development process.
iv. Interdisciplinary collaboration: Collaboration between AI developers, ethicists, social scientists, and domain experts is crucial to ensure ethical decision-making and identify potential biases or unintended consequences.

By addressing ethical considerations, data privacy concerns, and potential biases, organizations can build trust, ensure responsible AI deployment, and mitigate potential negative impacts associated with AI applications.

Recommendation for future research

Future research in the field of AI for the food industry can explore several areas to drive improvements and advancements. Research can focus on developing AI-based systems for real-time monitoring and early detection of contaminants, pathogens, and quality issues in food products. This includes exploring new sensor technologies, data fusion techniques, and ML algorithms to enhance food safety protocols. Further research can be done to optimize the food supply chain using AI. This includes developing AI models to predict demand, optimize inventory management, reduce food waste, and enhance logistics and distribution processes. Research can explore how AI can support sustainable practices in food production, such as precision agriculture, resource optimization, and eco-friendly farming techniques. AI can help optimize resource allocation, reduce environmental impact, and improve overall sustainability in the food industry.

References

Abdullah, A., Ali, M.A., 2021. Artificial intelligence-based food safety and quality control: a comprehensive review. Trends in Food Science and Technology 110, 942–957.

Agbai, C.-M., 2020. Application of artificial intelligence (AI) in food industry. GSC Biological and Pharmaceutical Sciences 13 (1), 171–178. https://doi.org/10.30574/gscbps.2020.13.1.0320.

Aggarwal, K.K., Billus, R., 2022. Artificial intelligence based marketing: a bibliometric analysis. International Journal of Bibliometrics in Business and Management 2 (2), 137–147. https://doi.org/10.1504/IJBBM.2022.125983.

Ahmad, B., Khalid, S., Zhang, D., Zhang, J., 2020. Artificial intelligence and machine learning in food and beverage industry: a review. Computers. Materials and Continua 64 (1), 349–366. https://doi.org/10.32604/cmc.2020.010086.

Albert-Weiss, D., Osman, A., 2022. Interactive deep learning for shelf life prediction of Muskmelons based on an active learning approach. Sensors 22 (2), 414. https://doi.org/10.3390/s22020414.

Andy, H., 2022. The Use of Artificial Intelligence in Food Processing. In: https://www.foodprocessing.com/on-the-plant-floor/automation/article/11290754/the-use-of-artificial-intelligence-in-food-processing.

Bandyopadhyay, K., Ghosh, S., Kumari Gope, R., 2021. Application of artificial intelligence in food industry—a review. International Journal of Engineering Applied Sciences and Technology 5 (11). https://doi.org/10.33564/ijeast.2021.v05i11.021.

Bendre, S., Shinde, K., Kale, N., Gilda, S., 2022. Artificial intelligence in food industry: a current panorama. Asian Journal of Pharmacy and Technology 242–250. https://doi.org/10.52711/2231-5713.2022.00040.

Carvalho, T.P., Soares, F.A.A.M.N., Vita, R., Francisco, R.D.P., Basto, J.P., Alcalá, S.G.S., 2019. A systematic literature review of machine learning methods applied to predictive maintenance. Computers and Industrial Engineering 137, 106024.

Elmasri, R., Park, S., 2021. Intelligent computational approaches for food recipe optimization: a review. Food Research International 139, 109–915.

Esquerre, C., Uysal, I., Dubois, J., 2018. Predictive modeling of food spoilage: recent advances and future perspectives. Comprehensive Reviews in Food Science and Food Safety 17 (5), 1174–1193.

Galitsky, B., 2020. AI-based chatbots for marketing: opportunities, challenges, and recommendations. Journal of the Academy of Marketing Science 48 (1), 17–37.

Garre, A., Ruiz, M.C., Hontoria, E., 2020. Application of machine learning to support production planning of a food industry in the context of waste generation under uncertainty. Operations Research Perspectives 7, 100147. https://doi.org/10.1016/j.orp.2020.100147.

Hemamalini, V., Rajarajeswari, S., Nachiyappan, S., Sambath, M., Devi, T., Singh, B.K., Raghuvanshi, A., 2022. Food quality inspection and grading using efficient image segmentation and machine learning-based system. Journal of Food Quality 1–6. https://doi.org/10.1155/2022/5262294.

Iorliam, I.B., Ikyo, B.A., Iorliam, A., Okube, E.O., Kwaghtyo, K.D., Shehu, Y.I., 2021. Application of machine learning techniques for Okra shelf life prediction. Journal of Data Analysis and Information Processing 9 (3), 136–150. https://doi.org/10.4236/jdaip.2021.93009.

Irfan, D., Tang, X., Narayan, V., Mall, P.K., Srivastava, S., Saravanan, V., 2022. Prediction of quality food sale in mart using the AI-based TOR method. Journal of Food Quality 2022. https://doi.org/10.1155/2022/6877520.

Jacobs, T., 2023. Unlocking the Value of Artificial Intelligence (AI) in Supply Chains and Logistics. https://throughput.world/blog/ai-in-supply-chain-and-logistics/. (Accessed 15 June 2023).

Kim, S., 2023. A study on the prediction of electrical energy in food storage using machine learning. Applied Science 13 (1), 346. https://doi.org/10.3390/app13010346.

Konur, S., Lan, Y., Thakker, D., Morkyani, G., Polovina, N., Sharp, J., 2021. Towards design and implementation of Industry 4.0 for food manufacturing. Neural Computing and Applications. https://doi.org/10.1007/s00521-021-05726-z.

Kumar, V., 2020. Artificial intelligence (AI) in marketing: a consensus study on AI's future impact. Journal of the Academy of Marketing Science 48 (1), 79−113.

Liu, G., He, Y., Liu, P., Chen, Z., Chen, X., Wan, L., Li, Y., Lu, J., 2020. Development of bioimplants with 2D, 3D, and 4D additive manufacturing materials. Engineering. ISSN: 20958099 6 (11), 1232−1243. https://doi.org/10.1016/j.eng.2020.04.015.

Ma, X., Zhang, P., 2020. Applications of artificial intelligence in food industry: a comprehensive review. Journal of Food Processing and Preservation 44 (11), e14991. https://doi.org/10.1111/jfpp.14991.

Mavani, N.R., Ali, J.M., Othman, S., Hussain, M.A., Hashim, H., Rahman, N.A., 2022. Application of artificial intelligence in food industry—a guideline. Food Engineering Review 14 (1), 134−175. https://doi.org/10.1007/s12393-021-09290-z.

Rahman, N.A., Hussain, M.A., Jahim, M.J., 2012. Production of fructose using recycle fixed-bed reactor and batch bioreactor. Journal of Food, Agriculture and Environment 10 (2), 268−273.

Rahman, M.S., Ghosh, T., Aurna, N.F., Kaiser, M.S., Anannya, M., Hosen, A.S.M.S., 2023. Machine learning and internet of things in industry 4.0: a review. Sensor 28, 100822. https://doi.org/10.1016/j.measen.2023.100822.

Rajnish, K., Elkady, G., Kantilal, R., Singh, A., Md, S.H., Dheeraj, M., Samrat, R., Komal, K.B., 2022. Machine learning and artificial intelligence in the food industry: a sustainable approach. Journal of Food Quality 2022, 1−9. https://doi.org/10.1155/2022/8521236.

Ren, l., 2021. Artificial intelligence and marketing analytics for demand prediction in food supply chains: a comprehensive review. International Journal of Production Economics 235.

Ribeiro, M.C., Barros, R.M., Gomes, T., Cabrita, I., 2020. Artificial intelligence in food industry: present and future. Trends in Food Science and Technology 97, 28−36.

Saha, S., Kumar, V., 2021. Reinforcement learning applications in supply chain management: a systematic review. Computers and Industrial Engineering 152.

Satwekar, A., Panda, A., Nandula, P., Sripada, S., Govindaraj, R., Rossi, M., 2023. Digital by design approach to develop a universal deep learning AI architecture for automatic chromatographic peak integration. Biotechnology and Bioengineering 120 (7), 1822−1843. https://doi.org/10.1002/bit.28406.

Sharma, S., Gahlawat, V.K., Rahul, K., Mor, R.S., Malik, M., 2021. Sustainable innovations in the food industry through artificial intelligence and big data analytics. Logistics 5 (4). https://doi.org/10.3390/logistics5040066.

Shekhawat, N.S., Tatavarthy, A., 2020. Artificial intelligence in marketing and advertising: transforming the future of customer engagement. International Journal of Information Management 54, 102−147. https://doi.org/10.1016/j.ijinfomgt.2020.102147.

Sonwani, E., Bansal, U., Alroobaea, R., Baqasah, A.M., Hedabou, M., 2022. An artificial intelligence approach toward food spoilage detection and analysis. Frontiers 9, 1−13. https://doi.org/10.3389/fpubh.2021.816226.

Torres-Sánchez, R., Martínez-Zafra, M.T., Castillejo, N., Guillamón-Frutos, A., Artés-Hernández, F., 2020. Real-time monitoring system for shelf life estimation of fruit and vegetables. Sensors 20 (7), 1860. https://doi.org/10.3390/s20071860.

Wang, Z., Li, L., Han, J., Ma, X., 2019. A review of the applications of artificial intelligence in the food industry. Engineering, Technology and Applied Science Research 9 (6), 4822–4827. https://doi.org/10.48084/etasr.3043.

Yanfi, Y., Heryadi, Y., Lukas, L., Suparta, W., Arifin, Y., 2022. Sentiment analysis of user review on Indonesian food and beverage group using machine learning techniques. In: 2022 IEEE Creative Communication and Innovative Technology, ICCIT 2022. Institute of Electrical and Electronics Engineers Inc., Indonesia https://doi.org/10.1109/ICCIT55355.2022.10118707.

Use of AI in conservation and for understanding climate change

13

Mehreen Khaleel[1], Naureen Murtaza[2], Qazi Hammad Mueen[3], Syed Aadam Ahmad[4] and Syed Fatima Qadri[1]

[1]*Wildlife Research and Conservation Foundation, Srinagar, Jammu and Kashmir, India;* [2]*Department of Environmental Sciences, Faculty of Engineering and Technology, Jamia Millia Islamia, New Delhi, India;* [3]*Department of Biological Sciences, Middle East Technical University, Üniversiteler Mahallesi, Çankaya, Ankara, Turkey;* [4]*Department of Information Technology, Cluster University Srinagar, Srinagar, Jammu and Kashmir, India*

Introduction

As our understanding of the environment and its diverse components becomes increasingly complex, the development and use of new technologies, such as artificial intelligence (AI), becomes imperative to deal with the intricate conceptual, theoretical, and practical aspects of environmental and ecological research, as well as conservation. AI refers to the simulation of human intelligence in machines that are programmed to think and learn like humans. It encompasses various techniques and algorithms that enable computers to perceive, reason, and take actions based on data inputs. It can be used to analyze vast amounts of data, optimize energy systems, enhance resource management, enable smart grids, support precision agriculture, and facilitate efficient transportation systems.

While there is no set definition for AI, the core essence of AI can be summarized as "AI = A + A," where A represents autonomy and adaptivity. Autonomy refers to the capability of performing tasks in intricate environments without constant human guidance. Adaptivity, on the other hand, pertains to the ability to enhance performance through learning from experiences. These two attributes lie at the heart of the definition of AI. Consequently, any AI application must possess these fundamental properties: the capacity to execute designated tasks in complex environments while acquiring knowledge and adapting to the environment (Ghosh and Singh, 2020). Today, AI is being used extensively in education, giving rise to adaptive learning (Verma, 2018), smart farming, smart healthcare, disease detection, AI-assisted security services, and climate change control and monitoring services (Ghosh and Singh, 2020). Fig. 13.1 provides a visual representation of the AI/machine learning (ML) landscape.

Artificial intelligence offers strong potential to find itself a special prominence in ecological studies, making it possible to employ previously unreachable methods of discovery and analysis to simplify our understanding of the environment and ecosystem. These developing tools and techniques make it possible to identify and visualize interactions between intersecting components in deeper and

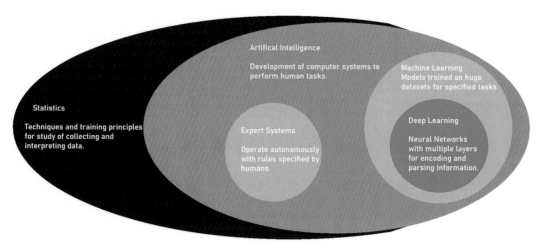

FIGURE 13.1

Venn diagram relationship of AI, ML, DL, expert systems and statistics. *AI*, artificial intelligence; *DL*, deep learning; *ML*, machine learning.

Modified from Bulletin of the American Meteorological Society.

subtler layers of the environment that may have been out of human limitations. It implies employing the use of complex technological systems and machinery as well as human-driven processes. The importance of AI in environmental research, thus, is reflected in the diverse ways in which it makes it possible to broaden the scope of research and its implications beyond human limits. Contextualizing the prominence of AI as an increasingly important aspect of environmental research and conservation, this chapter will review existing literature in four broad sub-fields: ecological modeling, biodiversity monitoring and conservation, climate change, and smart farming. Whereas there are other aspects of environmental research, conservation and protection that extensively employ the use of AI, these sub-fields are prioritized in this review since a more comprehensive body of research work is available pertaining to AI. The review of each sub-field will progress from an introduction to the field and its potential reliability on AI, a review of historical progress, its current use and application of AI, and future prospects and anticipated developments.

The first subsection of the chapter will give a brief description of the use of ecological modeling in environmental research studies and concisely review key studies, focusing on ecological modeling for river water quality, lake modeling, forest modeling, and integrated modeling. The second subsection will focus on the use of AI in biodiversity conservation and monitoring. This will include a review of essential studies that focus on the development of AI for ecological monitoring, its current and prospective use in data collection, and the emergence of ML and deep learning for scaling big data. The third subsection will be devoted to understanding the role of AI in mitigating climate change, its use in climate monitoring and change, current AI methodologies incorporated in climate research, the present use of AI in climate change research, and the future of AI and the challenges that come along with it. The last subsection will focus on aspects of AI use in smart farming, primarily reviewing the use of AI and the Internet of Things in agricultural sensors, precision agriculture, and open issues and challenges of smart farming.

Ecological modeling

Ecological or ecosystem models are mathematical representations of ecological systems. These are simplified models to understand the complex relationship of a real system. A model is developed as the result of systems analysis. Human influences now extensively dominate the alteration of ecosystems, leading to the extinction of species and environmental degradation on a global scale. Consequently, the imperative for ecosystem management (Odum, 2013), although lacking a precise definition, is increasing. To effectively manage complex systems, such as ecosystems, humans rely on the creation of models (Forrester, 1968) as a foundation for informed decision-making and functional comprehension. Therefore, ecological modeling plays a crucial role in the management of ecological systems, typically taking the form of mathematical representations of real ecosystems.

History

The initial development of ecological models, such as the Streeter—Phelps model for oxygen balance in streams and the Lotka—Volterra model for prey—predator relationships, occurred in the early 1920s (Lotka, 1920; Streeter and Phelps, 1958; Wangersky, 1978; Volterra, 1926). Further advancements in population dynamic models took place (Buckland et al., 2007; Newman et al., 2014; Raftery et al., 1995) during the 1950s and 60s, including the development of more intricate river models which, as an example, helped in understanding that the searching efficiency of fish for zooplankton can be influenced by the velocity of the river (Holling, 1959, 1966), marking the second generation of models.

Around 1970, the use of ecological models in environmental management significantly increased (Jørgensen and Mejer, 1979; Botkin et al., 1972; Lombardo and Franz, 1972). This period, referred to as the third generation of models, witnessed the emergence of eutrophication models and the development of complex river models. These models were often overly complex, facilitated by the ease of writing computer programs to handle intricate models. However, it became evident in the mid-1970s that limitations in modeling stemmed from data and our understanding of ecosystems and ecological processes rather than computer technology and mathematics. This period led to various recommendations, such as strictly following the procedural steps, finding an appropriate balance between data, problem, ecosystem, and knowledge when determining model complexity, and conducting sensitivity analyses to select model components.

Concurrently, ecologists adopted a more quantitative approach to environmental and ecological problems, likely driven by the demands of environmental management. The quantitative research outcomes from the late 1960s onward greatly influenced the quality of ecological models and were considered as significant as advancements in computer technology (For example Botkin et al. (1972)). The subsequent period, spanning from the mid-1970s to the mid-1980s, marked the fourth generation of models. These models were characterized by a strong ecological foundation, emphasizing realism and simplicity (Morioka and Chikami, 1986). Many models from this period were successfully validated, and some even demonstrated the ability to make accurate predictions.

However, limitations in modeling became apparent, highlighting the rigidity of the models compared to the dynamic nature of ecosystems. The hierarchical feedback mechanisms inherent in ecosystems were not adequately accounted for, rendering the models unable to predict adaptation and structural changes. Since the mid-1980s, researchers have proposed new approaches to address these shortcomings. These include fuzzy modeling (Salski, 1992, 2006; Van Broekhoven et al., 2007), examining catastrophic and chaotic behavior (Jørgensen, 2008) and applying goal functions to incorporate adaptation and structural changes (Jørgensen, 1986, 2002). Additionally, the application of objective and individual modeling, expert knowledge, and AI has introduced further advantages in modeling. Collectively, these recent developments are referred to as the fifth generation of modeling. Fig. 13.2 shows a year-wise overview of the development of ecological modeling.

Ecological modeling for river water quality

Rajaee et al. (2020) reviewed the use of AI techniques to model water quality in rivers. The review analyzed 51 journal papers published from 2000 to 2016 that focused on water quality modeling in rivers. The review found that there has been an increasing trend toward using AI models in water quality modeling in recent years. Artificial neural networks (ANNs) were the most popular AI model and were used to model a variety of water quality variables. Combining ANNs with other AI models, such as wavelet transform, genetic algorithm, fuzzy logic, and autoregressive integrated moving average (ARIMA) improved the accuracy of water quality prediction. The most successful hybrid model was wavelet-based ANN (WANN). WANN uses wavelet transform to preprocess the input time series data, which improves the accuracy of the ANN model. Some AI models have not been used for long-term prediction, and some have not been used to model other types of water quality variables. More research is needed to compare the different modeling approaches and to develop robust AI approaches for water quality modeling.

Pertinent to mention here, the review found that most of the studies were conducted in the United States. The authors recommended that future studies should use a variety of performance criteria to evaluate the models. The authors also recommended that future studies should use a wider range of water quality variables.

Lake modeling

Lake models describe the processes of eutrophication. They were initially developed for environmental management in the 1970s, and by the 1990s, various models with different levels of complexity were utilized for environmental management purposes (Stefan et al., 1989). Since the early 1980s, suitable eutrophication models have been available for almost any lake with an existing dataset. However, there is ample room for the development and application of better or additional lake models within the context of environmental management (Jayaweera and Asaeda, 1995).

In recent years, lake models have shifted their focus toward integrating hydrodynamic and ecological processes to broaden their scope. Specifically, new models have been developed to examine food webs in lakes (Van Donk et al., 1989), which had not been previously incorporated in lake modeling. Furthermore, case studies have employed structurally dynamic models that account for adaptations and shifts in species composition (Zhang et al., 2003).

Jørgensen (2010) reviewed lake models that had been published from 2005 to 2010 and concluded that four papers introduced lake models, which utilized a structurally dynamic approach. These models were found to be effective in considering changes in the structure, adaptation, and species composition of lakes and were further recommended for improving calibration, and prognoses, particularly when structural dynamic changes occurred. One of the models combined a three-dimensional model with a structurally dynamic model, which increased the usage of three-dimensional lake models.

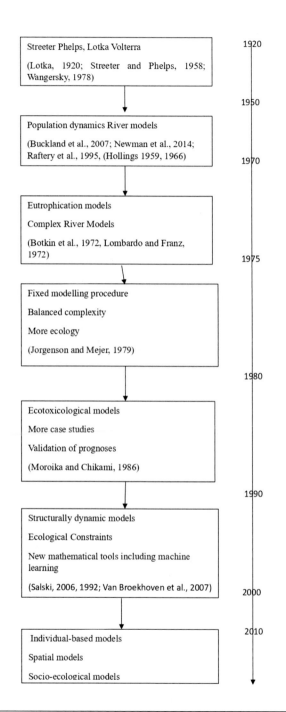

FIGURE 13.2

A year-wise overview of the development of ecological modeling flowchart.

Modifications were made to traditional eutrophication models to incorporate specific processes relevant to case studies. The modifications included multiple nutrient cycles, diverse phytoplankton (Lehman et al., 1975; Thomas et al., 2018) and zooplankton groups, nutrient recycling theory, competition between phytoplankton and macrophytes, woody habitats, cyanobacteria modeling, nitrogen fixation, and the role of mussels as filter feeders (Toma, 2013). Dynamic modeling proved valuable in addressing competition and capturing the hysteresis response under changing conditions. The modeling of multispecies fishery gained traction and showed promise for future management. Structurally dynamic models have proven effective in addressing cases involving adaptation and shifts in species composition. Additionally, there was an increasing interest in modeling the impact of climate change on lakes. The overall trend in modeling emphasized incorporating more scenario-specific details. Therefore, it was recommended to thoroughly study and consider relevant modeling experiences when tackling new modeling challenges.

A neural network is a computational model inspired by the brain that consists of interconnected artificial neurons and can learn from data to make predictions or perform tasks. Furthermore, "artificial neural network (ANN) is a model that is stimulated by the way biological nervous systems, such as the brain, process information" (Goethals et al., 2007). ANN has been proven beneficial in numerous factual tasks that are covenants with vastly collaborating and multifarious developments (Goethals et al., 2007). In a study conducted by Olomukoro and Odigie (2020), ANN was utilized to predict the benthic macroinvertebrate fauna of the Obueniyomo River, Nigeria. The dataset was divided into 75% for model testing and 25% for training. The variables were scaled between 0 and 1, and the model was implemented using R statistical software. A total of 39 physical and chemical predictors were used as inputs, and the model successfully predicted the presence or absence of macroinvertebrate fauna in the study stations. The visualized neural network model showed 25 output parameters that contributed to the prediction. Sensitivity analysis was conducted to determine the influence of the output parameters and identify the variables that significantly influenced the model's output. The study showcased the effective utilization of ANN in predicting benthic macroinvertebrates in a tropical rainforest river. The accuracy of the predictions was exceptional when the environmental variables remained relatively stable or exhibited minimal fluctuations. This underscores the reliability of ANN models in ecological studies and river management, as they can accurately forecast both common and rare benthos taxa. However, it is crucial to identify the conditions in which specific methods should be applied and optimize the parameters that determine the neural network architecture for accurate prediction of macroinvertebrates in aquatic ecosystems. By improving these strategies, researchers can enhance the reliability and applicability of ANN models in ecological studies and river management.

In their review, Agudelo et al. (2020) examined modeling techniques for studying interactions among multiple ecosystem services (ESs) in terrestrial landscapes. They found that logical and empirical models were commonly used but had limitations in capturing the capacity of diverse ecosystems to provide services and accounting for unpredictable inflection points influenced by ecosystem dynamics, human activities, overexploitation, and emerging technologies. To address these challenges, they advocated a transdisciplinary approach involving science and technology, fostering collaboration among experts to develop new models that better represent ES interactions. Regional cooperation and data generation were emphasized, as primary information is lacking in ES interaction studies. Bayesian networks were identified as a promising approach for modeling ES interactions due to their ability to represent uncertain relationships. The review provides a comprehensive overview of the current understanding of modeling ES interactions, highlighting the need for accurate and reliable models through a transdisciplinary approach.

Forest modeling

Forest models are useful for understanding ecosystem processes, which can be applied to aquatic and coastal systems. Forestry has faced modeling challenges, such as choosing between statistical models and process-based models. Two common types of forestry models are growth-yield models and gap models. Growth-yield models are favored by foresters for commercial timber species, while gap models are favored by ecologists studying ecological structure and function during succession. These models were originally focused on single-species stands or low-diversity stands.

The development of individual-based models (IBMs) provided a new approach to include ecological processes while maintaining individual characteristics. Some of the first IBMs were those developed by Shugart (1984). Trees, as countable objects, were a good candidate for process-based IBMs that consider the growth and death of individual trees. These individual tree stand models were originally used to research forest disease spread and succession but have had many applications within aquatic ecology.

Battaglia and Sands (1998) argued that process-based models are necessary for understanding the dynamics of complex forest systems. They concluded that "lumped-parameter process-based models and hybrid models provide the most immediate means through which our understanding of the biological processes underlying forest growth can be included in forest management systems."

Porté and Bartelink (2002) reviewed techniques for mixed forest growth models. They identified six modeling approaches and concluded that empirical models may be better predictors, given sufficient, high-quality data, but process-based models are necessary to model the dynamics of complex forest systems.

Integrated models

Integrated models (IMs) are valuable tools for decision-making in the sustainable management of regional, cultural, and ecological resources. However, a significant challenge in utilizing IMs effectively is ensuring consistency in incorporating disciplinary methodologies. There are two particular aspects that can potentially lead to inconsistencies: transforming model dimensions (units and goal functions) and constructing model dynamics. Socioeconomic–ecological systems are complex and interconnected, making IMs crucial for their sustainable management. Despite this, even the most advanced IM initiatives are not achieving their full potential. To effectively apply IMs to sustainable management, a deeper understanding of system dynamics and controllability is necessary. This understanding needs to be comprehensive, compatible, and commensurable. These complexities present opportunities for further development of IMs. Addressing the conceptual and technical barriers at the interfaces between disciplines is crucial for developing an integrated analytical framework. Each discipline operates within its own paradigm, which hinders effective communication, coordination, collaboration among researchers and stakeholders, and the dissemination of results. Moreover, individuals often remain unaware of being confined within their own paradigms. This long-standing issue is referred to as the incompatibility of discipline-specific languages and formalism.

The South Florida Everglades is another extensively studied aquatic system, which has experienced significant degradation due to human activities. The restoration of the system's sustainability has become a primary focus. Gentile and Harwell (2001) proposed a conceptual model that integrates ecological risk assessment and adaptive management principles to guide policy decisions at the regional scale. It involves a retrospective analysis to identify the causes of the current conditions and explicit consideration of societal preferences. The model illustrates the connections between management choices, societal preferences, and environmental stressors, serving as an effective communication tool.

In a similar vein, Volk et al. (2008) adopted a modeling approach that integrates ecological and socioeconomic assessments to develop a spatial decision support system for a river basin in North-Western Germany. Notably, their approach enables the transfer of scale-specific data from micro to meso to macro levels. The findings highlight the need for significant land management changes in the region to comply with the European Water Framework Directive.

Economic–ecological analyses typically involve the valuation of ecological services to align dimensions with financial currency. Turner et al. (2000) developed an integrated wetland research framework that combines economic valuation, stakeholder input, and multicriteria evaluation to assess policy consistency across different areas.

Chang et al. (2008) describe a dynamic decision support system for the management of coral reefs along the coast of Taiwan. Given the area's diverse coral species and high tourist activity, the model aims to develop sustainable management scenarios by integrating socioeconomic, environmental, and biological factors such as land development, wastewater treatment, local fish consumption, and tourist fees.

The application of ecological models for the management of aquatic systems has yielded a rich literature and a comprehensive toolbox. Langmead et al. (2009) adopted a DPSIR framework (driving forces, pressures, states, impacts, responses) to construct models for the Black Sea, incorporating socioeconomic drivers that impact the ecosystem. They address the dimensionality challenge by employing probabilistic dependencies through Bayesian belief networks as a common metric. Network approaches for model integration are further explored by Fath et al. (2007).

Nobre et al. (2009) developed an integrated economic and ecological decision-making approach in a case study of shellfish production in the East China Sea. The model's outputs align with standard economic theory and ecological economic theory, demonstrating that cultivated areas impose constraints on production and that reducing the production area can be an effective conservation technique.

Section summary

Ecological modeling has made significant progress in the past decade, with the emergence of new methodologies such as ANNs, individual-based models (IBM), species distribution models (SDMs), fuzzy-set models, spatial models, and stochastic models. These approaches have complemented the traditional compartmental model approach. The field of ecological modeling has become well-established and has contributed significantly to the understanding and management of ecological systems. Progress includes the establishment of standard procedures and protocols for modeling studies, a better balance between complexity and simplicity in models, improved description of processes and parameter values, and numerous case studies across various system types and ecological and environmental applications. Ecological modeling now has the capability to utilize a diverse range of model types to address different problems, ecosystems, and available data. It can account for adaptation, individuality, shifts in species composition, spatial distributions, and varying data quality. Despite these advancements, there are still areas that require further attention, some of which were identified as early as the 1970s. There is a need for conceptual breakthroughs to deepen the understanding and integration of ecosystem properties and to foster interdisciplinary modeling efforts. In general, there is a pressing need for a more comprehensive theory of modeling. As the field continues to mature, we are optimistic that such breakthroughs will occur in the near future.

Biodiversity monitoring and conservation

Biodiversity refers to the world's species, their phylogenetic histories, genetic variability within and among populations and species' distribution in regions ranging from local habitats to whole continents or oceans (National Research Council, 1999). Animal biodiversity is reducing at an unprecedented rate, accelerated by anthropogenic activities (Ceballos et al., 2020). The impact is so significant that it is causing the sixth mass species extinction event mainly due to the destruction of component populations (Ceballos et al., 2020). The biodiversity loss is presently not so well understood, with up to 17,000 species having "data-deficient" status (The IUCN Red List of Threatened Species, 2023). Biodiversity changes have significant implications for the sustainability of food provision, habitat protection, and regulation of ecosystems, carbon sequestration, tourism, and cultural identity (Tekwa et al., 2023). The rapid loss of biodiversity and resource depletion necessitates the urgent use of tools for efficient and rapid biodiversity assessment and population dynamics at a large scale. Recent international agreements have also highlighted the need to set up assessment and monitoring programs at regional and national levels.

However, there are many challenges to effective biodiversity change assessments, which go beyond economic incentives and investment in monitoring networks. There are significant knowledge gaps in components like sampling protocol design, metrics identification, estimation bias correction, quantification of uncertainties, attributing causes, projection of future pathways, and designing policies. These gaps have been responsible for the failure to achieve international targets to stop biodiversity loss earlier (Tekwa et al., 2023). For example, not even one of the 20 Aichi Biodiversity Targets, which were agreed upon by 196 nations for the 2011−2020 period, has been fully met (Silvestro et al., 2022). Therefore, there is now a desperate need to devise more realistic and effective policies for a sustainable future, which will help deliver the conservation targets under the post-2020 Global Biodiversity Framework.

Several tools and algorithms, like Spatial Conservation Prioritization, have been developed for the facilitation of systematic conservation planning (Moilanen and Wilson, 2009). While there are many benefits to the quantitative conservation prioritization methods, their real-world adoption and implementation are still in the early stages. Marxan, the most widely used method so far, operates by identifying a set of protected areas, which collectively enable certain conservation targets to be achieved using a simulated annealing algorithm under minimal expenses (Moilanen and Wilson, 2009). However, such methods fail to incorporate changes through time and collect a single initial gathering of biodiversity and cost data. They also do not explicitly include climate change, anthropogenic pressures, or sensitivities of different species to such changes (Silvestro et al., 2022). The conventional management and conservation of animal species are based on the manual collection of data by field workers, which is time-consuming, expensive, and labor-intensive (Witmer, 2005). This can be risky as human presence can pose threats to wildlife, their habitats and often, humans themselves.

The discipline of conservation science keeps evolving. It is imperative that different approaches are attempted in order to find what works best for biodiversity conservation. One such approach is the use of AI. Today, AI is more accessible than ever. Although it presents its own unique challenges, like not being as sensitive or accurate as humans at many conservation research tasks or needing a lot of data to train it for recognition of images and sounds, it seems to have enabled researchers a huge boost in monitoring (Kwok, 2019). AI is not only faster than humans in processing data but also does not experience fatigue-caused deteriorated performance and is also able to detect infrequent and complex patterns.

For biodiversity conservation, ecological monitoring is of utmost importance (Cord et al., 2017). With the increasing need for monitoring, the ways in which scientists collect, analyze, and process data have also increased (Allan et al., 2018). The development of AI techniques for ecological monitoring is rapidly increasing, allowing scientists to develop and process larger volumes of data than possible before with conventional methods (McClure et al., 2020; Pecl et al., 2019). AI, along with Citizen Science, which can be described as scientific projects involving volunteers with varying levels of expertize in research, has the potential to transform monitoring by accelerating the processing and analysis of big data sources (Weinstein, 2018).

While Citizen Science enhances the spatial and temporal scale of projects, AI enhances the scale of human data collection by being incorporated into devices meant for monitoring (Hochachka et al., 2012; McClure et al., 2020). AI algorithms can be incorporated into citizen science smartphone applications for recognition of geographic locations and providing incentives (Fang et al., 2019). Advances in sensor technologies have drastically increased data collection capacity. Many areas which were previously inaccessible can now be studied using high-resolution remote sensing (Gottschalk et al., 2007). Data are being collected using noninvasive devices like camera traps (Steenweg et al., 2017), acoustic sensors (Sugai et al., 2019), and consumer cameras (Hausmann et al., 2018). Acoustic loggers like AudioMoth are programmed for the identification and recording of animal calls using classification algorithms (Hill et al., 2018). AI has been demonstrated to be faster and more accurate than humans in the acoustic classification of environmental sounds and images in the Serengeti National Park in Tanzania using the Zooniverse platform for the wildebeest census (Torney et al., 2019). Similarly, drones have been shown to count wildlife more accurately and precisely than humans in South Australia (Hodgson et al., 2018). The data collected from these advanced computer model-based technologies and others like GIS-based animal tracking units, global positioning system (GPS), and X-ray fluorescence (XRF) can then be processed, interpreted, analyzed, and encrypted via AI and ML-based technologies. Wildlife morphology, behavior, distribution, abundance, and diversity relating to human invasion can be deduced (Kerry et al., 2022).

Theoretically, AI can replace the need for manual processing but hyper-efficient and complex social machines with high accuracy can be achieved by the integration of people power with AI power (Trouille et al., 2019). The central challenge to data analysis using AI is the huge volume of data generated by modern collection methods. Platforms like eMammal, Agouti, and Zooniverse act as collaborative portals for data collection. However, due to the large volume of data, such approaches become unsustainable (Tuia et al., 2022). In order to enhance the conservation measures in scale and accuracy, methods for automatic cataloging, searching, and converting data into relevant information are desperately required.

Deep learning (DL) and ML are considered as promising tools, which could help scale local studies to a global level (Tuia et al., 2022). ML deals with learning patterns from data. It has emerged as a powerful tool to bridge the gap between big data and actionable ecological insights (Christin et al., 2019; Pichler and Hartig, 2023). Another cutting-edge AI system called DL can identify, count, and describe the behaviors of 48 species in the 3.2 million-image Snapshot Serengeti dataset with more than 93.8% accuracy (Norouzzadeh et al., 2018). A significant part of the success of ML can be attributed to DL, which is based on ANNs that have exhibited better performance in most ML use cases. ML technology has the potential to speed up and enhance traditional ecological research, from data collection to image retrieval and population surveys (Tuia et al., 2022).

AI techniques like the AI -assisted semantic Internet of Things (AI-SIoT) using a wireless sensor network (WSN) are also being proposed for environmental monitoring systems. It can be used to measure weather parameters like rainfall, sunlight, fire, and gas leakage remotely. It has been proven to accurately collect data from a set time interval and accurately calculate the acquisition phase location (Zhang et al., 2021). Another novel framework called Conservation Area Prioritization Through AI, (CAPTAIN) has been proposed for biodiversity conservation under a limited budget by developing more interpretable prioritization maps (Silvestro et al., 2022).

While DL seems like a promising tool with implementations in various ecological fields, it is still very new, and DL solutions are not easy to develop. Various factors like training time, training datasets, development complexity, and computing power have to be considered (Christin et al., 2019). As the reliance of ecology on AI increases, ecologists will need to acquire new skills. This might seem difficult at first, but this challenge can be simply solved by collaboration between different disciplines (Carey et al., 2019).

Climate change

One of the greatest challenges faced by humanity is climate change. To tackle this catastrophic threat, AI is widely being used to monitor and detect areas requiring immediate conservation action. AI is currently being applied in climate change research and initiatives, such as climate modeling, carbon monitoring, renewable energy optimization, and assessing climate risk and vulnerability. These applications highlight the significant role AI can play in addressing the complex challenges posed by climate change.

Climate change is widely recognized by the scientific community as a severe and imminent threat. This is exemplified by a recent statement from over 11,000 scientists who emphasize their moral duty to warn humanity about catastrophic risks and to convey the gravity of the situation. In their declaration, they unequivocally describe the state of the planet as a climate emergency (Gardner and Bullock, 2021).

The term "emergency" underscores the severity and urgency of the issue. Climate change is a profoundly serious problem due to its projected adverse impacts, which encompass extreme weather events, droughts, wildfires, floods, and rising sea levels. Without effective and comprehensive mitigation measures, it is projected that the average global temperature will increase by 3°C by 2100, or possibly even higher (Schwalm et al., 2020), leading to increasingly catastrophic consequences. The urgency to address climate change and implement robust mitigation strategies has never been more crucial.

A brief history of the origin of AI usage in climate monitoring and change

In 1984, the National Oceanic and Atmospheric Administration's (NOAA) Environmental Research Laboratories (ERL) took a significant step by appointing a Special Advisor in AI. The objective was to explore the potential of AI in enhancing weather forecasting and other aspects of NOAA's work. At that time, the prevalent AI implementations primarily revolved around expert systems. These systems had already demonstrated their effectiveness in various domains and appeared promising for meteorology as well (Haupt et al., 2022). Additionally, in 1984, the Artificial Intelligence Research in the Environmental Sciences (AIRES) workshops were initiated, with the first editions taking place in 1986 and 1987. Further down the years, these workshops eventually evolved into the AMS AI conferences, starting from 1998, which were organized by the same group of pioneers. Although the AI methods have undergone significant changes over time, their application areas have remained relatively consistent (Haupt et al., 2022). In the initial phase of AI development for environmental sciences (ES), spanning the mid-1980s to the 1990s, numerous expert systems were designed and tested specifically for weather forecasting purposes (Haupt et al., 2022). Expert systems, being rule-based systems that used knowledge from human experts to make predictions had limitations, such as the inability to express all knowledge in words, the difficulty of taking advantage of rapidly improving computer capabilities, and running the AI systems often took as much time for reasonably skilled forecasters as generating forecasts in more traditional ways (Haupt et al., 2022). Fuzzy logic was developed as an alternative to expert systems. Fuzzy logic allows for uncertainty and ambiguity, which are inherent in weather forecasting. Fuzzy logic systems have been used to improve the accuracy of weather forecasts, especially for short-term forecasts (Haupt et al., 2022). As AI usage in climate sciences improved further, neural networks were also introduced. Neural networks are a type of AI that learns from data, and they became the dominant AI method for weather forecasting in the late 1990s (Haupt et al., 2022). Tree-based methods have become increasingly popular in recent years. Tree-based methods are a type of AI that uses decision trees to make predictions. Decision trees are made up of nodes and branches, and each node represents a decision that can be made. The branches represent the possible outcomes of each decision. Tree-based methods have been used to improve the accuracy of weather forecasts, especially for localized forecasts. The use of AI in weather forecasting has improved the accuracy of forecasts by up to 10%. AI is still a developing field, and researchers are constantly working to improve the accuracy of AI-based weather forecasts (Haupt et al., 2022).

Present usage of AI in climate change research

The integration of AI in climate change initiatives is already demonstrating significant and beneficial outcomes. However, accurately quantifying the full extent of this impact and understanding its precise influence poses a complex challenge. It also acknowledges previous efforts to document the potential positive outcomes resulting from the application of AI in addressing climate change.

Despite the scientific consensus on the fundamental aspects of climate change, numerous uncertainties persist within the environmental crisis. These uncertainties encompass the comprehension of historical and present events, as well as the accurate prediction of future outcomes. Leveraging its capacity to process vast volumes of unstructured, multidimensional data through sophisticated optimization techniques, AI is already facilitating the understanding of complex climate datasets and enabling the forecasting of future trends (Huntingford et al., 2019).

AI methodologies have been employed to forecast changes in global mean temperatures (Ise and Oba, 2019; Cifuentes et al., 2020), predict climatic and oceanic phenomena such as El Niño (Ham et al., 2019), cloud systems (Rasp et al., 2018), and tropical instability waves (Zheng et al., 2020), and enhance comprehension of various aspects of the weather system, including rainfall patterns in general (Sønderby et al., 2020; Larraondo et al., 2020) and in specific regions such as Malaysia (Ridwan et al., 2021), as well as their cascading effects such as water demand (Shrestha et al., 2020; Xenochristou and Kapelan, 2020).

Moreover, AI tools can assist in anticipating the occurrence of extreme weather events that are increasingly prevalent due to global climate change, such as damage caused by heavy rainfall (Choi et al., 2018) and wildfires (Jaafari et al., 2019), as well as other consequential phenomena like patterns of human migration (Robinson and Dilkina, 2018). In many instances, AI techniques can enhance or expedite existing forecasting and prediction systems, automating tasks such as labeling climate modeling data (Chattopadhyay et al., 2020), refining approximations for atmospheric simulations and extracting meaningful signals from climate observations (Gagne et al., 2020).

Artificial intelligence has diverse applications in natural resource management. AI can be utilized to monitor forests, mitigate deforestation, and provide early warnings for forest fires, aiding decision-makers in taking timely action. The implementation of AI in intelligent ecosystem restoration and climate change adaptation enables effective pollution reduction and the implementation of efficient conservation strategies. Furthermore, AI is widely being used to create energy-efficient technology and modules focusing mostly on maximizing renewable resources with a minimum carbon footprint. Effective management of urban water resources requires the integration of AI technologies. Additionally, the use of remote sensing and geographic information systems enhances urban land use and planning practices (Chen et al., 2023b).

Harnessing AI for energy efficiency

In today's global landscape, energy concerns have emerged as a critical issue. With the continuous expansion of the global economy and a growing population, the demand for energy has experienced an exponential surge (Chen et al., 2023a; Osman et al., 2023; Yang, 2022). This presents a significant challenge in achieving sustainable development and the judicious utilization of energy (Chen et al., 2023b). To address this challenge and mitigate the adverse environmental impacts, it is imperative to implement effective measures that enhance energy efficiency and reduce energy wastage (Cai et al., 2019; Nižetić et al., 2019). AI has emerged as a powerful tool in the energy sector, offering novel prospects and challenges in improving energy efficiency and realizing sustainable development (Baysan et al., 2019; Farghali et al., 2023).

In the energy sector, the implementation of AI can significantly enhance energy utilization efficiency. By predicting energy demand, optimizing energy production and consumption, and enabling intelligent control, AI technologies can curtail energy costs, reduce environmental pollution, and foster sustainable development (Khalilpourazari et al., 2021; Lee and Yoo, 2021). Consequently, the relationship between AI and energy efficiency has become a highly discussed topic in both the research community and the corporate world, attracting the attention of numerous scholars and corporations (Kumari et al., 2020; Ahmad et al., 2021). Furthermore, the judicious application of AI is believed to yield tangible improvements in energy efficiency, paving the way for a more promising future for human society.

AI has revolutionized the energy sector as a groundbreaking technological tool, offering new opportunities and challenges in enhancing energy efficiency and achieving sustainable development (Ahmed et al., 2022a; Yang et al., 2022; Farghali et al., 2023). Notably, AI has been successfully applied in various domains of energy efficiency, including fault detection and diagnosis, thermal comfort prediction and control, demand response, and energy storage optimization. These applications have shown promising results in improving energy efficiency, reducing energy waste, and fostering sustainable development (Chopra et al., 2022; Fang et al., 2023). However, it is important to note that the effectiveness of implementing AI in energy efficiency relies heavily on the accuracy of input data and the careful selection of AI algorithms (Arumugam et al., 2022; Ouadah et al., 2022). Thus, continued research and refinement are essential in leveraging the full potential of AI for energy efficiency.

Research conducted in Italy and Japan has shown widespread adoption of AItechnologies in energy management systems, yielding favorable outcomes (Zhao et al., 2022; Enholm et al., 2022; Yang, 2022). Similarly, studies in the United Kingdom indicate that although the use of AI in predictive maintenance is still in its early stages, it has shown good effectiveness (Ahmed et al., 2022a). In China and India, AI is utilized for fault detection and diagnosis, as well as for integrating renewable energy and demand response (Chai et al., 2022). The findings highlight that AI is a powerful tool for improving energy efficiency and promoting sustainable development, although further evaluation of its potential is necessary. Despite the potential benefits, some scholars argue that the high cost of AI technology poses a significant obstacle to its implementation in energy efficiency (Enholm et al., 2022; Yang, 2022; Zhao et al., 2022). The creation and deployment of AI-based systems require substantial investments that may exceed the financial capacity of certain organizations (Ahmed et al., 2022b). Additionally, the scarcity of data and qualified AI experts presents a significant challenge to the widespread adoption in energy efficiency (Chai et al., 2022). Moreover, most AI applications in various aspects of energy efficiency are still in their early stages, warranting further investigation into their effectiveness. Nevertheless, the increasing need to reduce energy consumption, mitigate environmental impact, and achieve sustainable development is expected to drive the utilization of AI technologies in energy efficiency.

Harnessing AI for environment monitoring, planning, and resource management
This topic analyzes the exploration of diverse methods and applications of AI in climate change, resource management, and environment monitoring, as illustrated in Fig. 13.3. In this field, various syntaxes are utilized to represent different parameters, including water-related issues, agriculture, energy, wildfire, and the implementation of specific AI techniques. Additionally, the analysis extends to encompass topics such as water management, agricultural practices, land management, and the profound impact that AI has on addressing specific challenges in these areas.

Deforestation prediction and monitoring
To forecast deforestation rates in the Amazon rainforest, Dominguez et al. (2022) utilized a dense neural network and an extended short-term memory network. By comparing prediction results and retraining the model with updated data, future forest loss can be estimated, enabling proactive measures. Similarly, Mayfield et al. (2017) achieved consistent predictive performance through Gaussian processes but were limited to determining deforestation risk without quantifying its extent or factors. Torres et al. (2020) demonstrated the effectiveness of a weightless neural network architecture

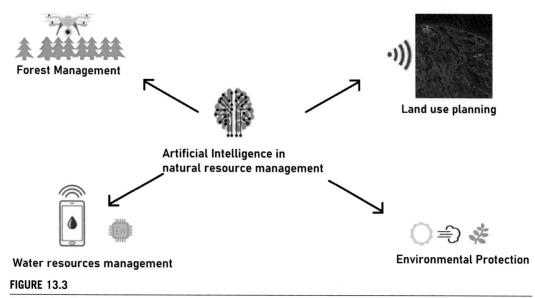

Forest Management

Land use planning

Artificial Intelligence in
natural resource management

Water resources management

Environmental Protection

FIGURE 13.3

Artificial intelligence (AI) usage in environment monitoring and conservation, and resource planning and management. Illustration of various domains of AI usage.

combined with unmanned aerial vehicles for deforestation monitoring and visual navigation assessment.

Ecosystem restoration planning

Yin et al. (2021) proposed an AI-assisted planning framework for ecological retreat configuration in the Changsha-Zhuzhou-Xiangtan urban area. By identifying environmental components and utilizing ML, retreat configurations help comprehend the impact of urban growth on ecological processes. Liu et al. (2022) suggested a neural network regression model optimized by a genetic algorithm for road planning in Zhejiang province, enhancing processing power and addressing local minima issues.

Water resource management

Urban water resource management necessitates advanced technological platforms to accommodate increasing demands due to climate change, urbanization, and population growth (Mrówczyńska et al., 2019). Adaptive intelligent dynamic water resource planning, a subset of AI technology, simplifies the process of enhancing water efficiency in metropolitan settings (Xiang et al., 2021; Liu et al., 2019) improved water quality evaluation accuracy by incorporating dynamic inertia weights into the moth flame algorithm. Afzaal et al. (2020) employed recurrent neural networks and long- and short-term memory to address the dynamic inputs of climate change in water resource management.

Land use planning

Urban land use planning significantly impacts socioeconomic, environmental, and ecological activities. Aerial imaging analysis combined with deep learning models enables cost-effective and time-saving investigations into physical surface materials and human land use. Alem and Kumar (2022) and Ghavami et al. (2017) demonstrated the feasibility of an intelligent planning support system based on a multiagent system and Bayesian learning methods for automated urban land use planning consultations. Carranza-García et al. (2019) found that convolutional neural networks outperformed other methods for land cover/land classification tasks.

AI in renewable energy

The utilization of AIin renewable energy encompasses various applications and methodologies. In the domain of wind energy, studies have evaluated the role of AI through correlation, statistical, physical, and neural network models to assess generated power and wind speed. Neural learning algorithms have been predominantly employed to estimate wind speed and power, exemplified by the proposed neural network (PNN) (Ashfaq et al., 2022).

Additionally, fuzzy logic and neural network approaches have been explored for wind power forecasting (Ashfaq et al., 2022). In the context of solar energy, AI has been employed for modeling weather data, sizing and simulation of photo voltaic systems, and building energy consumption calculations. The performance of AI models, such as the Back Propagation Neural Network (BPNN), has been compared to other approaches in estimating ambient temperature and predicting solar radiation (Ashfaq et al., 2022), and these have been applied in solar energy applications. Mashohor et al. (2008) proposed the use of genetic algorithms (GA) for solar tracking to enhance the performance of photovoltaic (PV) systems. The GA approach, illustrated in Fig. 13.4 for the GA-solar system, involves an initial population size of 50 and utilizes mutation and crossover probabilities of 0.001 and 0.7, respectively (Ashfaq et al., 2022). Various studies have employed the adaptive neuro-fuzzy inference system (ANFIS) method, achieving 98% accuracy in modeling PV power supply systems. Additionally, ANFIS has been utilized for hourly prediction of global radiation using satellite image data, modeling power supply systems, predicting solar radiation based on temperature, sunshine duration, and evaluating the performance of Solar Chimney Power Plants (SCPP) (Ashfaq et al., 2022).

Furthermore, geothermal energy applications have leveraged AI techniques, predominantly utilizing ANNs for tasks such as predicting system performance and optimizing the vertical ground

FIGURE 13.4

Photovoltaic panel for a genetic algorithm (GA)-solar system.

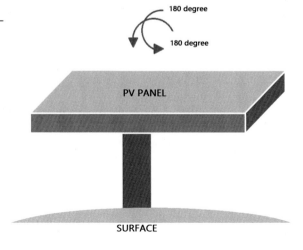

source heat pump (VGSHP) system (Ashfaq et al., 2022). Hybrid AI methods, including Adaptive Neuro Fuzzy Interference System (ANFIS), have been employed for analyzing and comparing the performance of VGSHP and AGDHS systems. The role of AI in geothermal energy extends to thermodynamic and thermo-economic optimizations as well (Ashfaq et al., 2022). Overall, AI plays a crucial role in enhancing renewable energy systems across the wind, solar, and geothermal domains, contributing to improved efficiency and optimized operations (Ashfaq et al., 2022).

Harnessing AI to counter-forest fires, flooding, and desertification

Applications of ML to Earth Observation data have increased substantially over the last decade, due to great advances in the field of deep learning, especially computer vision algorithms and increasing availability of high-resolution satellite imagery (Hoeser and Kuenzer, 2020), coupled with new-generation computers. Applications in climate impacts are relatively small in number and include the identification of desertification trends (Vinuesa et al., 2020), mapping fuel and detection of wildfires (Jain et al., 2020), and flood mapping (Bentivoglio et al., 2022). These applications address the challenge of system complexity by providing new data and models, which can fill gaps in scientific knowledge, and localization, by providing high-resolution data, which can be used to replace or improve physical models.

Flooding

Flooding poses a major natural hazard, causing extensive damage to property, infrastructure, and loss of life. Traditional flood mapping methods, such as remote sensing and ground-based surveys, are often time-consuming, expensive, and prone to inaccuracies, especially in areas with complex terrain or vegetation (Bentivoglio et al., 2022). However, DL methods, a type of ML, offer a promising solution. These methods excel at learning complex data relationships and have proven effectiveness in tasks like image classification and object detection. By leveraging deep learning, flood mapping can be significantly improved. Deep learning models enable fast and accurate mapping of floods, even in challenging areas with intricate terrain or dense vegetation (Bentivoglio et al., 2022). A notable case study conducted by Bentivoglio et al. (2022) showcased the potential of deep learning in flood mapping. Using a deep learning model called U-Net, the researchers classified satellite images of the Netherlands and successfully identified flooded regions, even in areas with complex terrain and vegetation (Bentivoglio et al., 2022). The study demonstrated the effectiveness of deep learning methods in accurately mapping floods. In addition, various AI models can significantly contribute to disaster relief efforts by employing satellite data to map floods and locate and assist refugee camps (Logar et al., 2020).

Wildfires

In the paper titled "*A Systematic Review of Applications of Machine Learning Techniques for Wildfire Management Decision Support*" by Bot and Borges (2022), the authors delve into the potential applications of ML in supporting wildfire management decisions. The study reveals that ML techniques exhibit a high level of accuracy in predicting wildfire spread, identifying vulnerable areas, optimizing fire suppression strategies, evaluating the effectiveness of suppression efforts, and enhancing firefighter safety. These findings carry significant implications for wildfire management, suggesting that ML can contribute to mitigating wildfire risks and saving lives.

The paper highlights several specific examples of how ML can enhance wildfire management practices. Firstly, ML algorithms can analyze historical data and weather forecasts to predict the areas where wildfires are likely to spread. This knowledge enables timely evacuations and more efficient deployment of fire suppression resources. Secondly, ML can analyze factors such as vegetation type, slope, and fuel load to identify areas at a higher risk of wildfires. This information helps in developing proactive fire management plans and targeting fire prevention efforts in vulnerable areas. Thirdly, ML algorithms can consider variables like weather conditions, fuel load, and terrain to devise optimal fire suppression strategies, minimizing the loss of life and property. Additionally, ML can track the progression of wildfires and the resources used for their containment, facilitating the assessment of suppression efforts and enabling improvements for future responses. Many of the ML methods used in fire-susceptibility mapping have also been used to examine landscape controls—the relative importance of weather, vegetation, topography, and structural and anthropogenic variables, on fire activity—which may facilitate hypothesis formation and testing or model building (Jain et al., 2020; Tien Bui et al., 2017) employed particle swarm optimization neuro-fuzzy modeling to predict forest fires in Vietnam, achieving optimal parameter values and causation insights. Additionally, Tien Bui et al. (2016) developed fire sensitivity maps to aid in forest fire planning and management. Lastly, ML can assist in identifying high-risk areas for firefighters and devising strategies to reduce their exposure to danger, ultimately enhancing firefighter safety.

Desertification

Desertification, a severe ecological and environmental issue in arid regions, necessitates low-cost and accurate methods for quantitative monitoring. In a recent study conducted in Mongolia, six machine-learning techniques were employed using Google Earth Engine and Landsat images to monitor desertification dynamics from 1990 to 2020 (Meng et al., 2021). By analyzing spatiotemporal distributions and changes in desertification, researchers utilized gravity center change and intensity analysis models. The results indicated that the maximum entropy method provided the most precise assessment, achieving an accuracy of 96%. The study observed a significant increase in desertified land area from 1990 to 2005, followed by a slight decrease (Meng et al., 2021). Lightly and moderately desertified lands exhibited the highest change intensities and sensitivity to environmental factors. While both natural and human factors influence desertification dynamics, precipitation played a dominant role in Mongolia (Meng et al., 2021). The comprehensive analysis presented in this study offers valuable insights into the desertification status and trends in Mongolia, providing a foundation for formulating preventive measures and guiding desertification prevention and control efforts. Research findings have revealed an upward trend of approximately 1900 Km^2 per year in the total area of desertified land in Mongolia, with the light and moderate desertification areas experiencing the most rapid expansion. This escalating desertification directly impacts Mongolia's ecosystem and neighboring regions (Meng et al., 2021).

Future of AI in climate change and challenges

As a result of climate change and global warming, the utilization of renewable energies is increasing worldwide (Ashfaq et al., 2022). Renewable energy provides a viable solution to today's issues (Ashfaq et al., 2022). When combined with AI, renewable energy has the potential to revolutionize the energy sector and promote sustainability on both national and global levels (Jha et al., 2017). Through pattern recognition, predictive capabilities, autonomous task execution, supply optimization, and improved decision-making, AI can enhance the efficiency of the renewable energy industry (Ashfaq et al., 2022). The rapid advancements in AI technology offer faster forecasting and smarter connections between critical elements, providing valuable insights into energy processes (Ashfaq et al., 2022).

Artificial intelligence represents a new era in data-driven societies and global development. It introduces innovative approaches to leverage traditional production factors and establish modern frameworks for addressing present challenges (Ashfaq et al., 2022). While acknowledging the key breakthroughs in AI's history, recent advancements in neural networks, cognitive computing, and ML have brought unexpected competencies in energy management, optimization, and monitoring (Jha et al., 2017).

Like any technology, it has the potential to either improve or worsen the world we live in. However, when it comes to combating climate change, AI and ML can make significant contributions across various domains by playing a crucial role in automatic monitoring by utilizing remote sensing techniques to detect deforestation, collect data on buildings, and assess the damage caused by disasters (Rolnick et al., 2023). Additionally, it can expedite the process of scientific discovery by suggesting new materials for batteries, construction, carbon capture, and combining climate models with energy scheduling models.

There are abundant opportunities to develop cost-effective, autonomous, energy-efficient, and user-friendly Internet of Things (IoT)-based solutions for agriculture by utilizing long-range, ZigBee, Wi-Fi, 5G, and emerging narrow-band IoT communication technologies. These solutions can be built with a robust architecture and require low maintenance (Cicioğlu and Çalhan, 2021). One potential approach is to combine crop simulation models with remote sensing to gather crop phenotype data. Integrating phenotypic and genotypic data at the plot level using AI models can further address the complex challenges in agriculture (Khaki and Wang, 2019).

Other challenges facing AI in the future are ethical in nature. Policy considerations are essential when integrating AI systems, as they can entail risks and unintended consequences (Kaack et al., 2022). The European Commission's High-Level Expert Group on AI has outlined seven requirements for trustworthy AI, applicable to climate change and other areas. In the context of climate strategies, several key issues emerge. Firstly, there is a fundamental asymmetry regarding data on AI's climate impacts, with energy consumption estimates being relatively accessible compared to impact data on specific use cases (Kaack et al., 2022). Proactive policy approaches should address both energy use and application-specific impact, given the rapid evolution of the AI sector. Equity is a central concern, as AI-driven approaches may amplify inequities, such as through the digital divide or algorithmic biases. Additionally, the use of AI may reshape power dynamics between public and private entities based on data control, analytical capabilities, and access conditions (Kaack et al., 2022). Considering that many climate strategies involve the public sector, it is important for public entities to factor in these considerations when deciding on in-house AI capacities. Moreover, critical infrastructure, particularly in the energy sector, plays a crucial role in climate change mitigation (Kaack et al., 2022). AI applications in critical infrastructure should incorporate safety and security considerations. Addressing these policy-relevant considerations is vital to ensure responsible and effective use of AI in climate change strategies.

Artificial intelligence has the potential to drive positive change in cities and societies and contribute to achieving multiple sustainable development goals (Vinuesa et al., 2020). However, it is crucial to advance the implementation of appropriate policies and regulations to mitigate the harm caused by AI to vulnerable urban and social groups, as well as the environment.

The integration of AI in energy systems raises crucial questions about system optimization, parameterization, and personalized consumption and production (Ellabban et al., 2014). The adoption of AI in the energy industry requires an examination of social, economic, and technical considerations, particularly in envisioning intelligent agents driven by smart grid technology and AI algorithms (Ashfaq et al., 2022). Various AI technologies, including organized data management, data mining capabilities, and ML strategies, can contribute to an AI ecosystem or a system of systems within energy sectors, enabling an AI-powered Smart Energy Grid. To fully realize these advancements, AI must be integrated at all levels of energy systems (Ashfaq et al., 2022).

Section summary

Looking ahead, the future of AI in addressing climate change is promising. AI technologies have the potential to play pivotal roles in various aspects of climate change mitigation and adaptation. One of the key areas where AI can make a significant impact is in optimizing energy systems. By leveraging AI algorithms and data analytics, it becomes possible to optimize energy production, distribution, and consumption patterns. This can lead to improved energy efficiency, reduced greenhouse gas emissions, and the integration of renewable energy sources into the grid. AI can also enhance the management of smart grids, enabling more reliable and resilient energy networks.

AI's predictive capabilities and data analysis techniques can greatly assist in climate modeling and forecasting. Climate models powered by AI algorithms can simulate complex climate systems and provide more accurate projections of future climate scenarios. This information is crucial for policymakers, enabling them to make informed decisions and develop effective climate change strategies.

In the realm of sustainable transportation, AI can revolutionize the way we travel. AI-powered systems can optimize transportation routes, manage traffic flow, and promote the use of electric and autonomous vehicles. These advancements can significantly reduce emissions from the transportation sector and enhance overall transportation efficiency.

AI also holds immense potential in the field of agriculture and food systems. By leveraging AI technologies, farmers can optimize irrigation, fertilization, and pest control, resulting in reduced resource use and increased crop yields. AI can also assist in monitoring and managing ecosystem health, contributing to the preservation of biodiversity and natural habitats.

Furthermore, AI can aid in the development of climate-resilient infrastructure and urban planning. By analyzing data and predicting the impact of climate change on cities, AI can inform the design of buildings, transportation networks, and urban landscapes that are better adapted to changing climate conditions. This can enhance the resilience of communities and reduce the vulnerability of urban areas to climate-related risks.

In conclusion, AI has a bright future in the fight against climate change. Its potential applications span across various sectors, from energy systems and transportation to agriculture and urban planning. By harnessing the power of AI, we can unlock innovative solutions, optimize resource management, and make significant strides toward a more sustainable and resilient future. However, it is essential to ensure that the deployment of AI is guided by ethical principles, responsible practices, and a strong commitment to the well-being of both present and future generations. With the right approach, AI can be a game-changer in tackling the complex challenges posed by climate change and paving the way for a greener and more sustainable world.

Use of AI in smart farming through the Internet of Things
Introduction to AI in smart farming through the Internet of Things

Countries such as South Korea, China, and North America are making significant investments in advanced technologies in agriculture to meet the growing demands of their populations (Jha et al., 2019). In this context, AI is playing a crucial role in precision agriculture to reduce the environmental impact caused by the use of fertilizers and chemicals. Precision farming leverages state-of-the-art sensors and predictive analytics to collect real-time data on various factors such as soil conditions, crop maturity, air quality, weather patterns, equipment, labor availability, and prices (Ghosh and Singh, 2020). By adopting a data-driven approach, precision agriculture aims to increase agricultural yields, enhance decision-making processes, and minimize the negative environmental consequences associated with farming practices (Raj et al., 2021; Das et al., 2018). Another innovative technology, the IoT, holds great promise in transforming agriculture by enhancing user perception and enabling the automation of tasks (Gonzalez et al., 2021). Furthermore, AI technology has the potential to seamlessly integrate with the expanding opportunities provided by the IoT and renewable energy sources within the energy industry. Fig. 13.5 illustrates AI/ML, IoT devices, and renewable energy systems.

The IoT has revolutionized various domains, including healthcare, retail, traffic, security, smart homes, smart cities, and agriculture, by focusing on communication, automation, and cost-saving (Jha et al., 2019). In the agricultural sector, IoT deployment is seen as an ideal solution for the continuous monitoring of Controlled Environment Agriculture (CEA) systems. The global demand for food is projected to increase by 35%–56% between 2010 and 2050, while the population at risk of hunger is expected to change by −91% to +8% during the same period (van Dijk et al., 2021). Therefore, sustainable agriculture practices are crucial to support the world's population growth, mitigating climate change, and preserving water and natural resources (Gonzalez et al., 2021).

Considering the challenges posed by climate change, traditional agricultural methods alone may not suffice to meet the increasing food demands. Thus, field-based sensors, drones, advanced tractors, and hydroponic farming are being utilized to enhance crop production at lower costs (Pratyush Reddy et al., 2020). Consequently, there has been a significant increase in the usage and implementation of IoT-based agriculture solutions in recent years (Gonzalez et al., 2021).

Using IoT for precision agriculture

Artificial intelligence holds immense potential for leveraging big data, which can be easily accessible through the use of unmanned aircraft systems (UAS), to improve the efficiency and resiliency of food production systems (Jung et al., 2021). In agriculture, IoT is employed at various stages of the production chain, offering benefits such as enhanced monitoring and data collection (Medela et al., 2013). The integration of AI into precision agriculture has further contributed to the profitability and sustainability of modern farming practices (Wei et al., 2020; Costa et al., 2021). AI technology plays a vital role in disease detection, optimizing the use of fertilizers and pesticides, pest identification, and yield prediction (Bacco et al., 2018).

The combination of remote sensing, simulation models, and AI-based UAS enables efficient and reliable crop phenotyping (Ashapure et al., 2019) as illustrated in Fig. 13.6. However, the current limitations of UAS, such as limited battery life and flight time, make extensive spatial coverage

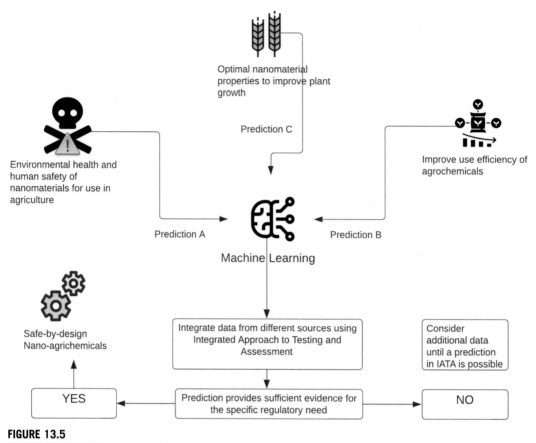

FIGURE 13.5

Machine learning over datasets and real-time data acquired through Internet of Things (IoT) sensors.

Modified from Zhang, X., Shu, K., Rajkumar, S., Sivakumar, V., 2021. Research on deep integration of application of artificial intelligence in environmental monitoring system and real economy. Environmental Impact Assessment Review 86, 01959255, 106499. https://doi.org/10.1016/j.eiar.2020.106499.

impractical, and data processing costs increase exponentially with larger data volumes required for larger areas (Singh and Frazier, 2018). Satellite data, with its coarser spatial and temporal resolution, have been widely used in precision agriculture applications, but there is a need to adapt it for scale-appropriate precision agriculture (Herbei et al., 2016).

Crop management

Advancements in computer vision, ML, and DL have facilitated accurate and rapid identification of crop diseases, and overall crop management while robotics and AI technologies enhance productivity and maximize human potential in agriculture (Barile et al., 2019). The advancement of UAVs has brought about advancements in precision agriculture and geodesy (Colomina and Molina, 2014). Precision spraying technology addresses the challenge of chemical transfer to unintended areas by reducing the

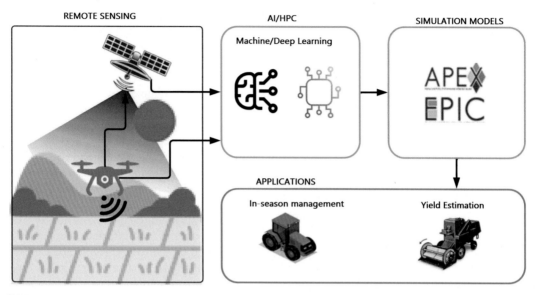

FIGURE 13.6

The integration of remote sensing data, crop simulation models, and artificial intelligence (AI) in digital agriculture advances crop management and marketing projections through in-season prescriptive tools and yield forecasting.

quantity of herbicides required and minimizing environmental impact, crop damage, chemical residues, and expenses (Creech et al., 2015; Balafoutis et al., 2017). By employing deep learning and convolutional neural networks, robots equipped with computer vision and AI have the potential to significantly reduce chemical usage in crop spraying and herbicide costs (Swaminathan et al., 2023). Moreover, farmers now have easy access to spatial data collection in precision agriculture thanks to the utilization of UAVs. According to Zhang and Kovacs (2012), UAVs possess the ability to monitor changes in agricultural fields with exceptional spatial and temporal precision. The gathered imagery can then undergo processing and analysis using ML algorithms, which are increasingly employed to classify large volumes of intricate data, particularly high-resolution UAV imagery. These algorithms have demonstrated their superiority over conventional supervised classification algorithms in terms of classification accuracy and time efficiency (Belgiu and Drăgut, 2016; Gašparović et al., 2017). By integrating UAVs into precision agriculture, farmers can enhance crop yields and reduce treatment costs by efficiently employing fertilizers and herbicides. Furthermore, UAVs are remarkably effective in detecting even minor abnormalities in agricultural fields, as emphasized by Seelan et al. (2003). Not only are these spatial data collection methods more efficient than traditional terrestrial approaches, but they are also more cost-efficient and time-efficient, as asserted by Gašparović et al. (2017).

Crop productivity

Artificial intelligence can drive positive change in cities and societies and contribute to achieving multiple sustainable development goals (Vinuesa et al., 2020). The energy and environment management sectors are increasingly recognizing and incorporating the numerous advantages offered by the IoT. Through the use of smart devices, farmers can now anticipate climatic changes within their fields, leading to a reduction in waste and better control of agricultural processes based on factors such as weather conditions, relative humidity, soil moisture, visible and UV rays, and other external variables (Gonzalez et al., 2021). For instance, Carnegie Mellon University has implemented wireless sensor technology in a plant nursery (Zubairi, 2009). In another example, a real-time rice crop monitoring system has been developed to enhance productivity (Satyanarayana and Mazaruddin, 2013). This system, described in Rajesh (2011), collects rainfall and temperature data, analyzes it, and employs the findings to mitigate the risk of crop loss and boost productivity.

Artificial intelligence provides systems that are proven to be scalable, stable, and accurate to provide real-time data for precision agriculture, and AI -supported precision agriculture eliminates randomness, provides precise and required amounts of fertilizers and pesticides, and can increase food productivity by utilizing the limited available arable land for farming (Leal Filho et al., 2022). There are various communication technologies available, such as long-range, ZigBee, wireless fidelity, 5G, and emerging narrow-band IoT, which offer great potential for creating cost-effective, autonomous, energy-efficient, and user-friendly solutions for agriculture based on the IoT (Cicioğlu and Çalhan, 2021). One possible approach is to combine crop simulation models and remote sensing to gather crop phenotype data, and employing AI models to integrate phenotypic and genotypic data at the plot level can further contribute to addressing the complex challenges in agriculture (Khaki and Wang, 2019; Ma et al., 2018).

Soil and irrigation management are crucial factors influencing crop productivity and yield. Inadequate management practices in these areas can result in crop losses and diminished quality. Bralts et al. (1993) developed a rule-based expert system that enables the estimation of design and performance parameters for micro-irrigation systems. Sicat et al. (2005) proposed a fuzzy-based recommendation system that utilizes farmers' knowledge to determine the suitable crop to be cultivated on a specific type of soil. Additionally, rainfall is a significant factor affecting crop production and Manek and Singh (2016) compared various neural network architectures to predict rainfall.

Advances in ML technology, particularly in handling big data, offer unique opportunities for the development of accurate, large-scale prediction and prescriptive models (Halevy et al., 2009). Crop simulation models, which utilize input variables such as crop management information, weather, and soil data, have become powerful tools for estimating crop productivity and linking physiology, genetics, and phenomics. Research is focused on upscaling these models from field-scale to large regions, with the main challenge being the calibration of model inputs beyond the field scale (Pearl, 2019). Integrating remotely sensed crop phenotypic data with crop simulation models holds promise, as it combines the deterministic nature of ML methodologies with the ability of simulation models to handle nonexperienced scenarios. Benchmarks are being developed to help solve complex agricultural problems by integrating phenotypic and genotypic data at the plot level (Jung et al., 2021).

Weed detection

Effective weed management is essential for maintaining crop quality and productivity, and AI applications are employed to address this issue. Pasqual (1994) developed a rule-based expert system specifically designed for the recognition and removal of weeds in crops such as oats, barley, triticale, and wheat. On the other hand, Shi et al. (2008) proposed and developed an approach that combines image analysis techniques with neural networks to identify weeds in real time. This method enables the timely detection and management of weeds in agricultural fields. Furthermore, in the context of weed management, usage of UAVs has increased resulting in successful weed mapping through the utilization of low-cost RGB cameras. López-Granados et al. (2016) and Pérez-Ortiz et al. (2016) also highlighted the significant potential of low-cost UAVs in weed management, particularly when combined with near-real-time approaches facilitated by automatic classification methods. The increasing adoption of ML algorithms, notably, and the RF classification algorithm has shown effective performance in automatically classifying UAV imagery, particularly when employing the object-based classification approach for weed mapping (De Castro et al., 2018; López-Granados et al., 2016; Pérez-Ortiz et al., 2016).

Precision fertilizers

Precision fertilizer application utilizes models and data inputs to optimize fertilizer usage based on soil nutrient levels and field segmentation, resulting in increased crop yields, balanced soil nutrients, and reduced atmospheric emissions (Elbeltagi et al., 2022). By integrating genome analysis, editing techniques, precision agriculture, and AI technologies, it becomes possible to develop crops suited to specific land conditions and maximize plant production, while minimizing the impact of chemicals on the soil and reducing chemical fertilizer usage (Ghosh and Singh, 2020; Joseph et al., 2021).

Reducing the impact of chemicals on the soil can lead to a decrease in the use of chemical fertilizers in agriculture, promoting more environmentally friendly farming practices. In the districts of Hafizabad and Sheikhupura, Elahi et al. (2019) estimated target values for agrochemicals used in rice farming while maintaining current rice yields. The results indicated that 52.6% of pesticide inputs and 43.6% of pure nitrogen fertilizer inputs could be reduced, yielding a favorable and significant impact. Putra et al. (2020) developed a model to simulate the availability and loss of oil palm nutrients based on the nutrient data stored and released through fertilizer application. This approach enables effective determination of nutrient balance and maintenance through targeted fertilizer application.

In the realm of cotton cultivation, Du et al. (2021) devised a water and fertilizer control system that leverages soil conductivity thresholds to enhance the utilization of water and fertilizer. By considering soil conductivity and moisture content, this system achieved a 10.89% reduction in water and fertilizer usage.

The utilization of deep learning and convolutional neural networks is increasingly prevalent in agricultural remote sensing applications (Kussul et al., 2017). According to Swaminathan et al. (2023), the implementation of robots equipped with computer vision and AI for weed monitoring and spraying could potentially reduce chemical usage on crops by 80% and decrease herbicide costs by 90%. In precision fertilization, a fertilizer application model is employed to determine the appropriate fertilizer input based on soil nutrient levels. This model facilitates the precise application of fertilizer using a variable rate applicator, which divides the field into a grid (Elbeltagi et al., 2022). Precision fertilizer application offers several benefits, including minimized fertilizer usage, increased crop yields, improved soil nutrient balance, and reduced atmospheric emissions. By employing genome analysis and editing techniques, precision agriculture and AI technologies have the potential to cultivate crops that are well-suited to the land and optimize plant production (Joseph et al., 2021).

Pest detection and management

The smart sprayer, a technological innovation, integrates weed recognition, a mapping system, and a unique rapid and precise spraying mechanism. It also employs a newly developed algorithm to generate visual maps, further enhancing its effectiveness in targeted pest management. Partel et al. (2019) utilized an embedded graphics processing unit within a smart sprayer to achieve precise weed control of artificial and amaranth weeds, achieving an accuracy rate of 59%−71%. This approach holds the potential to significantly decrease pesticide costs, minimize crop damage, mitigate the risk of excessive herbicide residues, and potentially reduce environmental impacts.

In a separate study, Facchinetti et al. (2021) employed a "Rover" sprayer vehicle that accurately distinguished color differences between salad crops and the ground, resulting in a 55% reduction in pesticide spraying. Another noteworthy system, the I2PDM, is an intelligent integrated pest management wireless sensor network that employs sensor nodes to collect images, pest data, and species information, which are then stored in a database for analysis. This system generates models that can be translated into numerical information, aiding in pest management decisions (Rustia et al., 2020). Applying this technique to a tomato field, Rustia et al. (2020) successfully reduced the pesticide dose from 235 to 204 L/time (16%), indicating the effective reduction of pests through insecticide spraying.

The main application domains of IoT in agriculture

Various sub-domains have been identified for IoT-based agricultural monitoring solutions. These sub-domains include soil monitoring, air monitoring, temperature monitoring, water monitoring, disease monitoring, location monitoring, environmental conditions monitoring, pest monitoring, and fertilization monitoring. These different areas of monitoring enable comprehensive data collection and analysis to support efficient and effective agricultural practices (Farooq et al., 2020). The selected mainstream application domains by Farooq et al. (2020) are monitoring, tracking, and controlling.

The applications of IoT-based agricultural monitoring can be broadly classified into several categories, such as: irrigation monitoring and controlling, precision farming, soil monitoring, temperature monitoring, humidity monitoring, animal monitoring and tracking, water monitoring and controlling, disease monitoring, air monitoring, and fertilization monitoring (Farooq et al., 2020).

Open issues and challenges in smart farming

There are many open issues and challenges that are associated with the implementation of IoT applications. Some of the challenges identified in the literature include:

(a) *Security*

Security issues pose significant challenges that need to be addressed (Farooq et al., 2020). Risks include data loss, physical interference, and unauthorized access to devices (Soumyalatha, 2016). There are also limitations to current internet architecture concerning mobility, manageability, and scalability. Ensuring robust security measures, encryption, and access control is crucial to protect sensitive data in IoT-based agricultural systems (Bandyopadhyay and Sen, 2011).

(b) *Cost and reliability*

Cost and reliability are important considerations when implementing IoT in agriculture. The setup costs, including hardware expenses for IoT devices, sensors, base station infrastructure, and gateways, as well as ongoing subscription fees for device management and information exchange, need to be carefully managed (Farooq et al., 2020). The IoT network's complexity and diversity introduce the risk of software or hardware failures, which can have significant consequences. Even power outages can disrupt operations and cause inconvenience (Soumyalatha, 2016). Managing heterogeneous applications, environments, and devices also constitutes a major challenge (Bandyopadhyay and Sen, 2011). Addressing security and privacy concerns on limited-capability devices is crucial and requires convincing solutions. Additionally, developing and implementing technological architectures that prioritize privacy must serve as the foundation for all future advancements (van Kranenburg and Bassi, 2012).

(c) *Scalability*

The deployment of a large number of IoT devices and sensors in the agriculture field necessitates the implementation of an intelligent IoT management system. This system is essential for effectively identifying and controlling each node within the network (Farooq et al., 2020). By employing such a management system, agricultural stakeholders can ensure efficient monitoring, control, and coordination of the IoT devices and sensors, enabling optimized agricultural operations and resource utilization.

(d) *Interoperability*

The vast number of IoT devices, along with the need for interoperability, poses significant challenges. Interoperability encompasses various aspects, including semantic, syntactic, technical, and organizational policy (Farooq et al., 2020). Currently, with devices from different manufacturers being interconnected in IoT systems, there is a lack of international standardization for compatibility in terms of tagging and monitoring (Naveen, 2016). This lack of standardization hinders seamless integration and communication between devices, making it essential to address interoperability issues to ensure smooth interoperability among IoT devices from different vendors.

(e) *Integration challenges of IoT and cloud computing in agriculture*

The objective of the cloud-based IoT paradigm is to analyze and integrate real-world data into IoT objects. This approach involves interacting with a vast number of geographically dispersed end devices (Farooq et al., 2020). While the cloud platform offers advanced techniques that benefit farmers, there are limitations that can lead to technology loss. Challenges such as Internet connectivity issues, low-power communication devices, and integration challenges can hinder the adoption and effectiveness of cloud-based IoT solutions for farmers. Additionally, monitoring and controlling processes can be impeded by connectivity issues and latency problems faced by IoT communication devices, making data sharing more difficult (Soumyalatha, 2016). Overcoming these limitations is crucial to ensure reliable and seamless integration of IoT devices into the cloud-based paradigm, enabling efficient monitoring, control, and data analysis for improved agricultural practices.

Section summary

In conclusion, the expansion of the global economy and population has led to increased energy demand, resulting in environmental damage and climate change. AI technology emerges as a promising tool to address these issues and mitigate their adverse effects. AI contributes to climate change mitigation by predicting energy demand, enhancing energy efficiency, and optimizing renewable energy production. It also improves weather prediction, aids in sustainable architecture and urban planning, reduces agrochemical use in agriculture, optimizes industrial processes, enhances environmental protection efforts, and contributes to the development of resilient cities. AI's potential in the energy sector is further amplified by its integration with the IoT and renewable energy systems. By increasing energy efficiency and providing decision-makers with accurate data, AI plays a crucial role in combating climate change. Moreover, in the field of agriculture, the integration of AI in precision agriculture revolutionizes farming practices, improves resource efficiency, and increases food productivity while promoting sustainability on limited arable land. As we move forward, it is important to focus on multidisciplinary collaboration, standard data protocols, and the quality of raw data to fully leverage the potential of AI in addressing future challenges and meeting global needs.

Conclusion

The advent of technology and AI, while dominating modern methods of scientific inquiry, has had quite an impact on the field of environmental research, with pertinent implications for the integration of and progress in all the reviewed areas. Historically, the use of the precursors of AI has helped develop monumental works that have shaped our understanding of the environment and ecosystems, as is seen in the development of ecological models from the 1950s until now. Whereas the use of AI in biodiversity conservation, research and monitoring have been relatively novel, methods of inquiry are increasingly becoming more dependent on the use of AI and machine-learning-driven tools. Similar usage has been seen in the field of Climate Change research and mitigation, with climate projections and models being developed for a wide range of uses, including climate policy, while involving a wide range of components such as optimization of energy systems, making climate resilient infrastructure, transportation, and climate forecasting. The combination of AI and IoT, by intensifying the use of technology in agriculture, has enabled the integration of the two, with smart farming and precision agriculture developing at a rapid pace. The progress due to AI has been imminent, and future usage of the technology in the reviewed fields should be of interest for ecologists in the near future.

Climate change crisis, biodiversity conservation and monitoring, ecological modeling, and smart farming hold utmost importance for sustainability of the planet. These interconnected components are essential for addressing the challenges we face and are deeply intertwined with human dependence. The preservation of biodiversity and the environment is intricately linked to the issue of climate change, which is a pressing concern. Moreover, the climate crisis itself has a profound impact on smart farming, which represents a crucial step toward improving agricultural yields and ensuring food security for all. Currently, the methodologies employed for these components rely predominantly on ML, simulations, or traditional methods, some of which require further refinement. Therefore, it is vital to emphasize the role of AI in ecological modeling, biodiversity monitoring and conservation. By harnessing the power of AI, we can revolutionize our efforts in combating the climate crisis and addressing global hunger.

Author contributions

MK conceived, compiled, and reviewed manuscript, NM wrote the Ecological Modeling section, SFQ wrote the Biodiversity monitoring and conservation section, SAA wrote the Climate Change and Smart Farming Section, and QHM conceived, compiled, and reviewed the manuscript and bibliography.

References

Afzaal, H., Farooque, A.A., Abbas, F., Acharya, B., Esau, T., 2020. Computation of evapotranspiration with artificial intelligence for precision water resource management. Applied Sciences 10 (5). https://doi.org/10.3390/app10051621.

Agudelo, C.A.R., Bustos, S.L.H., Moreno, C.A.P., 2020. Modeling interactions among multiple ecosystem services. A critical review. Ecological Modelling 429. https://doi.org/10.1016/j.ecolmodel.2020.109103.

Ahmad, T., Zhang, D., Huang, C., Zhang, H., Dai, N., Song, Y., Chen, H., 2021. Artificial intelligence in sustainable energy industry: status Quo, challenges and opportunities. Journal of Cleaner Production 289, 125834. https://doi.org/10.1016/j.jclepro.2021.125834.

Ahmed, Q.W., Garg, S., Rai, A., Ramachandran, M., Jhanjhi, N.Z., Masud, M., Baz, M., 2022a. AI-based resource allocation techniques in wireless sensor internet of things networks in energy efficiency with data optimization. Electronics 11 (13), 2071. https://doi.org/10.3390/electronics11132071.

Ahmed, S., Alshater, M.M., Ammari, A.E., Hammami, H., 2022b. Artificial intelligence and machine learning in finance: a bibliometric review. Research in International Business and Finance 61. https://doi.org/10.1016/j.ribaf.2022.101646.

Alem, A., Kumar, S., 2022. Transfer learning models for land cover and land use classification in remote sensing image. Applied Artificial Intelligence 36 (1). https://doi.org/10.1080/08839514.2021.2014192.

Allan, B.M., Nimmo, D.G., Ierodiaconou, D., VanDerWal, J., Koh, L.P., Ritchie, E.G., 2018. Futurecasting ecological research: the rise of technoecology. Ecosphere 9 (5). https://doi.org/10.1002/ecs2.2163.

Arumugam, K., Swathi, Y., Sanchez, D.T., Mustafa, M., Phoemchalard, C., Phasinam, K., Okoronkwo, E., 2022. Towards applicability of machine learning techniques in agriculture and energy sector. Materials Today: Proceedings 51, 2260−2263. https://doi.org/10.1016/j.matpr.2021.11.394.

Ashapure, A., Jung, J., Yeom, J., Chang, A., Maeda, M., Maeda, A., Landivar, J., 2019. A novel framework to detect conventional tillage and no-tillage cropping system effect on cotton growth and development using multi-temporal UAS data. ISPRS Journal of Photogrammetry and Remote Sensing 152, 49−64. https://doi.org/10.1016/j.isprsjprs.2019.04.003.

Ashfaq, A., Kamran, M., Rehman, F., Sarfaraz, N., Ilyas, H.U., Riaz, H.H., 2022. Role of artificial intelligence in renewable energy and its scope in future. In: 2022 5th International Conference on Energy Conservation and Efficiency, ICECE 2022 − Proceedings. Institute of Electrical and Electronics Engineers Inc., Pakistan. https://doi.org/10.1109/ICECE54634.2022.9758957

Bacco, M., Berton, A., Ferro, E., Gennaro, C., Gotta, A., Matteoli, S., Paonessa, F., Ruggeri, M., Virone, G., Zanella, A., 2018. Smart farming: opportunities, challenges and technology enablers. In: 2018 IoT Vertical and Topical Summit on Agriculture − Tuscany, IOT Tuscany 2018. Institute of Electrical and Electronics Engineers Inc., Italy, pp. 1−6. https://doi.org/10.1109/IOT-TUSCANY.2018.8373043, 9781538669303.

Balafoutis, A., Beck, B., Fountas, S., Vangeyte, J., Wal, T., Soto, I., Gómez-Barbero, M., Barnes, A., Eory, V., 2017. Precision agriculture technologies positively contributing to GHG emissions mitigation, farm productivity and economics. Sustainability 9 (8). https://doi.org/10.3390/su9081339.

Bandyopadhyay, D., Sen, J., 2011. Internet of things: applications and challenges in technology and standardization. Wireless Personal Communications 58 (1), 49−69. https://doi.org/10.1007/s11277-011-0288-5.

Barile, S., Piciocchi, P., Bassano, C., Spohrer, J., Pietronudo, M.C., 2019. Re-defining the role of artificial intelligence (AI) in wiser service systems. Advances in Intelligent Systems and Computing 787, 159−170. https://doi.org/10.1007/978-3-319-94229-2_16.

Battaglia, M., Sands, P.J., 1998. Process-based forest productivity models and their application in forest management. Forest Ecology and Management 102 (1), 13−32. https://doi.org/10.1016/S0378-1127(97)00112-6.

Baysan, S., Kabadurmus, O., Cevikcan, E., Satoglu, S.I., Durmusoglu, M.B., 2019. A simulation-based methodology for the analysis of the effect of lean tools on energy efficiency: an application in power distribution industry. Journal of Cleaner Production 211, 895−908. https://doi.org/10.1016/j.jclepro.2018.11.217.

Belgiu, M., Drăgut, L., 2016. Random forest in remote sensing: a review of applications and future directions. ISPRS Journal of Photogrammetry and Remote Sensing 114, 24−31. https://doi.org/10.1016/j.isprsjprs.2016.01.011.

Bentivoglio, R., Isufi, E., Jonkman, S.N., Taormina, R., 2022. Deep learning methods for flood mapping: a review of existing applications and future research directions. Hydrology and Earth System Sciences 26 (16), 4345−4378. https://doi.org/10.5194/hess-26-4345-2022.

Bot, K., Borges, J.G., 2022. A systematic review of applications of machine learning techniques for wildfire management decision support. Inventions 7 (1). https://doi.org/10.3390/inventions7010015.

Botkin, D.B., Janak, J.F., Wallis, J.R., 1972. Some ecological consequences of a computer model of forest growth. Journal of Ecology 60 (3), 849. https://doi.org/10.2307/2258570.

Bralts, V.F., Driscoll, M.A., Shayya, W.H., Cao, L., 1993. An expert system for the hydraulic analysis of microirrigation systems. Computers and Electronics in Agriculture 9 (4), 275−287. https://doi.org/10.1016/0168-1699(93)90046-4.

Buckland, S.T., Newman, K.B., Fernández, C., Thomas, L., Harwood, J., 2007. Embedding population dynamics models in inference. Statistical Science 22 (1), 44−58. https://doi.org/10.1214/088342306000000673.

Cai, W., Lai, K.H., Liu, C., Wei, F., Ma, M., Jia, S., Jiang, Z., Lv, L., 2019. Promoting sustainability of manufacturing industry through the lean energy-saving and emission-reduction strategy. Science of the Total Environment 665, 23−32. https://doi.org/10.1016/j.scitotenv.2019.02.069.

Carey, C.C., Ward, N.K., Farrell, K.J., Lofton, M.E., Krinos, A.I., McClure, R.P., Subratie, K.C., Figueiredo, R.J., Doubek, J.P., Hanson, P.C., Papadopoulos, P., Arzberger, P., 2019. Enhancing collaboration between ecologists and computer scientists: lessons learned and recommendations forward. Ecosphere 10 (5). https://doi.org/10.1002/ecs2.2753.

Carranza-García, M., García-Gutiérrez, J., Riquelme, J.C., 2019. A framework for evaluating land use and land cover classification using convolutional neural networks. Remote Sensing 11 (3). https://doi.org/10.3390/rs11030274.

Ceballos, G., Ehrlich, P.R., Raven, P.H., 2020. Vertebrates on the brink as indicators of biological annihilation and the sixth mass extinction. Proceedings of the National Academy of Sciences of the United States of America 117 (24), 13596−13602. https://doi.org/10.1073/pnas.1922686117.

Chai, S.Y., Hayat, A., Flaherty, G.T., 2022. Integrating artificial intelligence into haematology training and practice: opportunities, threats and proposed solutions. British Journal of Haematology 198 (5), 807−811. https://doi.org/10.1111/bjh.18343.

Chang, Y.C., Hong, F.W., Lee, M.T., 2008. A system dynamic based DSS for sustainable coral reef management in Kenting coastal zone, Taiwan. Ecological Modelling 211 (1−2), 153−168. https://doi.org/10.1016/j.ecolmodel.2007.09.001.

Chattopadhyay, A., Hassanzadeh, P., Pasha, S., 2020. Predicting clustered weather patterns: a test case for applications of convolutional neural networks to spatio-temporal climate data. Scientific Reports 10 (1). https://doi.org/10.1038/s41598-020-57897-9.

Chen, L., Huang, L., Hua, J., Chen, Z., Wei, L., Osman, A.I., Fawzy, S., Rooney, D.W., Dong, L., Yap, P.S., 2023a. Green construction for low-carbon cities: a review. Environmental Chemistry Letters 21 (3), 1627−1657. https://doi.org/10.1007/s10311-022-01544-4.

Chen, L., Chen, Z., Zhang, Y., Liu, Y., Osman, A.I., Farghali, M., Hua, J., Al-Fatesh, A., Ihara, I., Rooney, D.W., Yap, P.-S., 2023b. Artificial intelligence-based solutions for climate change: a review. Environmental Chemistry Letters. https://doi.org/10.1007/s10311-023-01617-y.

Choi, C., Kim, J., Kim, J., Kim, D., Bae, Y., Kim, H.S., 2018. Development of heavy rain damage prediction model using machine learning based on big data. Advances in Meteorology 2018, 1–11. https://doi.org/10.1155/2018/5024930.

Chopra, R., Magazzino, C., Shah, M.I., Sharma, G.D., Rao, A., Shahzad, U., 2022. The role of renewable energy and natural resources for sustainable agriculture in ASEAN countries: do carbon emissions and deforestation affect agriculture productivity? Resources Policy 76. https://doi.org/10.1016/j.resourpol.2022.102578.

Christin, S., Hervet, É., Lecomte, N., Ye, H., 2019. Applications for deep learning in ecology. Methods in Ecology and Evolution 10 (10), 1632–1644. https://doi.org/10.1111/2041-210x.13256.

Cicioğlu, M., Çalhan, A., 2021. Smart agriculture with internet of things in cornfields. Computers and Electrical Engineering 90. https://doi.org/10.1016/j.compeleceng.2021.106982.

Cifuentes, J., Marulanda, G., Bello, A., Reneses, J., 2020. Air temperature forecasting using machine learning techniques: a review. Energies 13 (16). https://doi.org/10.3390/en13164215.

Colomina, I., Molina, P., 2014. Unmanned aerial systems for photogrammetry and remote sensing: a review. ISPRS Journal of Photogrammetry and Remote Sensing 92, 79–97. https://doi.org/10.1016/j.isprsjprs.2014.02.013.

Cord, A.F., Brauman, K.A., Chaplin-Kramer, R., Huth, A., Ziv, G., Seppelt, R., 2017. Priorities to advance monitoring of ecosystem services using earth observation. Trends in Ecology & Evolution 32 (6), 416–428. https://doi.org/10.1016/j.tree.2017.03.003.

Costa, L., Archer, L., Ampatzidis, Y., Casteluci, L., Caurin, G.A.P., Albrecht, U., 2021. Determining leaf stomatal properties in citrus trees utilizing machine vision and artificial intelligence. Precision Agriculture 22 (4), 1107–1119. https://doi.org/10.1007/s11119-020-09771-x.

Creech, C.F., Henry, R.S., Werle, R., Sandell, L.D., Hewitt, A.J., Kruger, G.R., 2015. Performance of post-emergence herbicides applied at different carrier volume rates. Weed Technology 29 (3), 611–624. https://doi.org/10.1614/WT-D-14-00101.1.

Das, S., Ghosh, I., Banerjee, G., Sarkar, U., 2018. Artificial intelligence in agriculture: a literature survey. International Journal of Scientific Research in Computer Science Applications and Management Studies 7, 1–6.

De Castro, A.I., Torres-Sánchez, J., Peña, J.M., Jiménez-Brenes, F.M., Csillik, O., López-Granados, F., 2018. An automatic random forest-OBIA algorithm for early weed mapping between and within crop rows using UAV imagery. Remote Sensing 10 (2). https://doi.org/10.3390/rs10020285.

Dominguez, D., de del Villar, L.J., Pantoja, O., González-Rodríguez, M., 2022. Forecasting Amazon rain-forest deforestation using a hybrid machine learning model. Sustainability 14 (2). https://doi.org/10.3390/su14020691.

Du, C., Zhang, L., Ma, X., Lou, X., Shan, Y., Li, H., Zhou, R., 2021. A cotton high-efficiency water-fertilizer control system using wireless sensor network for precision agriculture. Processes 9 (10). https://doi.org/10.3390/pr9101693.

Elahi, E., Weijun, C., Zhang, H., Abid, M., 2019. Use of artificial neural networks to rescue agrochemical-based health hazards: a resource optimisation method for cleaner crop production. Journal of Cleaner Production 238, 117900. https://doi.org/10.1016/j.jclepro.2019.117900.

Elbeltagi, A., Kushwaha, N.L., Srivastava, A., Zoof, A.T., 2022. Artificial intelligent-based water and soil management. Deep Learning for Sustainable Agriculture 129–142. https://doi.org/10.1016/B978-0-323-85214-2.00008-2.

Ellabban, O., Abu-Rub, H., Blaabjerg, F., 2014. Renewable energy resources: current status, future prospects and their enabling technology. Renewable and Sustainable Energy Reviews 39, 748–764. https://doi.org/10.1016/j.rser.2014.07.113.

Enholm, I.M., Papagiannidis, E., Mikalef, P., Krogstie, J., 2022. Artificial intelligence and business value: a literature review. Information Systems Frontiers 24 (5), 1709–1734. https://doi.org/10.1007/s10796-021-10186-w.

Facchinetti, D., Santoro, S., Galli, L.E., Fontana, G., Fedeli, L., Parisi, S., Bonacchi, L.B., Šušnjar, S., Salvai, F., Coppola, G., Matteucci, M., Pessina, D., 2021. Reduction of pesticide use in fresh-cut salad production through artificial intelligence. Applied Sciences 11 (5). https://doi.org/10.3390/app11051992.

Fang, F., Tambe, M., Dilkina, B., Plumptre, A.J., 2019. Engaging Citizen Scientists in Data Collection for Conservation. Cambridge University Press (CUP), pp. 194−209. https://doi.org/10.1017/9781108587792.011.

Fang, B., Yu, J., Chen, Z., Osman, A.I., Farghali, M., Ihara, I., Hamza, E.H., Rooney, D.W., Yap, P.S., 2023. Artificial intelligence for waste management in smart cities: a review. Environmental Chemistry Letters. https://doi.org/10.1007/s10311-023-01604-3.

Farghali, M., Osman, A.I., Mohamed, I.M.A., Chen, Z., Chen, L., Ihara, I., Yap, P.S., Rooney, D.W., 2023. Strategies to save energy in the context of the energy crisis: a review. Environmental Chemistry Letters. https://doi.org/10.1007/s10311-023-01591-5.

Farooq, M.S., Riaz, S., Abid, A., Umer, T., Zikria, Y.B., 2020. Role of IoT technology in agriculture: a systematic literature review. Electronics 9 (2). https://doi.org/10.3390/electronics9020319.

Fath, B.D., Scharler, U.M., Ulanowicz, R.E., Hannon, B., 2007. Ecological network analysis: network construction. Ecological Modelling 208 (1), 49−55. https://doi.org/10.1016/j.ecolmodel.2007.04.029.

Forrester, J.W., 1968. Industrial dynamics—after the first decade. Management Science 14 (7), 398−415. https://www.jstor.org/stable/2628888.

Gagne, D.J., Christensen, H.M., Subramanian, A.C., Monahan, A.H., 2020. Machine learning for stochastic parameterization: generative adversarial networks in the Lorenz '96 model. Journal of Advances in Modeling Earth Systems 12 (3). https://doi.org/10.1029/2019MS001896.

Gardner, C.J., Bullock, J.M., 2021. In the climate emergency, conservation must become survival ecology. Frontiers in Conservation Science 2. https://doi.org/10.3389/fcosc.2021.659912.

Gašparović, M., Seletković, A., Berta, A., Balenović, I., 2017. The evaluation of photogrammetry-based DSM from low-cost UAV by LiDAR-based DSM. South-East European Forestry 8 (2), 117−125. https://doi.org/10.15177/seefor.17-16.

Gentile, J.H., Harwell, M.A., 2001. Strategies for assessing cumulative ecological risks. Human and Ecological Risk Assessment: An International Journal 7 (2), 239−246. https://doi.org/10.1080/20018091094358.

Ghavami, S.M., Taleai, M., Arentze, T., 2017. An intelligent spatial land use planning support system using socially rational agents. International Journal of Geographical Information Science 31 (5), 1022−1041. https://doi.org/10.1080/13658816.2016.1263306.

Ghosh, S., Singh, A., 2020. The scope of Artificial Intelligence in mankind: a detailed review. Journal of Physics: Conference Series 1531 (1). https://doi.org/10.1088/1742-6596/1531/1/012045.

Goethals, P.L.M., Dedecker, A.P., Gabriels, W., Lek, S., De Pauw, N., 2007. Applications of artificial neural networks predicting macroinvertebrates in freshwaters. Aquatic Ecology 41 (3), 491−508. https://doi.org/10.1007/s10452-007-9093-3.

Gonzalez, C.J., Espinosa, A., Ponte, D., Gibeaux, S., 2021. Smart-IoT platform to monitor microclimate conditions in tropical regions. IOP Conference Series: Earth and Environmental Science 835 (1). https://doi.org/10.1088/1755-1315/835/1/012011.

Gottschalk, T.K., Huettmann, F., Ehlers, M., 2007. Review article: thirty years of analysing and modelling avian habitat relationships using satellite imagery data: a review. International Journal of Remote Sensing 26 (12), 2631−2656. https://doi.org/10.1080/01431160512331338041.

Halevy, A., Norvig, P., Pereira, F., 2009. The unreasonable effectiveness of data. IEEE Intelligent Systems 24 (2), 8−12. https://doi.org/10.1109/MIS.2009.36.

Ham, Y.G., Kim, J.H., Luo, J.J., 2019. Deep learning for multi-year ENSO forecasts. Nature 573 (7775), 568−572. https://doi.org/10.1038/s41586-019-1559-7.

Haupt, S.E., Gagne, D.J., Hsieh, W.W., Krasnopolsky, V., McGovern, A., Marzban, C., Moninger, W., Lakshmanan, V., Tissot, P., Williams, J.K., 2022. The history and practice of AI in the environmental sciences.

Bulletin of the American Meteorological Society 103 (5), E1351–E1370. https://doi.org/10.1175/BAMS-D-20-0234.1.

Hausmann, A., Toivonen, T., Slotow, R., Tenkanen, H., Moilanen, A., Heikinheimo, V., Di Minin, E., 2018. Social media data can be used to understand tourists' preferences for nature-based experiences in protected areas. Conservation Letters 11 (1). https://doi.org/10.1111/conl.12343.

Herbei, M.V., Popescu, C.A., Bertici, R., Smuleac, A., Popescu, G., 2016. Processing and use of satellite images in order to extract useful information in precision agriculture. Bulletin of University of Agricultural Sciences and Veterinary Medicine Cluj-Napoca — Agriculture 73 (2). https://doi.org/10.15835/buasvmcn-agr:12442.

Hill, A.P., Prince, P., Piña Covarrubias, E., Doncaster, C.P., Snaddon, J.L., Rogers, A., 2018. AudioMoth: evaluation of a smart open acoustic device for monitoring biodiversity and the environment. Methods in Ecology and Evolution 9 (5), 1199–1211. https://doi.org/10.1111/2041-210X.12955.

Hochachka, W.M., Fink, D., Hutchinson, R.A., Sheldon, D., Wong, W.K., Kelling, S., 2012. Data-intensive science applied to broad-scale citizen science. Trends in Ecology & Evolution 27 (2), 130–137. https://doi.org/10.1016/j.tree.2011.11.006.

Hodgson, J.C., Mott, R., Baylis, S.M., Pham, T.T., Wotherspoon, S., Kilpatrick, A.D., Raja Segaran, R., Reid, I., Terauds, A., Koh, L.P., 2018. Drones count wildlife more accurately and precisely than humans. Methods in Ecology and Evolution 9 (5), 1160–1167. https://doi.org/10.1111/2041-210X.12974.

Hoeser, T., Kuenzer, C., 2020. Object detection and image segmentation with deep learning on earth observation data: a review-part I: evolution and recent trends. Remote Sensing 12 (10). https://doi.org/10.3390/rs12101667.

Holling, C.S., 1959. Some characteristics of simple types of predation and parasitism. The Canadian Entomologist 91 (7), 385–398. https://doi.org/10.4039/ent91385-7.

Holling, C.S., 1966. The functional response of invertebrate predators to prey density. Memoirs of the Entomological Society of Canada 98 (S48), 5–86. https://doi.org/10.4039/entm9848fv.

Huntingford, C., Jeffers, E.S., Bonsall, M.B., Christensen, H.M., Lees, T., Yang, H., 2019. Machine learning and artificial intelligence to aid climate change research and preparedness. Environmental Research Letters 14 (12). https://doi.org/10.1088/1748-9326/ab4e55.

Ise, T., Oba, Y., 2019. Forecasting climatic trends using neural networks: an experimental study using global historical data. Frontiers in Robotics and AI 6. https://doi.org/10.3389/frobt.2019.00032.

Jaafari, A., Zenner, E.K., Panahi, M., Shahabi, H., 2019. Hybrid artificial intelligence models based on a neuro-fuzzy system and metaheuristic optimization algorithms for spatial prediction of wildfire probability. Agricultural and Forest Meteorology 266–267, 198–207. https://doi.org/10.1016/j.agrformet.2018.12.015.

Jain, P., Coogan, S.C.P., Subramanian, S.G., Crowley, M., Taylor, S., Flannigan, M.D., 2020. A review of machine learning applications in wildfire science and management. Environmental Reviews 28 (4), 478–505. https://doi.org/10.1139/er-2020-0019.

Jayaweera, M., Asaeda, T., 1995. Impacts of environmental scenarios on chlorophyll-a in the management of shallow, eutrophic lakes following biomanipulation: an application of a numerical model. Ecological Engineering 5 (4), 445–468. https://doi.org/10.1016/0925-8574(95)00020-8.

Jha, S.K., Bilalovic, J., Jha, A., Patel, N., Zhang, H., 2017. Renewable energy: present research and future scope of Artificial Intelligence. Renewable and Sustainable Energy Reviews 77, 297–317. https://doi.org/10.1016/j.rser.2017.04.018.

Jha, K., Doshi, A., Patel, P., Shah, M., 2019. A comprehensive review on automation in agriculture using artificial intelligence. Artificial Intelligence in Agriculture 2, 1–12. https://doi.org/10.1016/j.aiia.2019.05.004.

Jørgensen, S.E., 1986. Structural dynamic model. Ecological Modelling, Scope and Limit in the Application of Ecological Models to Environmental Management\3-I\2-IV 31, 90051–90057. https://doi.org/10.1016/0304-3800.

Jørgensen, S.E., 2002. Integration of ecosystem theories: a pattern. Ecology & Environment. https://doi.org/10.1007/978-94-010-0381-0.

Jørgensen, S.E., 2008. Overview of the model types available for development of ecological models. Ecological Modelling 215 (1–3), 3–9. https://doi.org/10.1016/j.ecolmodel.2008.02.041.

Jørgensen, S.E., 2010. A review of recent developments in lake modelling. Ecological Modelling 221 (4), 689–692. https://doi.org/10.1016/j.ecolmodel.2009.10.022.

Jørgensen, S.E., Mejer, H., 1979. A holistic approach to ecological modelling. Ecological Modelling 7 (3), 169–189. https://doi.org/10.1016/0304-3800(79)90068-1.

Joseph, A., Chandra, J., Siddharthan, S., 2021. Genome analysis for precision agriculture using artificial intelligence: a survey. Lecture Notes in Networks and Systems 132, 221–226. https://doi.org/10.1007/978-981-15-5309-7_23.

Jung, J., Maeda, M., Chang, A., Bhandari, M., Ashapure, A., Landivar-Bowles, J., 2021. The potential of remote sensing and artificial intelligence as tools to improve the resilience of agriculture production systems. Current Opinion in Biotechnology 70, 15–22. https://doi.org/10.1016/j.copbio.2020.09.003.

Kaack, L.H., Donti, P.L., Strubell, E., Kamiya, G., Creutzig, F., Rolnick, D., 2022. Aligning artificial intelligence with climate change mitigation. Nature Climate Change 12 (6), 518–527. https://doi.org/10.1038/s41558-022-01377-7.

Kerry, R.G., Montalbo, F.J.P., Das, R., Patra, S., Mahapatra, G.P., Maurya, G.K., Nayak, V., Jena, A.B., Ukhurebor, K.E., Jena, R.C., Gouda, S., Majhi, S., Rout, J.R., 2022. An overview of remote monitoring methods in biodiversity conservation. Environmental Science and Pollution Research 29 (53), 80179–80221. https://doi.org/10.1007/s11356-022-23242-y.

Khaki, S., Wang, L., 2019. Crop yield prediction using deep neural networks. Frontiers in Plant Science 10. https://doi.org/10.3389/fpls.2019.00621.

Khalilpourazari, S., Khalilpourazary, S., Özyüksel Çiftçioğlu, A., Weber, G.W., 2021. Designing energy-efficient high-precision multi-pass turning processes via robust optimization and artificial intelligence. Journal of Intelligent Manufacturing 32 (6), 1621–1647. https://doi.org/10.1007/s10845-020-01648-0.

Kumari, A., Gupta, R., Tanwar, S., Kumar, N., 2020. Blockchain and AI amalgamation for energy cloud management: challenges, solutions, and future directions. Journal of Parallel and Distributed Computing 143, 148–166. https://doi.org/10.1016/j.jpdc.2020.05.004.

Kussul, N., Lavreniuk, M., Skakun, S., Shelestov, A., 2017. Deep learning classification of land cover and crop types using remote sensing data. IEEE Geoscience and Remote Sensing Letters 14 (5), 778–782. https://doi.org/10.1109/lgrs.2017.2681128.

Kwok, R., 2019. AI empowers conservation biology. Nature 567 (7746), 133–134. https://doi.org/10.1038/d41586-019-00746-1.

Langmead, O., McQuatters-Gollop, A., Mee, L.D., Friedrich, J., Gilbert, A.J., Gomoiu, M.T., Jackson, E.L., Knudsen, S., Minicheva, G., Todorova, V., 2009. Recovery or decline of the northwestern Black Sea: a societal choice revealed by socio-ecological modelling. Ecological Modelling 220 (21), 2927–2939. https://doi.org/10.1016/j.ecolmodel.2008.09.011.

Larraondo, P.R., Renzullo, L.J., Van Dijk, A.I.J.M., Inza, I., Lozano, J.A., 2020. Optimization of deep learning precipitation models using categorical binary metrics. Journal of Advances in Modeling Earth Systems 12 (5). https://doi.org/10.1029/2019MS001909.

Leal Filho, W., Wall, T., Rui Mucova, S.A., Nagy, G.J., Balogun, A.L., Luetz, J.M., Ng, A.W., Kovaleva, M., Safiul Azam, F.M., Alves, F., Guevara, Z., Matandirotya, N.R., Skouloudis, A., Tzachor, A., Malakar, K., Gandhi, O., 2022. Deploying artificial intelligence for climate change adaptation. Technological Forecasting and Social Change 180. https://doi.org/10.1016/j.techfore.2022.121662.

Lee, J., Yoo, H.-J., 2021. An overview of energy-efficient hardware accelerators for on-device deep-neural-network training. IEEE Open Journal of the Solid-State Circuits Society 1, 115–128. https://doi.org/10.1109/ojsscs.2021.3119554.

Lehman, J.T., Botkin, D.B., Likens, G.E., 1975. The assumptions and rationales of a computer model of phytoplankton population dynamics. Limnology and Oceanography 20 (3), 343−364. https://doi.org/10.4319/lo.1975.20.3.0343.

Liu, D., Zhang, G., Li, H., Fu, Q., Li, M., Faiz, M.A., Ali, S., Li, T., Imran Khan, M., 2019. Projection pursuit evaluation model of a regional surface water environment based on an Ameliorative Moth-Flame Optimization algorithm. Ecological Indicators 107. https://doi.org/10.1016/j.ecolind.2019.105674.

Liu, D., Liu, C., Tang, Y., Gong, C., 2022. A GA-BP neural network regression model for predicting soil moisture in slope ecological protection. Sustainability 14 (3). https://doi.org/10.3390/su14031386.

Logar, T., Bullock, J., Nemni, E., Bromley, L., Quinn, J.A., Luengo-Oroz, M., 2020. PulseSatellite: a tool using human-AI feedback loops for satellite image analysis in humanitarian contexts. Proceedings of the AAAI Conference on Artificial Intelligence 34 (09), 13628−13629. https://doi.org/10.1609/aaai.v34i09.7101.

Lombardo, P.S., Franz, D.D., 1972. Mathematical Model of Water Quality in Rivers and Impoundments. Hydrocomp Inc.

López-Granados, F., Torres-Sánchez, J., Serrano-Pérez, A., de Castro, A.I., Mesas-Carrascosa, F.-J., Peña, J.-M., 2016. Early season weed mapping in sunflower using UAV technology: variability of herbicide treatment maps against weed thresholds. Precision Agriculture 17 (2), 183−199. https://doi.org/10.1007/s11119-015-9415-8.

Lotka, A.J., 1920. Analytical note on certain rhythmic relations in organic systems. Proceedings of the National Academy of Sciences 6 (7), 410−415. https://doi.org/10.1073/pnas.6.7.410.

Ma, W., Qiu, Z., Song, J., Li, J., Cheng, Q., Zhai, J., Ma, C., 2018. A deep convolutional neural network approach for predicting phenotypes from genotypes. Planta 248 (5), 1307−1318. https://doi.org/10.1007/s00425-018-2976-9.

Manek, A.H., Singh, P.K., 2016. Comparative study of neural network architectures for rainfall prediction. In: Proceedings − 2016 IEEE International Conference on Technological Innovations in ICT for Agriculture and Rural Development, TIAR 2016. Institute of Electrical and Electronics Engineers Inc., India, pp. 171−174. https://doi.org/10.1109/TIAR.2016.7801233.

Mashohor, S., Samsudin, K., Noor, A.M., Rahman, A.R.A., 2008. Evaluation of genetic algorithm based solar tracking system for photovoltaic panels. In: 2008 IEEE International Conference on Sustainable Energy Technologies, ICSET 2008. Malaysia, pp. 269−273. https://doi.org/10.1109/ICSET.2008.4747015.

Mayfield, H., Smith, C., Gallagher, M., Hockings, M., 2017. Use of freely available datasets and machine learning methods in predicting deforestation. Environmental Modelling & Software 87, 17−28. https://doi.org/10.1016/j.envsoft.2016.10.006.

McClure, E.C., Sievers, M., Brown, C.J., Buelow, C.A., Ditria, E.M., Hayes, M.A., Pearson, R.M., Tulloch, V.J.D., Unsworth, R.K.F., Connolly, R.M., 2020. Artificial intelligence meets citizen science to supercharge ecological monitoring. Patterns 1 (7). https://doi.org/10.1016/j.patter.2020.100109.

Medela, A., Cendón, B., González, L., Crespo, R., Nevares, I., 2013. IoT multiplatform networking to monitor and control wineries and vineyards. In: 2013 Future Network and Mobile Summit, FutureNetworkSummit 2013. IEEE Computer Society, Spain.

Meng, X., Gao, X., Li, S., Li, S., Lei, J., 2021. Monitoring desertification in Mongolia based on Landsat images and Google Earth Engine from 1990 to 2020. Ecological Indicators 129. https://doi.org/10.1016/j.ecolind.2021.107908.

Moilanen, A., Wilson, K.A., 2009. Spatial Conservation Prioritization: Quantitative Methods and Computational Tools, Oxford Biology. Oxford University Press.

Morioka, T., Chikami, S., 1986. Basin-wide ecological fate model for management of chemicals hazard. Ecological Modelling 31 (1−4), 267−281. https://doi.org/10.1016/0304-3800(86)90068-2.

Mrówczyńska, M., Sztubecka, M., Skiba, M., Bazan-Krzywoszańska, A., Bejga, P., 2019. The use of artificial intelligence as a tool supporting sustainable development local policy. Sustainability 11 (15). https://doi.org/10.3390/su11154199.

National Research Council, 1999. What is biodiversity? In: Perspectives on Biodiversity: Valuing its Role in an Everchanging World. National Academies Press.

Naveen, S., 2016. Study of IoT: Understanding IoT Architecture, Applications, Issues and Challenges.

Newman, K.B., Buckland, S.T., Morgan, B.J.T., King, R., Borchers, D.L., Cole, D.J., Besbeas, P., Gimenez, O., Thomas, L., 2014. Modelling Population Dynamics Using Closed-Population Abundance Estimates. Springer Science and Business Media LLC, pp. 123–145. https://doi.org/10.1007/978-1-4939-0977-3_6.

Niżetić, S., Djilali, N., Papadopoulos, A., Rodrigues, J.J.P.C., 2019. Smart technologies for promotion of energy efficiency, utilization of sustainable resources and waste management. Journal of Cleaner Production 231, 565–591. https://doi.org/10.1016/j.jclepro.2019.04.397.

Nobre, A.M., Musango, J.K., de Wit, M.P., Ferreira, J.G., 2009. A dynamic ecological–economic modeling approach for aquaculture management. Ecological Economics 68 (12), 3007–3017. https://doi.org/10.1016/j.ecolecon.2009.06.019.

Norouzzadeh, M.S., Nguyen, A., Kosmala, M., Swanson, A., Palmer, M.S., Packer, C., Clune, J., 2018. Automatically identifying, counting, and describing wild animals in camera-trap images with deep learning. Proceedings of the National Academy of Sciences of the United States of America 115 (25), E5716–E5725. https://doi.org/10.1073/pnas.1719367115.

Odum, E., 2013. Ecological Vignettes: Ecological Approaches to Dealing with Human Predicaments. Taylor and Francis, United States, pp. 1–269. https://doi.org/10.4324/9781315079370.

Olomukoro, J.O., Odigie, J.O., 2020. Ecological modelling using artificial neural network for macroinvertebrate prediction in a tropical rainforest river. International Journal of Environment and Waste Management 26 (3), 325. https://doi.org/10.1504/ijewm.2020.10028733.

Osman, A.I., Chen, L., Yang, M., Msigwa, G., Farghali, M., Fawzy, S., Rooney, D.W., Yap, P.S., 2023. Cost, environmental impact, and resilience of renewable energy under a changing climate: a review. Environmental Chemistry Letters 21 (2), 741–764. https://doi.org/10.1007/s10311-022-01532-8.

Ouadah, A., Zemmouchi-Ghomari, L., Salhi, N., 2022. Selecting an appropriate supervised machine learning algorithm for predictive maintenance. International Journal of Advanced Manufacturing Technology 119 (7–8), 4277–4301. https://doi.org/10.1007/s00170-021-08551-9.

Partel, V., Charan Kakarla, S., Ampatzidis, Y., 2019. Development and evaluation of a low-cost and smart technology for precision weed management utilizing artificial intelligence. Computers and Electronics in Agriculture 157, 339–350. https://doi.org/10.1016/j.compag.2018.12.048.

Pasqual, G.M., 1994. Development of an expert system for the identification and control of weeds in wheat, triticale, barley and oat crops. Computers and Electronics in Agriculture 10 (2), 117–134. https://doi.org/10.1016/0168-1699(94)90016-7.

Pearl, J., 2019. The seven tools of causal inference, with reflections on machine learning. Communications of the ACM 62 (3), 54–60. https://doi.org/10.1145/3241036.

Pecl, G.T., Stuart-Smith, J., Walsh, P., Bray, D.J., Kusetic, M., Burgess, M., Frusher, S.D., Gledhill, D.C., George, O., Jackson, G., Keane, J., Martin, V.Y., Nursey-Bray, M., Pender, A., Robinson, L.M., Rowling, K., Sheaves, M., Moltschaniwskyj, N., 2019. Redmap Australia: challenges and successes with a large-scale citizen science-based approach to ecological monitoring and community engagement on climate change. Frontiers in Marine Science 6. https://doi.org/10.3389/fmars.2019.00349.

Pérez-Ortiz, M., Peña, J.M., Gutiérrez, P.A., Torres-Sánchez, J., Hervás-Martínez, C., López-Granados, F., 2016. Selecting patterns and features for between- and within- crop-row weed mapping using UAV-imagery. Expert Systems with Applications 47, 85–94. https://doi.org/10.1016/j.eswa.2015.10.043.

Pichler, M., Hartig, F., 2023. Machine learning and deep learning—a review for ecologists. Methods in Ecology and Evolution 14 (4), 994–1016. https://doi.org/10.1111/2041-210X.14061.

Porté, A., Bartelink, H.H., 2002. Modelling mixed forest growth: a review of models for forest management. Ecological Modelling 150 (1–2), 141–188. https://doi.org/10.1016/S0304-3800(01)00476-8.

Pratyush Reddy, K.S., Roopa, Y.M., Kovvada Rajeev, L.N., Nandan, N.S., 2020. IoT based smart agriculture using machine learning. In: Proceedings of the 2nd International Conference on Inventive Research in Computing Applications, ICIRCA 2020. Institute of Electrical and Electronics Engineers Inc., India, pp. 130–134. https://doi.org/10.1109/ICIRCA48905.2020.9183373.

Putra, D.P., Bimantio, M.P., Sahfitra, A.A., Suparyanto, T., Pardamean, B., 2020. Simulation of availability and loss of nutrient elements in land with android-based fertilizing applications. In: Proceedings of 2020 International Conference on Information Management and Technology, ICIMTech 2020. Institute of Electrical and Electronics Engineers Inc., Indonesia, pp. 312–317. https://doi.org/10.1109/ICIMTech50083.2020.9211268.

Raftery, A.E., Givens, G.H., Zeh, J.E., 1995. Inference from a deterministic population dynamics model for bowhead whales. Journal of the American Statistical Association 90 (430), 402–416. https://doi.org/10.1080/01621459.1995.10476529.

Raj, M., Gupta, S., Chamola, V., Elhence, A., Garg, T., Atiquzzaman, M., Niyato, D., 2021. A survey on the role of Internet of Things for adopting and promoting Agriculture 4.0. Journal of Network and Computer Applications 187. https://doi.org/10.1016/j.jnca.2021.103107.

Rajaee, T., Khani, S., Ravansalar, M., 2020. Artificial intelligence-based single and hybrid models for prediction of water quality in rivers: a review. Chemometrics and Intelligent Laboratory Systems 200, 103978. https://doi.org/10.1016/j.chemolab.2020.103978.

Rajesh, D., 2011. Application of spatial data mining for agriculture. International Journal of Computer Applications 15 (2), 7–9. https://doi.org/10.5120/1922-2566.

Rasp, S., Pritchard, M.S., Gentine, P., 2018. Deep learning to represent subgrid processes in climate models. Proceedings of the National Academy of Sciences 115 (39), 9684–9689. https://doi.org/10.1073/pnas.1810286115.

Ridwan, W.M., Sapitang, M., Aziz, A., Kushiar, K.F., Ahmed, A.N., El-Shafie, A., 2021. Rainfall forecasting model using machine learning methods: case study Terengganu, Malaysia. Ain Shams Engineering Journal 12 (2), 1651–1663. https://doi.org/10.1016/j.asej.2020.09.011.

Robinson, C., Dilkina, B., 2018. A machine learning approach to modeling human migration. In: Proceedings of the 1st ACM SIGCAS Conference on Computing and Sustainable Societies, COMPASS 2018. Association for Computing Machinery, Inc, United States. https://doi.org/10.1145/3209811.3209868.

Rolnick, D., Donti, P.L., Kaack, L.H., Kochanski, K., Lacoste, A., Sankaran, K., Ross, A.S., Milojevic-Dupont, N., Jaques, N., Waldman-Brown, A., Luccioni, A.S., Maharaj, T., Sherwin, E.D., Mukkavilli, S.K., Kording, K.P., Gomes, C.P., Ng, A.Y., Hassabis, D., Platt, J.C., Creutzig, F., Chayes, J., Bengio, Y., 2023. Tackling climate change with machine learning. ACM Computing Surveys 55 (2). https://doi.org/10.1145/3485128.

Rustia, D.J.A., Lin, C.E., Chung, J.Y., Zhuang, Y.J., Hsu, J.C., Lin, T.T., 2020. Application of an image and environmental sensor network for automated greenhouse insect pest monitoring. Journal of Asia-Pacific Entomology 23 (1), 17–28. https://doi.org/10.1016/j.aspen.2019.11.006.

Salski, A., 1992. Fuzzy knowledge-based models in ecological research. Ecological Modelling 63 (1–4), 103–112. https://doi.org/10.1016/0304-3800(92)90064-L.

Salski, A., 2006. Ecological applications of fuzzy logic. Ecological Informatics: Scope, Techniques and Applications 3–14. https://doi.org/10.1007/3-540-28426-5_1.

Satyanarayana, G.V., Mazaruddin, S., 2013. Wireless sensor based remote monitoring system for agriculture using ZigBee and GPS. In: Presented at the Conference on Advances in Communication and Control Systems (CAC2S 2013). Atlantis Press, pp. 110–114.

Schwalm, C.R., Glendon, S., Duffy, P.B., 2020. RCP8.5 tracks cumulative CO_2 emissions. Proceedings of the National Academy of Sciences 117 (33), 19656–19657. https://doi.org/10.1073/pnas.2007117117.

Seelan, S.K., Laguette, S., Casady, G.M., Seielstad, G.A., 2003. Remote sensing applications for precision agriculture: a learning community approach. IKONOS Fine Spatial Resolution Land Observation 88 (1), 157–169. https://doi.org/10.1016/j.rse.2003.04.007.

Shi, Y., Yuan, H., Liang, A., Zhang, C., 2008. Analysis and testing of weed real-time identification based on neural network. IFIP International Federation for Information Processing 1095–1101. https://doi.org/10.1007/978-0-387-77253-0_43.

Shrestha, M., Manandhar, S., Shrestha, S., 2020. Forecasting water demand under climate change using artificial neural network: a case study of Kathmandu Valley, Nepal. Water Supply 20 (5), 1823–1833. https://doi.org/10.2166/ws.2020.090.

Shugart, H.H., 1984. A Theory of Forest Dynamics: The Ecological Implications of Forest Succession Models. Springer-Verlag, p. 278.

Sicat, R.S., Carranza, E.J.M., Nidumolu, U.B., 2005. Fuzzy modeling of farmers' knowledge for land suitability classification. Agricultural Systems 83 (1), 49–75. https://doi.org/10.1016/j.agsy.2004.03.002.

Silvestro, D., Goria, S., Sterner, T., Antonelli, A., 2022. Improving biodiversity protection through artificial intelligence. Nature Sustainability 5 (5), 415–424. https://doi.org/10.1038/s41893-022-00851-6.

Singh, K.K., Frazier, A.E., 2018. A meta-analysis and review of unmanned aircraft system (UAS) imagery for terrestrial applications. International Journal of Remote Sensing 39 (15–16), 5078–5098. https://doi.org/10.1080/01431161.2017.1420941.

Sønderby, C.K., Espeholt, L., Heek, J., Dehghani, M., Oliver, A., Salimans, T., Agrawal, S., Hickey, J., Kalchbrenner, N., 2020. Metnet: a neural weather model for precipitation forecasting. arXiv. https://arxiv.org.

Soumyalatha, 2016. Study of IoT: understanding IoT architecture, applications, issues and challenges. In: 1st International Conference on Innovations in Computing & Net-Working (ICICN16), p. 478.

Steenweg, R., Hebblewhite, M., Kays, R., Ahumada, J., Fisher, J.T., Burton, C., Townsend, S.E., Carbone, C., Rowcliffe, J.M., Whittington, J., Brodie, J., Royle, J.A., Switalski, A., Clevenger, A.P., Heim, N., Rich, L.N., 2017. Scaling-up camera traps: monitoring the planet's biodiversity with networks of remote sensors. Frontiers in Ecology and the Environment 15 (1), 26–34. https://doi.org/10.1002/fee.1448.

Stefan, H.G., Jr, Dortch, M.S., 1989. Army Engineer Waterways Experiment Station Vicksburg MS Environmental LabFormulation of Water Quality Models for Streams, Lakes and Reservoirs: Modeler's Perspective. https://apps.dtic.mil/sti/citations/ADA211198.

Streeter, H.W., Phelps, E.B., 1958. A Study of the Pollution and Natural Purification of the Ohio River. United States Public Health Service, Washington.

Sugai, L.S.M., Silva, T.S.F., Ribeiro, J.W., Llusia, D., 2019. Terrestrial passive acoustic monitoring: review and perspectives. BioScience 69 (1), 5–11. https://doi.org/10.1093/biosci/biy147.

Swaminathan, B., Palani, S., Vairavasundaram, S., Kotecha, K., Kumar, V., 2023. IoT-driven artificial intelligence technique for fertilizer recommendation model. IEEE Consumer Electronics Magazine 12 (2), 109–117. https://doi.org/10.1109/mce.2022.3151325.

Tekwa, E., Gonzalez, A., Zurell, D., O'Connor, M., 2023. Detecting and attributing the causes of biodiversity change: needs, gaps and solutions. Philosophical Transactions of the Royal Society B: Biological Sciences 378 (1881). https://doi.org/10.1098/rstb.2022.0181.

The IUCN Red List of Threatened Species, 2023. https://www.iucnredlist.org. (Accessed 27 June 2023).

Thomas, M.K., Fontana, S., Reyes, M., Kehoe, M., Pomati, F., 2018. The predictability of a lake phytoplankton community, over time-scales of hours to years. Ecology Letters 21 (5), 619–628. https://doi.org/10.1111/ele.12927.

Tien Bui, D., Pham, B.T., Nguyen, Q.P., Hoang, N.D., 2016. Spatial prediction of rainfall-induced shallow landslides using hybrid integration approach of Least-Squares Support Vector Machines and differential evolution optimization: a case study in Central Vietnam. International Journal of Digital Earth 9 (11), 1077–1097. https://doi.org/10.1080/17538947.2016.1169561.

Tien Bui, D., Bui, Q.T., Nguyen, Q.P., Pradhan, B., Nampak, H., Trinh, P.T., 2017. A hybrid artificial intelligence approach using GIS-based neural-fuzzy inference system and particle swarm optimization for forest fire susceptibility modeling at a tropical area. Agricultural and Forest Meteorology 233, 32–44. https://doi.org/10.1016/j.agrformet.2016.11.002.

Toma, J.J., 2013. Limnological study of Dokan, Derbendikhan and Duhok lakes, Kurdistan region of Iraq. Open Journal of Ecology 03 (01), 23–29. https://doi.org/10.4236/oje.2013.31003.

Torney, C.J., Lloyd-Jones, D.J., Chevallier, M., Moyer, D.C., Maliti, H.T., Mwita, M., Kohi, E.M., Hopcraft, G.C., 2019. A comparison of deep learning and citizen science techniques for counting wildlife in aerial survey images. Methods in Ecology and Evolution 10 (6), 779–787. https://doi.org/10.1111/2041-210X.13165.

Torres, V.A.M.F., Jaimes, B.R.A., Ribeiro, E.S., Braga, M.T., Shiguemori, E.H., Velho, H.F.C., Torres, L.C.B., Braga, A.P., 2020. Combined weightless neural network FPGA architecture for deforestation surveillance and visual navigation of UAVs. Engineering Applications of Artificial Intelligence 87. https://doi.org/10.1016/j.engappai.2019.08.021.

Trouille, L., Lintott, C.J., Fortson, L.F., 2019. Citizen science frontiers: efficiency, engagement, and serendipitous discovery with human–machine systems. Proceedings of the National Academy of Sciences 116 (6), 1902–1909. https://doi.org/10.1073/pnas.1807190116.

Tuia, D., Kellenberger, B., Beery, S., Costelloe, B.R., Zuffi, S., Risse, B., Mathis, A., Mathis, M.W., van Langevelde, F., Burghardt, T., Kays, R., Klinck, H., Wikelski, M., Couzin, I.D., van Horn, G., Crofoot, M.C., Stewart, C.V., Berger-Wolf, T., 2022. Perspectives in machine learning for wildlife conservation. Nature Communications 13 (1). https://doi.org/10.1038/s41467-022-27980-y.

Turner, R.K., van den Bergh, J.C.J.M., Söderqvist, T., Barendregt, A., van der Straaten, J., Maltby, E., van Ierland, E.C., 2000. Ecological-economic analysis of wetlands: scientific integration for management and policy. Ecological Economics 35 (1), 7–23. https://doi.org/10.1016/S0921-8009(00)00164-6.

Van Broekhoven, E., Adriaenssens, V., De Baets, B., 2007. Interpretability-preserving genetic optimization of linguistic terms in fuzzy models for fuzzy ordered classification: an ecological case study. International Journal of Approximate Reasoning 44 (1), 65–90. https://doi.org/10.1016/j.ijar.2006.03.003.

van Dijk, M., Morley, T., Rau, M.L., Saghai, Y., 2021. A meta-analysis of projected global food demand and population at risk of hunger for the period 2010–2050. Nature Food 2 (7), 494–501. https://doi.org/10.1038/s43016-021-00322-9.

Van Donk, E., Gulati, R.D., Grimm, M.P., Grimm, M.P., 1989. Food web manipulation in Lake Zwemlust: positive and negative effects during the first two years. Hydrobiological Bulletin 23 (1), 19–34. https://doi.org/10.1007/BF02286424.

van Kranenburg, R., Bassi, A., 2012. IoT challenges. Communications in Mobile Computing 1 (1). https://doi.org/10.1186/2192-1121-1-9.

Verma, M., 2018. Artificial intelligence and its scope in different areas with special reference to the field of education. Open Submission 3, 5–10.

Vinuesa, R., Azizpour, H., Leite, I., Balaam, M., Dignum, V., Domisch, S., Felländer, A., Langhans, S.D., Tegmark, M., Fuso Nerini, F., 2020. The role of artificial intelligence in achieving the Sustainable Development Goals. Nature Communications 11 (1). https://doi.org/10.1038/s41467-019-14108-y.

Volk, M., Hirschfeld, J., Dehnhardt, A., Schmidt, G., Bohn, C., Liersch, S., Gassman, P.W., 2008. Integrated ecological-economic modelling of water pollution abatement management options in the Upper Ems River Basin. Ecological Economics 66 (1), 66–76. https://doi.org/10.1016/j.ecolecon.2008.01.016.

Volterra, V., 1926. Fluctuations in the abundance of a species considered mathematically. Nature 118 (2972), 558–560. https://doi.org/10.1038/118558a0.

Wangersky, P.J., 1978. Lotka-Volterra population models. Annual Review of Ecology and Systematics 9 (1), 189–218. https://doi.org/10.1146/annurev.es.09.110178.001201.

Wei, M.C.F., Maldaner, L.F., Ottoni, P.M.N., Molin, J.P., 2020. Carrot yield mapping: a precision agriculture approach based on machine learning. AI 1 (2), 229–241. https://doi.org/10.3390/ai1020015.

Weinstein, B.G., 2018. A computer vision for animal ecology. Journal of Animal Ecology 87 (3), 533–545. https://doi.org/10.1111/1365-2656.12780.

Witmer, G.W., 2005. Wildlife population monitoring: some practical considerations. Wildlife Research 32 (3). https://doi.org/10.1071/wr04003.

Xenochristou, M., Kapelan, Z., 2020. An ensemble stacked model with bias correction for improved water demand forecasting. Urban Water Journal 17 (3), 212–223. https://doi.org/10.1080/1573062X.2020.1758164.

Xiang, X., Li, Q., Khan, S., Khalaf, O.I., 2021. Urban water resource management for sustainable environment planning using artificial intelligence techniques. Environmental Impact Assessment Review 86, 106515. https://doi.org/10.1016/j.eiar.2020.106515.

Yang, C.-H., 2022. How artificial intelligence technology affects productivity and employment: firm-level evidence from Taiwan. Research Policy 51 (6). https://doi.org/10.1016/j.respol.2022.104536.

Yang, M., Chen, L., Msigwa, G., Tang, K.H.D., Yap, P.S., 2022. Implications of COVID-19 on global environmental pollution and carbon emissions with strategies for sustainability in the COVID-19 era. Science of the Total Environment 809. https://doi.org/10.1016/j.scitotenv.2021.151657.

Yin, X., Li, J., Kadry, S.N., Sanz-Prieto, I., 2021. Artificial intelligence assisted intelligent planning framework for environmental restoration of terrestrial ecosystems. Environmental Impact Assessment Review 86. https://doi.org/10.1016/j.eiar.2020.106493.

Zhang, C., Kovacs, J.M., 2012. The application of small unmanned aerial systems for precision agriculture: a review. Precision Agriculture 13 (6), 693–712. https://doi.org/10.1007/s11119-012-9274-5.

Zhang, J., Jørgensen, S.E., Tan, C.O., Beklioglu, M., 2003. A structurally dynamic modelling-Lake Mogan, Turkey as a case study. Ecological Modelling 164, 51–56. https://doi.org/10.1016/S0304-3800(03)00051-6.

Zhang, X., Shu, K., Rajkumar, S., Sivakumar, V., 2021. Research on deep integration of application of artificial intelligence in environmental monitoring system and real economy. Environmental Impact Assessment Review 86, 106499. https://doi.org/10.1016/j.eiar.2020.106499.

Zhao, P., Gao, Y., Sun, X., 2022. How does artificial intelligence affect green economic growth?—evidence from China. Science of the Total Environment 834, 155306. https://doi.org/10.1016/j.scitotenv.2022.155306.

Zheng, G., Li, X., Zhang, R.H., Liu, B., 2020. Purely satellite data-driven deep learning forecast of complicated tropical instability waves. Science Advances 6 (29). https://doi.org/10.1126/sciadv.aba1482.

Zubairi, J.A., 2009. Applications of modern high performance networks. 12. In: Applications of Modern High Performance Networks. Bentham Science Publishers Ltd., United States. https://doi.org/10.2174/978160 80507721090101

Artificial intelligence in marine biology

14

Gulustan Dogan[1], Doorva Vaidya[1], Megdalia Bromhal[1] and Nelofar Banday[2]

[1]*University of North Carolina Wilmington, Computer Science Department, Wilmington, NC, United States;* [2]*Sher-e-Kashmir University of Agricultural Sciences and Technology of Kashmir, Srinagar, Jammu and Kashmir, India*

Introduction

Concerning marine science, artificial intelligence (AI) offers scientists an opportunity to streamline research and promote newfound discoveries in the fields of marine biology and oceanography. Numerous applications of AI in marine science now exist because AI has emerged as a powerful tool in advancing research and exploration. For example, AI can make predictions from the vast amount of data available from sources such as buoys, sensors, satellites, and underwater images and analyze the information. Oceanographic forecasting, species monitoring, coral reef assessment, and much more are possible from the marriage between AI and this abundant marine and oceanographic information.

Some applications of AI in marine science include identifying and classifying marine species with computer vision. The algorithms can analyze the visual data from the images and videos captured from underwater cameras or drones, and then these algorithms can identify different species. Thus, AI aids marine biologists in studying biodiversity, conservation, and ecological studies. Additionally, AI can take acoustic data and analyze these underwater recordings of mammal vocalizations. Machine learning (ML) algorithms can then identify species-specific vocalizations, track behavioral patterns, and estimate population sizes. In this way, AI could help scientists collect information on the effects of human activities on the oceans and the creatures that live there.

Due to an overall warming planet, coral reefs must increasingly deal with rising sea temperatures, and unfortunately, most coral reefs as of now cannot thrive—or sometimes live at all—with such stressors. In this area, AI can monitor and assess coral reef health by analyzing satellite and underwater imagery. The algorithms can detect coral bleaching, track coral growth, and classify different coral species. This enables scientists to analyze reef health, biodiversity, and composition on a large scale, greatly contributing to the conservation of coral reef ecosystems.

Furthermore, the combination of robotics and AI has brought forth marine robots using AI, and autonomous underwater robots (AUVs) hold the focus of scientific research currently due to their multitude of possible applications. AUVs can identify species, collect data, and explore underwater ecosystems, thereby enriching research and helping scientists understand the ocean. Especially for remote marine environments where human divers could not easily collect data, AUVs provide valuable opportunities.

A Biologist's Guide to Artificial Intelligence. **https://doi.org/10.1016/B978-0-443-24001-0.00014-2**

In this paper, we will explore an overview of applications of AI in marine science and oceanography, particularly regarding computer vision, ML, predictive modeling, and AI-enabled marine robotics. Even with this short list of examples, it is clear AI is a useful tool and will continue to help marine scientists expand what is known about the ocean and its creatures (Dong et al., 2022; Salman, 2023). We will discuss the challenges and future directions of AI in marine science, with further integration of AI into marine science.

Marine biology, a quick overview

Marine biology is the scientific study of organisms, ecosystems, and processes occurring in the world's oceans, seas, and other saltwater environments (Karleskint et al., 2012). It is a branch of biology that focuses on understanding marine life and the interactions between organisms and their marine habitats. Marine biologists investigate a wide range of topics, including the diversity and distribution of marine species, their adaptations to the marine environment, their ecological roles and interactions, and the functioning of marine ecosystems. They study various organisms, from microorganisms and plankton to fish, marine mammals, and even deep-sea organisms. Marine biology encompasses various sub-disciplines, such as marine ecology, marine physiology, marine genetics, marine microbiology, and marine conservation. It involves fieldwork, laboratory experiments, and data analysis to gather knowledge about marine organisms and their habitats.

Some key areas of study within marine biology include:

- **Biodiversity:** Identifying and cataloging marine species, studying their evolutionary relationships, and understanding patterns of species diversity in different marine habitats. Scientists estimate that there are 10−100 million species yet to be discovered which emphasizes the importance of this field (Bouchet, 2006).
- **Ecology:** Investigating the interactions between organisms and their environment, including predator−prey relationships, competition, symbiotic associations, and the flow of energy and nutrients within marine ecosystems (Tait et al., 1998).
- **Oceanography:** Examining physical and chemical aspects of the ocean, such as water temperature, salinity, currents, and the impacts of climate change on marine ecosystems (Lalli and Parsons, 1997).
- **Conservation and management:** Working toward the conservation and sustainable management of marine resources, protecting vulnerable species and habitats, and addressing issues such as over-fishing, habitat destruction, pollution, and climate change impacts (Lotze, 2021).

Marine biology plays a crucial role in understanding and addressing the challenges and complexities of marine environments. It contributes to our knowledge of marine biodiversity, helps to inform conservation and management strategies, and sheds light on the interconnectedness between marine ecosystems and the well-being of our planet.

Big data and marine biology

Big data in marine biology refers to the large and complex datasets that are generated from various sources in the field of marine biology. With advancements in technology and data collection techniques, researchers now have access to vast amounts of data related to marine ecosystems, species distribution, oceanographic parameters, and more. Big data is being generated by all spheres of life (Hamadani et al., 2023), and marine biology is no exception. The storage of these data is also a challenge, and scientists are evaluating new ways of storing these data (Hamadani et al., 2022). The application of big data analytics in marine biology allows scientists to analyze and interpret these massive datasets to gain valuable insights into the dynamics of marine ecosystems, species interactions, and environmental changes. It provides a comprehensive view of marine biodiversity, helping researchers understand patterns and trends over time and space. The growth of data collection systems in Marine Sciences has led also to the development of OBI or Ocean Biodiversity Informatics (OBI) (Costello and Vanden Berghe, 2006). OBI refers to the application of computer technologies in managing marine biodiversity information. It encompasses various tasks such as data capture, storage, search, retrieval, visualization, mapping, modeling, analysis, and publication. OBI enables scientists and researchers to efficiently organize and utilize the wealth of information related to oceanic ecosystems and their diverse species. By leveraging advanced computational tools, OBI plays a crucial role in enhancing our understanding of marine biodiversity and supporting conservation efforts.

Data generation

It is now becoming increasingly possible to generate big data in marine biology is generated through various sources and techniques. This is only expected to increase in the future, and a brief list of ways in which big data is being generated in marine biology today is given below:

- **Sensor networks:** Sensor networks deployed in the ocean collect vast amounts of data on various environmental parameters. These sensors include devices for measuring temperature, salinity, dissolved oxygen, pH, and other water quality parameters. They may also include acoustic sensors for studying marine mammal populations or underwater cameras for visual observations. These sensor networks provide continuous and high-resolution data, generating large datasets over time while collecting broad-ranging data about biotic, abiotic, static, and dynamic ocean variables (Goddijn-Murphy et al., 2021). Acoustic technologies are used to monitor marine mammal populations, fish abundance, and behavior. Hydrophones and other underwater acoustic sensors capture sounds and signals emitted by marine organisms. These acoustic data provide insights into the distribution, migration, and communication patterns of marine species. Fig. 14.1 gives a quick overview of the ML methodology starting at data collection.
- **Satellite remote sensing:** Satellites equipped with sensors capture imagery and data related to oceanographic parameters such as sea surface temperature, chlorophyll concentration, ocean color, as well as ocean biota (Chauhan and Raman, 2017). These remote sensing techniques provide a synoptic view of large oceanic areas, allowing for the generation of extensive datasets on a global scale (Hogan et al., 2022). Remote sensing is being used for marine life and sea birds as well (Goddijn-Murphy et al., 2021).
- **Autonomous underwater vehicles (AUVs) and Gliders:** AUVs and gliders are autonomous underwater vehicles that can be deployed for extended periods to collect data on water properties, currents, and other variables. Equipped with various sensors, these vehicles traverse the ocean, collecting data at different depths and locations. The continuous data collection capability of AUVs and gliders contributes significantly to big data generation (Stutters et al., 2008).

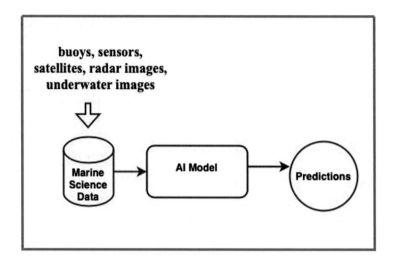

FIGURE 14.1

Marine science artificial intelligence (AI) model.

- **Citizen science and crowdsourcing:** Citizen science initiatives engage the public in data collection. Marine biology projects often involve citizen scientists collecting data on species sightings, coastal observations, or marine debris. These crowdsourced datasets contribute to big data generation, providing valuable information on species distribution, phenology, and ecological trends. For example, globally citizen scientists are coming together to understand the impact of plastics on our environment (Zettler et al., 2017).
- **Genomic sequencing:** The advancement of genomic sequencing technologies has led to the generation of large genomic datasets in marine biology (Ribeiro et al., 2017). Researchers sequence the genomes of marine organisms to study their genetic diversity, evolutionary relationships, and adaptations to different environments. Genomic data contribute to our understanding of marine biodiversity and aid in conservation efforts. In recent years, a novel method called environmental DNA (eDNA) meta-barcoding has emerged as an innovative approach to identify individuals, species, and communities, monitor their movements and distributions, and investigate biological diversity over time and space (Taberlet et al., 2018).

Data repositories and monitoring programs: Numerous data repositories and monitoring programs collect and store data from various sources, such as research expeditions, long-term monitoring stations, and collaborations. These repositories include databases of species occurrences, oceanographic data, satellite imagery, and more. The aggregation of these datasets creates comprehensive and accessible resources for big data analysis.

Overall, the combination of advanced technologies, collaborative efforts, and data-driven research in marine biology generates large and diverse datasets. These datasets enable scientists to explore and uncover patterns, relationships, and trends that contribute to our understanding of marine ecosystems and inform conservation and management strategies. It should be noted that AI algorithms work with data, and without enough data, it is impossible to build AI models.

Data preprocessing

Having enough data is the first step, but data have to be preprocessed to be ready to use in AI models. Data preprocessing is an important step in getting the collected data ready for AI models. Data processing includes cleaning the mistaken fields, handling missing fields, and merging multiple datasets. AI techniques can help automatically identify and correct these issues, making the data more accurate and reliable. Another challenge is dealing with missing data, which can affect predictions. Researchers use different methods to estimate missing values based on the available data. In marine science, data often come from different sources, and merging these data are necessary to get a complete picture of the oceans. AI algorithms can help combine datasets by matching variables or using advanced techniques, which helps researchers gain better insights into marine science.

Data quality and integrity

Algorithms such as clustering methods, statistical techniques, or anomaly detection algorithms can analyze real-time data streams and identify erroneous or outlier data points. This ensures that only accurate and reliable data are used in forecasting models, maintaining data integrity and improving the quality of predictions, which are essential to marine science fields such as oceanographic forecasting, climate modeling, ecosystem monitoring, and water quality assessment (Hogan et al., 2022). Furthermore, high-quality data practices and documentation are important for the reproducibility of marine science research.

Artificial intelligence in marine science

Artificial intelligence and ML have revolutionized marine science by enhancing our ability to analyze and interpret complex oceanic data. Through AI and ML algorithms, vast amounts of marine data, such as satellite imagery, oceanographic measurements, and underwater sensor data, can be efficiently processed, leading to improved understanding of marine ecosystems and their dynamics. These technologies enable the identification of patterns, correlations, and anomalies in data, facilitating the prediction of marine phenomena, such as harmful algal blooms or ocean currents. AI and ML also contribute to species identification, habitat mapping, and ecosystem modeling, aiding conservation efforts and sustainable resource management. By harnessing the power of AI and ML, marine scientists can make informed decisions and advance our knowledge of the oceans, leading to better protection and preservation of marine environments. Computer vision in marine science utilizes image and video processing techniques to extract valuable insights from underwater data, enabling researchers to study marine life, monitor habitats, and assess environmental changes. It plays a crucial role in advancing our knowledge of the oceans and informing conservation efforts.

Deep learning computer vision techniques analyze images and videos captured by underwater cameras or drones to automatically identify and classify various marine species, aiding in biodiversity assessments and conservation efforts. Additionally, deep learning algorithms enable the monitoring of marine mammal populations by processing acoustic data, identifying species-specific vocalizations, and tracking behavioral patterns, providing valuable insights for research, conservation, and management strategies. Deep learning methods have also been used to identify plant diseases from pictures (Aoki et al., 2022).

One area that deep learning techniques can be used is oceanic forecasting. Salman (2023) mention in their paper that AI models, including CNNs or recurrent neural networks (RNNs), can be trained on historical data and weather patterns to identify precursors and patterns indicative of extreme oceanographic events, aiding in their prediction. Similarly, image recognition algorithms or RNNs can be used to quickly analyze satellite imagery, radar data, and buoy observations to provide up-to-date information on current ocean conditions, supporting nowcasting and short-term forecasting. An overview of a CNN is given in (Fig. 14.2).

Besides deep learning methods, ML algorithms can be used in various domains as well. In a study done by Dogan et al., buoy data have been used to predict ocean wave conditions (Dogan et al., 2021).

Species identification

Artificial intelligence is used to develop automated systems for species identification and classification. Computer vision algorithms can analyze images and videos of marine organisms, such as fish, corals, and plankton, enabling quicker and more accurate identification compared to manual methods.

Computer vision models can be trained to detect and localize specific marine species or objects of interest within images or videos. Through deep learning techniques like convolutional neural networks (CNNs), these models learn to recognize patterns and distinctive features that characterize different species. By identifying and localizing the presence of marine species in images, researchers can efficiently collect data for population assessments and ecological studies.

Using geospatial software such as ArcGIS, most previous studies of detecting marine animals using satellite imagery have detected animals by manually scanning through imagery. However, this manual detection process is time-consuming and is a limiting factor for generating near-real-time detections. Replacing these techniques with a semi-automated ML approach has the potential to rapidly accelerate our ability to process large volumes of imagery using effective models like CNNs (Khan et al., 2023).

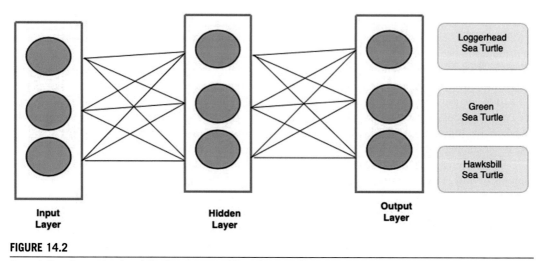

FIGURE 14.2

Convolutional neural network for image recognition of sea turtles.

Previous studies on marine animal detection using satellite imagery have primarily relied on manual scanning methods, often utilizing geospatial software like ArcGIS. However, this manual process is time-consuming. To overcome this limitation, a semi-automated ML approach using effective models like CNNs holds great potential. Using AI image recognition techniques can expedite the processing of large volumes of imagery.

In a study of the applications of deep learning, a subset of ML, in marine science, De La Houssaye et al. (2019) found that Keras and TensorFlow worked quite effectively in predicting ocean sediment data. Both Keras and TensorFlow are open-sourced software libraries for ML, and an increasing number of AUVs run on data from these libraries (Allken et al., 2019; Logares et al., 2021). However, as Logares et al. (2021) notes, science has a long way to go to better apply deep learning to marine science. Granted, ML is excitingly revolutionizing every branch of science, yet there is still much to consider and add to the learning networks to better help scientists understand marine and oceano-graphic sciences (Logares et al., 2021).

Computer vision techniques also aid in biodiversity assessments by automatically processing large volumes of visual data. These algorithms can identify and quantify the presence of various species in images or videos, providing valuable information about species richness, abundance, and distribution. Biodiversity assessments using computer vision contribute to understanding ecosystem health, tracking changes over time, and informing conservation and management decisions.

Monitoring of marine life

Monitoring marine life using AI involves the use of advanced algorithms and technologies to analyze data from various sources such as underwater images, videos, acoustic recordings (Souza et al., 2023), and environmental sensors. AI can identify and classify different species, track their movements and behavior, and provide valuable insights into population dynamics and habitat changes. By automating data collection and analysis, AI enables researchers to better understand and conserve marine eco-systems, contributing to the sustainable management of our oceans (Ditria et al., 2022). Monitoring and prediction systems are not just important for domestic livestock but also for marine life (Merz, 2001).

Mammal monitoring

Artificial intelligence algorithms can process acoustic data, such as underwater recordings of marine mammals (Hamadani and Ganai, 2022), to identify and track species, estimate population sizes, and monitor their behavior (Caruso et al., 2020). This technology aids in conservation efforts and helps understand the impact of human activities on marine mammal populations. Bio-logging is also becoming increasingly useful for mammal monitoring and tracking (Ropert-Coudert and Wilson, 2005).

Machine learning algorithms, such as support vector machines (SVMs), random forests, or neural networks, can be trained on acoustic data to classify and identify marine mammal vocalizations. These algorithms learn patterns and features from labeled data, enabling automated species identification and classification. Hidden Markov Models (HMMs) are statistical models that can analyze temporal patterns in acoustic data. They are often used to recognize and classify different vocalizations emitted by marine mammals. HMMs can identify unique features in vocalizations and track the presence of specific species or behavioral states. Clustering algorithms, such as k-means clustering or hierarchical clustering, can group similar vocalizations based on acoustic features. These algorithms aid in identifying and separating different species' vocalizations, enabling population estimation and tracking changes in vocal behavior.

Coral reef monitoring

Artificial intelligence algorithms are utilized to analyze satellite imagery and underwater images to assess the health and condition of coral reefs (González-Rivero et al., 2020). These algorithms can detect coral bleaching, monitor coral growth, and track changes in reef structure, assisting in the conservation and management of these fragile ecosystems.

Artificial intelligence algorithms can analyze underwater images and videos of coral reefs to automatically classify and identify different coral species, substrate types, and benthic organisms. ML models, trained on labeled image datasets, can learn to recognize unique visual features and patterns associated with different coral species. This automated image analysis allows for rapid and large-scale assessment of reef composition and biodiversity.

The research conducted by (Nunes et al., 2020) emphasizes the importance of using maps and time series analysis to visualize the community structure and health status of coral reefs. These visualizations provide valuable information about the proportion of species, ecological groups, and endangered species present in coral reef ecosystems, as well as the intensity of bleaching, diseases, and sedimentation affecting their health. Sorting these visualizations by location or specific components of the coral reef community enables researchers to gain a better understanding of the spatial and temporal dynamics of these ecosystems, facilitating targeted conservation efforts.

Machine learning techniques can analyze other environmental data such as sea surface temperature, nutrient levels, and water quality parameters to predict and monitor potential stressors on coral reefs. By integrating data from multiple sources, including satellites, buoys, and sensors, these models can also identify environmental factors influencing coral health, bleaching events, or coral mortality. These forecasts help with predicting harmful algal blooms, sea surface temperature anomalies, and extreme weather events like hurricanes, supporting coastal management and maritime operations. In one study, (González-Rivero et al., 2020) uses CNNs for coral reef monitoring, for example.

Marine robots

In connection with the ever-improving study and creation of robotics, scientists can combine robotics with AI, and these can be specialized to work in marine environments (Molina-Molina et al., 2021). These AI-enabled robots could potentially be one of the most capable assistants to scientists on dives or possibly replace scientist divers altogether in some instances (Christensen et al., 2022). They can be trained to identify weak and diseased coral, different species of marine wildlife, and so much more (Saad et al., 2020). Ideally, these robots would be autonomous underwater vehicles (AUVs). In one study, AUVs used sonar to create a 3D representation of the seafloor to find manganese crust deposits rich in cobalt (Neettiyath et al., 2019). Because cobalt is a magnetic metallic element commonly used in alloys, its applications span from airplane engine construction to a blue color dye, and as such, scientists worry about expanding the supply of cobalt on the earth's surface (Cobalt Statistics and Information, 2023). In this case, the AUVs used in this study introduced a new way to search for cobalt on the ocean floor, instead of costly and sometimes dangerous deep-ocean drills (Neettiyath et al., 2019).

The study of underwater robotics still contains many challenges and puzzles that have yet to be solved. Indeed, there are many elements that the robot must deal with that must be considered and planned for in marine environments. For example, how will the robot respond to low visibility in murky water? How will it recalibrate from waves or a boat's wake? Scientists must design responsive, applicable solutions to these problems and more. Even though much is still unknown, the progress is slow, AI-enabled marine robotics holds multitudes of possibilities. With so much yet undiscovered, AI robotics could help scientists understand the ocean in ways nearly impossible to fathom today.

Ocean monitoring and prediction

Artificial intelligence can integrate data from various sources, including satellite observations, buoys, and sensors, to monitor ocean conditions in real time. It helps in predicting harmful algal blooms, identifying marine pollution, and forecasting oceanographic events like storms or temperature changes. By integrating various sources of data, such as satellite imagery, oceanographic sensors, and historical records, AI models can accurately predict oceanographic parameters like temperature, currents, and sea level rise (Saad et al., 2020). These predictions enable us to anticipate and manage natural hazards such as storms, tsunamis, and harmful algal blooms.

Additionally, AI-powered systems can monitor and analyze oceanic conditions in real time, providing valuable insights into the health of marine ecosystems, the impact of climate change (Rau et al., 2012), and the effectiveness of conservation measures. Ultimately, AI-driven ocean monitoring and prediction contribute to the sustainable management and preservation of our marine resources.

Conservation and marine protected areas

Artificial intelligence can assist in designing and managing marine protected areas by analyzing ecological data, predicting habitat suitability, and optimizing conservation strategies. It aids in understanding species distribution, identifying critical habitats, and assessing the effectiveness of conservation measures. Conservation and the effective management of marine protected areas (MPAs) are vital for safeguarding marine biodiversity and ecosystems. AI plays a significant role in these efforts by enabling advanced monitoring, analysis, and decision-making. AI algorithms can analyze large datasets from various sources, including satellite imagery, acoustic sensors, and underwater cameras, to assess the health of marine ecosystems, detect illegal fishing activities, and identify areas of high ecological importance.

By integrating AI-based predictive models, managers can optimize MPA design and zoning, ensuring maximum protection for vulnerable species and habitats. AI-powered tools also facilitate real-time surveillance (Aoki et al., 2022) and adaptive management, allowing for prompt responses to environmental changes and threats. Ultimately, AI empowers conservationists and policymakers with valuable insights and tools for effective conservation and the sustainable management of our marine resources.

Climate change impact assessment

Artificial intelligence studies the oceans to gain insights about climate change by analyzing vast amounts of oceanic data collected from various sources. These data include satellite imagery, oceanographic measurements, underwater sensor data, and historical climate records. AI can analyze climate models and historical data to assess the impacts of climate change on marine ecosystems. It helps predict shifts in species distributions, changes in oceanographic conditions, and potential ecological disruptions.

Artificial intelligence algorithms process and analyze this data to identify patterns, correlations, and anomalies, enabling scientists to understand the complex interactions between oceanic processes and climate change. AI can also simulate and model oceanic systems to predict future scenarios and assess the impacts of climate change on marine ecosystems, such as changes in ocean temperature, sea level rise, ocean acidification, and shifts in species distributions. This AI-driven analysis provides valuable insights into the dynamics of the oceans and enhances our understanding of climate change's effects on marine environments.

Mariculture and AI

Mariculture, which refers to the cultivation of marine organisms in controlled environments, can greatly benefit from the integration of AI technologies (Phillips, 2009). Here are some ways in which AI can be applied to enhance mariculture practices:

Monitoring and optimization: AI can be used to monitor and optimize mariculture systems. By analyzing real-time data from sensors measuring parameters such as water quality, temperature, dissolved oxygen levels, and feed consumption, AI models can identify trends, patterns, and anomalies. This information can help optimize feeding schedules, water flow, and environmental conditions, leading to improved growth rates, health, and overall productivity of cultured organisms.

Disease detection and prevention: AI can assist in the early detection of diseases in mariculture systems. By analyzing sensor data, including behavior patterns, feeding habits, and environmental conditions, AI algorithms can identify potential signs of disease or stress in cultured organisms. This enables timely intervention and preventive measures to minimize disease outbreaks and reduce economic losses.

Feed optimization: AI can optimize feed management in mariculture operations. By analyzing feeding patterns, growth rates, and environmental conditions, AI models can predict optimal feeding schedules and adjust feed composition based on the nutritional needs of cultured organisms. This helps reduce feed waste, improve feed conversion efficiency, and minimize environmental impacts associated with excess feed discharge.

Water quality management: AI can play a crucial role in managing and maintaining optimal water quality in mariculture systems. By analyzing sensor data from multiple sources, AI models can predict and manage factors such as oxygen levels, pH, salinity, and nutrient concentrations. This helps prevent water quality fluctuations that may negatively impact the health and growth of cultured organisms.

Stock management and inventory control: AI can assist in stock management and inventory control in mariculture operations. By analyzing growth rates, mortality rates, and environmental conditions, AI models can predict optimal stocking densities and determine the optimal time for harvesting or transferring cultured organisms. This helps ensure efficient use of space, minimize resource wastage, and improve overall production efficiency.

Decision support systems: AI-powered decision support systems can assist mariculturists in making informed decisions. By integrating data from various sources, such as weather forecasts, market trends, and historical data, these systems provide insights and recommendations for optimizing production strategies, market timing, and risk management.

Overall, the integration of AI in mariculture can lead to improved productivity, sustainability, and profitability. By leveraging AI technologies, mariculturists can make data-driven decisions, optimize resource utilization, and minimize environmental impacts, contributing to the growth and development of sustainable mariculture practices.

Challenges and future directions

The application of AI in marine science holds immense promise for understanding and managing our oceans. However, several challenges and limitations must be addressed for the continued progress of this field.

Data availability

One significant challenge in applying AI to marine science is the limited availability of high-quality data. Although a lot of data are being generated on daily bases, the challenges of its consolidation, validation, and generalization remain valid to this day. Also, given the magnitude of the ocean surface, the data are insufficient. While there is a vast amount of data coming from sources such as buoys, sensors, and satellite imagery, certain marine environments or remote areas may have sparse data coverage. This is especially true for inaccessible areas. In fact, more than 80% of the ocean area remains unexplored. To overcome this challenge, researchers must focus on improving data collection methods, expanding data sharing initiatives, and exploring innovative approaches, such as citizen science, to augment the available data. Collaborative efforts between scientists, policymakers, and technology developers are essential in addressing the issue of data scarcity.

Model robustness in diverse marine environments

Marine environments exhibit high variability and complexity, posing challenges for AI models that are trained on specific data distributions. To make AI models effective across diverse marine environments, it is crucial to ensure their robustness and generalizability. Future research should focus on developing models that can adapt to different environmental conditions, leveraging techniques such as transfer learning, domain adaptation, and ensemble methods to improve model performance across various marine settings. Additionally, incorporating domain experts' knowledge and integrating interdisciplinary approaches can help enhance model robustness and ensure accurate predictions.

Future directions

Artificial intelligence in marine science has come a long way, but there is still a long way to go, and many areas warrant further exploration, enhancing the accuracy and interpretability of models by advancing AI algorithms, deep learning structures, and more. Moreover, autonomous underwater vehicles and remote sensing technologies could offer innovations in methods of collecting data, which will improve scientists' reach and understanding of the oceans. Also, the study of the oceans involves marine scientists as much as it does computer scientists, statisticians, and policymakers, so collaborations will foster problem-solving habits.

Ethical concerns with AI in marine science, as well, must be considered for responsible and reliable research to result. The risk of AI disturbing ecosystems due to malfunctioning hardware or inaccurate predictions is the main concern. Algorithmic bias, where AI models unintentionally leave out certain species or such in their data collection or identification, is an ethical area of interest. To begin resolving these issues, AI can provide transparency with explainability techniques. Understanding why AI models make the predictions they do can also help scientists to identify and resolve issues before they affect the marine environment. Thus, additional research on how environmental factors play into AI's predictions is required.

Conclusion

In conclusion, the integration of AI in marine science has revolutionized many fields such as marine species monitoring, coral reef assessment, oceanographic forecasting, and AI-enabled marine robotics. AI algorithms have enhanced species identification, provided insights into reef health, enabled accurate predictions of ocean conditions, and facilitated scientific exploration. The continued advancement of AI technologies holds great promise for expanding our understanding of the ocean and supporting conservation and management efforts. Because so much is unknown about the ocean, implementing AI in this region of academia will undoubtedly reveal many more of the ocean's mysteries and help scientists better understand the practicalities of this knowledge.

References

Allken, V., Handegard, N.O., Rosen, S., Schreyeck, T., Mahiout, T., Malde, K., O'Driscoll, R., 2019. Fish species identification using a convolutional neural network trained on synthetic data. ICES Journal of Marine Science 76 (1), 342–349. https://doi.org/10.1093/icesjms/fsy147.

Aoki, L.R., Rappazzo, B., Beatty, D.S., Domke, L.K., Eckert, G.L., Eisenlord, M.E., Graham, O.J., Harper, L., Hawthorne, T.L., Hessing-Lewis, M., Hovel, K.A., Monteith, Z.L., Mueller, R.S., Olson, A.M., Prentice, C., Stachowicz, J.J., Tomas, F., Yang, B., Duffy, J.E., Gomes, C., Harvell, C.D., 2022. Disease surveillance by artificial intelligence links eelgrass wasting disease to ocean warming across latitudes. Limnology & Oceanography 67 (7), 1577–1589. https://doi.org/10.1002/lno.12152.

Bouchet, P., 2006. The Exploration of Marine Biodiversity: Scientific and Technological Challenges, pp. 31–62.

Caruso, F., Dong, L., Lin, M., Liu, M., Gong, Z., Xu, W., Alonge, G., Li, S., 2020. Monitoring of a nearshore small dolphin species using passive acoustic platforms and supervised machine learning techniques. Frontiers in Marine Science 7. https://doi.org/10.3389/fmars.2020.00267.

Chauhan, P., Raman, M., 2017. Satellite remote sensing for ocean biology: an Indian perspective. Proceedings of the National Academy of Sciences, India, Section A: Physical Sciences 87 (4), 629–640. https://doi.org/10.1007/s40010-017-0439-5.

Christensen, L., de Gea Fernández, J., Hildebrandt, M., Koch, C.E.S., Wehbe, B., 2022. Recent advances in AI for navigation and control of underwater robots. Current Robotics Reports 3 (4), 165–175. https://doi.org/10.1007/s43154-022-00088-3.

Cobalt Statistics and Information, 2023. Cobalt Statistics and Information | U.S. Geological Survey.

Costello, M.J., Vanden Berghe, E., 2006. 'Ocean biodiversity informatics': a new era in marine biology research and management. Marine Ecology Progress Series 316, 203–214. https://doi.org/10.3354/meps316203.

De La Houssaye, B., Flaming, P., Nixon, Q., Acton, G., 2019. Machine learning and deep learning applications for International Ocean Discovery Program Geoscience Research. SMU Data Science Review 2 (3), 9.

Ditria, E.M., Buelow, C.A., Gonzalez-Rivero, M., Connolly, R.M., 2022. Artificial intelligence and automated monitoring for assisting conservation of marine ecosystems: a perspective. Frontiers in Marine Science 9. https://doi.org/10.3389/fmars.2022.918104.

Dogan, G., Ford, M., James, S., 2021. Predicting ocean-wave conditions using buoy data supplied to a hybrid RNN-LSTM neural network and machine learning models. In: Proceedings of the 2021 IEEE International Conference on Machine Learning and Applied Network Technologies, ICMLANT 2021. Institute of Electrical and Electronics Engineers Inc., United States. https://doi.org/10.1109/ICMLANT53170.2021.9690528. http://ieeexplore.ieee.org/xpl/mostRecentIssue.jsp?punumber=9690525.

Dong, C., Xu, G., Han, G., Bethel, B.J., Xie, W., Zhou, S., 2022. Recent developments in artificial intelligence in oceanography. Ocean-Land-Atmosphere Research 2022. https://doi.org/10.34133/2022/9870950.

Goddijn-Murphy, L., O'Hanlon, N.J., James, N.A., Masden, E.A., Bond, A.L., 2021. Earth observation data for seabirds and their habitats: an introduction. Remote Sensing Applications: Society and Environment 24, 100619. https://doi.org/10.1016/j.rsase.2021.100619.

González-Rivero, M., Beijbom, O., Rodriguez-Ramirez, A., Bryant, D.E.P., Ganase, A., Gonzalez-Marrero, Y., Herrera-Reveles, A., Kennedy, E.V., Kim, C.J.S., Lopez-Marcano, S., Markey, K., Neal, B.P., Osborne, K., Reyes-Nivia, C., Sampayo, E.M., Stolberg, K., Taylor, A., Vercelloni, J., Wyatt, M., Hoegh-Guldberg, O., 2020. Monitoring of coral reefs using artificial intelligence: a feasible and cost-effective approach. Remote Sensing 12 (3). https://doi.org/10.3390/rs12030489.

Hamadani, A., Ganai, N.A., 2022. Development of a multi-use decision support system for scientific management and breeding of sheep. Scientific Reports 12 (1). https://doi.org/10.1038/s41598-022-24091-y.

Hamadani, A., Ganai, N.A., Farooq, S.F., Bhat, B.A., 2022. Big data management: from hard drives to DNA drives. Indian Journal of Animal Sciences 90 (2), 134−140. https://doi.org/10.56093/ijans.v90i2.98761.

Hamadani, A., Ganai, N.A., Bashir, J., 2023. Artificial neural networks for data mining in animal sciences. Bulletin of the National Research Centre 47 (1). https://doi.org/10.1186/s42269-023-01042-9.

Hogan, R., Ford, M., Battaglia, M., Sturdivant, J., Dogan, G., Bresnahan, 2022. Predicting water quality estimates using satellite images in coastal and estuarine environments. Journal of Computing Sciences in Colleges 38 (5), 87−95.

Karleskint, G., Turner, R., Small, J., 2012. Introduction to Marine Biology. Cengage Learning.

Khan, C.B., Goetz, K.T., Cubaynes, H.C., Robinson, C., Murnane, E., Aldrich, T., Sackett, M., Clarke, P.J., LaRue, M.A., White, T., Leonard, K., Ortiz, A., Lavista Ferres, J.M., 2023. A biologist's guide to the galaxy: leveraging artificial intelligence and very high-resolution satellite imagery to monitor marine mammals from space. Journal of Marine Science and Engineering 11 (3), 595. https://doi.org/10.3390/jmse11030595.

Lalli, C.M., Parsons, T.R., 1997. Biological Oceanography: An Introduction. In: 2nd ed (Ed.).

Logares, R., Alos, J., Catalan, I.A., Solana, A.C., del Campo, F.J., Ercilla, G., et al., 2021. Oceans of Big Data and Artificial Intelligence. CSIC Scientific Challenges towards 2030, pp. 163−179. In press.

Lotze, H.K., 2021. Marine biodiversity conservation. Current Biology 31 (19), R1190−R1195. https://doi.org/10.1016/j.cub.2021.06.084.

Merz, C.R., 2001. An overview of the coastal ocean monitoring and prediction system (COMPS). Oceans Conference Record (IEEE) 2, 1183−1187.

Molina-Molina, J.C., Salhaoui, M., Guerrero-González, A., Arioua, M., 2021. Autonomous marine robot based on AI recognition for permanent surveillance in marine protected areas. Sensors 21 (8), 2664. https://doi.org/10.3390/s21082664.

Neettiyath, U., Thornton, B., Sangekar, M., Nishida, Y., Ishii, K., Sato, T., Bodenmann, A., Ura, T., 2019. An AUV based method for estimating hectare-scale distributions of deep sea cobalt-rich manganese crust deposits. In: OCEANS 2019 - Marseille, OCEANS Marseille 2019. Institute of Electrical and Electronics Engineers Inc., Japan. https://doi.org/10.1109/OCEANSE.2019.8867481. http://ieeexplore.ieee.org/xpl/mostRecentIssue.jsp?punumber=8846157.

Nunes, J.A.C.C., Cruz, I.C.S., Nunes, A., Pinheiro, H.T., 2020. Speeding up coral reef conservation with AI-aided automated image analysis. Nature Machine Intelligence 2 (6), 292. https://doi.org/10.1038/s42256-020-0192-3.

Phillips, M., 2009. Mariculture overview. In: Encyclopedia of Ocean Sciences. Elsevier Ltd, Thailand, pp. 537−544. https://doi.org/10.1016/B978-012374473-9.00752-9. http://www.sciencedirect.com/science/referenceworks/9780123744739.

Rau, G.H., McLeod, E.L., Hoegh-Guldberg, O., 2012. The need for new ocean conservation strategies in a high-carbon dioxide world. Nature Climate Change 2 (10), 720−724. https://doi.org/10.1038/nclimate1555.

Ribeiro, ngela M., Foote, A.D., Kupczok, A., Frazão, B., Limborg, M.T., Piñeiro, R., Abalde, S., Rocha, S., da Fonseca, R.R., 2017. Marine genomics: news and views. Marine Genomics 31, 1−8. https://doi.org/10.1016/j.margen.2016.09.002.

Ropert-Coudert, Y., Wilson, R.P., 2005. Trends and perspectives in animal-attached remote sensing. Frontiers in Ecology and the Environment 3 (8), 437−444. https://doi.org/10.1890/1540-9295(2005)003[0437:tapiar]2.0.co;2.

Saad, A., Stahl, A., Våge, A., Davies, E., Nordam, T., Aberle, N., Ludvigsen, M., Johnsen, G., Sousa, J., Rajan, K., 2020. Advancing ocean observation with an AI-driven mobile robotic explorer. Oceanography 33 (3), 50−59. https://doi.org/10.5670/oceanog.2020.307.

Salman, A., 2023. Editorial: application of machine learning in oceanography and marine sciences. Frontiers in Marine Science 10. https://doi.org/10.3389/fmars.2023.1207337.

Souza Jr., P.M., Olsen, Z., Brandl, S.J., 2023. Paired passive acoustic and gillnet sampling reveal the utility of bioacoustics for monitoring fish populations in a turbid estuary. ICES Journal of Marine Science 80 (5), 1240−1255.

Stutters, L., Liu, H., Tiltman, C., Brown, D.J., 2008. Navigation technologies for autonomous underwater vehicles. IEEE Transactions on Systems, Man, and Cybernetics - Part C: Applications and Reviews 38 (4), 581−589. https://doi.org/10.1109/TSMCC.2008.919147.

Taberlet, P., Bonin, A., Zinger, L., Coissac, E., 2018. Environmental DNA: for biodiversity research and monitoring. In: Environmental DNA: For Biodiversity Research and Monitoring. Oxford University Press, France, pp. 1−253. https://doi.org/10.1093/oso/9780198767220.001.0001. http://www.oxfordscholarship.com/view/10.1093/oso/9780198767220.001.0001/oso-9780198767220.

Tait, R., Victor, F., Dipper, 1998. Elements of Marine Ecology. Butterworth-Heinemann.

Zettler, E.R., Takada, H., Monteleone, B., Mallos, N., Eriksen, M., Amaral-Zettler, L.A., 2017. Incorporating citizen science to study plastics in the environment. Analytical Methods 9 (9), 1392−1403. https://doi.org/10.1039/C6AY02716D.

Advances in robotics for biological sciences

15

Shabia Shabir[1] and Henna Hamadani[2]

[1]*Islamic University of Science and Technology (IUST), Awantipora, Jammu and Kashmir, India;* [2]*Sher-e-Kashmir University of Agricultural Sciences and Technology of Kashmir, Srinagar, Jammu and Kashmir, India*

Introduction

In artificial intelligence (AI), a robot is a machine that understands its environment and can perform tasks automatically. The term "robot" is used to name an AI agent that can receive input through receptors and provide output through actuators. Some of the unique advantages of robots are accuracy, automation, and the capability of operating under inaccessibility or harsh conditions (Dahiya et al., 2023). In the field of biological sciences, robotics has become an important tool that enables researchers to explore and understand the complexity of organisms and their environments. The history of robotics in biology is intertwined with the general development of robotics, a brief timeline of robotics in biology is given in Fig. 15.1.

As AI robots become more sophisticated, they can perform even broader tasks, from simple tasks to complex medical procedures such as surgery, rehabilitation, and other medical applications. They can provide greater precision and accuracy than human surgeons (Borse et al., 2022). Building powerful systems for laboratory experiments is another important contribution robots make to the biological sciences (Wolf et al., 2022).

Apart from this, in medical science, the conduct of the surgeries is hampered by a number of laparoscopy's intrinsic difficulties even after the surgeon has gathered years of experience. These hazards include an unsteady platform for the video camera, the straight laparoscopic instruments' restricted degrees of freedom, two-dimensional imagery, and poor ergonomics for the surgeon. It is hypothesized that robotics and telerobotics provide remedies for these annoying laparoscopic surgical drawbacks (Ballantyne, 2002). Thus, these technologies have improved the efficiency and reproducibility of research methods, enabling researchers to tackle larger projects, conduct more efficient experiments, and more efficiently analyze large amounts of data. The field of AI robotics is growing rapidly, as researchers develop new ways to give robots more intelligence and autonomy. As they become more sophisticated, they have the ability to perform a wider range of tasks in biology. This could lead to new discoveries in medicine, ecology, and other fields.

A Biologist's Guide to Artificial Intelligence. https://doi.org/10.1016/B978-0-443-24001-0.00015-4

255

FIGURE 15.1

Brief timeline of developments of robotics in biology.

Principles and features of robotics

The principles applied to the development of AI robots in biology are designed with the aim to ensure that the robots are safe, effective, and ethical. These principles determine the features of robotic systems and thus guide the researchers working in the field of robotics in biology, to develop efficient, adaptable, and capable human interactive machines. Some of them are listed below

1. **Physical nature:** AI robots must adopt a physical nature that allows them to interact with the living environment around them as part of their physiology. This could lead to a humanoid body or a prosthetic body with a specific goal in mind.
2. **Less energy consumption:** Designing lightweight constructions and low-power sensors and actuators can all contribute to the consumption of less energy and resource. It entails having sensors that can pick up on sound, touch, light, and smell. Fig. 15.2 shows the AI agent and environment interaction.
3. **Power of sensing:** For AI robots to effectively interact with the world, they need to be able to sense. Advanced sensory skills, such as vision, touch, and automatic hearing, need to be incorporated in order to observe and understand their environment. The better the observation, the better will be the data collected, in turn, the wiser shall be the decisions.
4. **Learning ability:** To thrive over time, AI robots need to be able to learn from their environment. Such robots are the learning agents that perform much better due to the training process. This is performed using previous experience, which makes it suitable for predictive analysis and lets them learn from their own experience, as well as data collected from other robots.
5. **Adaptability and safety:** AI robots must be able to adapt to their environment. This includes the ability to learn new ideas and change behavior in changing circumstances. They need to be safe when interacting with humans and the environment. This includes a secure system and the ability to identify and manage potential hazards.

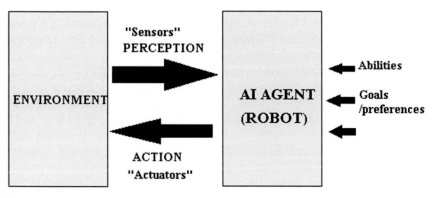

FIGURE 15.2

Interaction of artificial intelligence (AI) agent (robot) with the environment.

Advancements and contributions—A review

Robotics has made significant advancements in the field of biological science. The integration of robotics in biological sciences continues to expand the capabilities and possibilities for researchers. Some of the recent work reflects functions and the extent to which robotics have changed the scenario by integrating with biological science (Mazzolai, 2016). Over years, a number of breakthroughs in robotics have advanced medical sciences and aided surgeries and rehabilitation. Some major milestones are given in Fig. 15.3. Other important advances are also listed under.

1. **Robotics in laboratory and field automation:** Due to increasing interest in drug discovery, genomics, and synthetic biology, the level of automation in pharmaceutical laboratories and production has been increased by integrating with robotics. This has transformed the process of laboratory workflows and experimentation processes and thus automated repetitive tasks. This has, in turn, resulted in improved efficiency, reduced human error, and has enabled researchers to process larger volumes of data (Wolf et al., 2022). Total lab automation (TLA), which is the integration of robotics and automation technologies to streamline and optimize laboratory processes, is now being achieved. It involves the automation of various tasks, such as sample handling, specimen preparation, testing, and result analysis. TLA aims to improve efficiency, accuracy, and throughput in clinical laboratories. TLA minimizes manual intervention, reducing the risk of errors and enabling high-speed processing of samples. Automation reduces the potential for human error, improving the accuracy and reliability of test results and enables faster processing of samples (Al Naam et al., 2022), leading to quicker diagnostic results and prompt patient care. Automation also allows laboratories to handle a large volume of samples simultaneously, increasing the overall testing capacity, and ensures consistency in sample processing and test procedures, reducing variability between different laboratory technicians. TLA systems are often integrated with LIS, enabling seamless data management and streamlined workflow. The detection of multidisease resistant organisms (MDROs) is yet another critical aspect of infectious disease management. Robotics has played a significant role in enhancing the speed and accuracy of MDRO detection, particularly through the use of automated molecular diagnostic platforms. Robotic systems can process multiple samples simultaneously, enabling the rapid detection of MDROs in a high-throughput manner. Automation reduces the potential for human error and ensures consistent and precise testing, leading to more reliable detection of MDROs. Additionally robotic platforms can perform multiplex assays, simultaneously detecting multiple MDROs or genetic markers associated with resistance, saving time and resources. Such systems when integrated with laboratory information systems allow for efficient sample tracking and streamlined data management. Automation also minimizes the handling of infectious samples, reducing the risk of laboratory-acquired infections and improving safety for laboratory personnel.

2. **Simulating robotics:** In order to emulate the form, function, and locomotion of animals, robotics has played an important role which enables researchers to gain insights into the principles of biological systems and apply them to solve complex problems. For example, insect or bird models of robots can be used to study animal behavior, environmental interactions, and attachment theories. The main aim is to create robots that mimic the shape, behavior, and movement of living things. By analyzing this, researchers learn more about the fundamentals of behavior, perception,

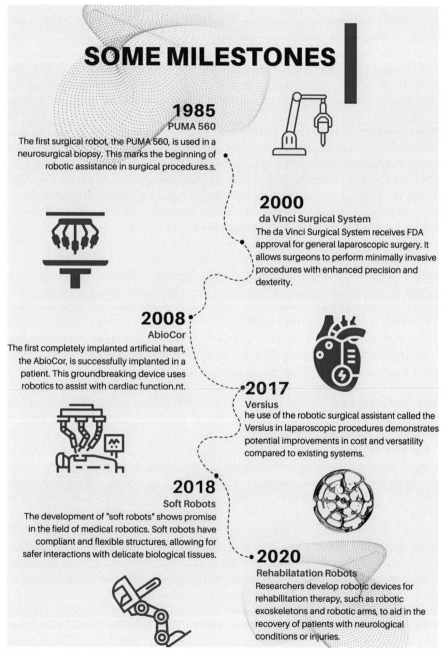

SOME MILESTONES

1985
PUMA 560
The first surgical robot, the PUMA 560, is used in a neurosurgical biopsy. This marks the beginning of robotic assistance in surgical procedures.s.

2000
da Vinci Surgical System
The da Vinci Surgical System receives FDA approval for general laparoscopic surgery. It allows surgeons to perform minimally invasive procedures with enhanced precision and dexterity.

2008
AbioCor
The first completely implanted artificial heart, the AbioCor, is successfully implanted in a patient. This groundbreaking device uses robotics to assist with cardiac function.nt.

2017
Versius
he use of the robotic surgical assistant called the Versius in laparoscopic procedures demonstrates potential improvements in cost and versatility compared to existing systems.

2018
Soft Robots
The development of "soft robots" shows promise in the field of medical robotics. Soft robots have compliant and flexible structures, allowing for safer interactions with delicate biological tissues.

2020
Rehabilatation Robots
Researchers develop robotic devices for rehabilitation therapy, such as robotic exoskeletons and robotic arms, to aid in the recovery of patients with neurological conditions or injuries.

FIGURE 15.3

Some major breakthroughs in robotics with reference to the medical field.

and movement. These bio-engineered robots allow scientists to study the basic functions of biological systems and apply them to complex issues in areas such as environmental monitoring and agriculture (Plum et al., 2020).

3. **Computer vision for visually impaired children:** Based on artificial vision recognition, an environment suited to the specific cognitive abilities has been provided to visually impaired children in which the educational content fosters spatial awareness of movement and verbal abilities and promotes it in a logical and structured way (Cavazos-Carrizales and Ruiz-Sanchez, 2022).

4. **Robotic underwater vehicle:** These automatic vehicles designed for underwater exploration are equipped with sensors and imaging systems that allow them to collect data on marine ecosystems, biodiversity, temperature, salinity, and oceanographic or chemical parameters, thus revolutionizing the study. They have been instrumental in mapping and monitoring coral reefs, studying marine life, and investigating underwater geological features Such robots have the advantage of operating in challenging environments and gathering data with high precision and efficiency. A robotic system using Yolov4, with the ability to detect the classes of objects in an image, combined with channel attention has been provided, wherein an integrated channel detection methodology facilitates underwater biological detection (Li et al., 2022).

5. **Intelligent drones:** These unmanned aerial vehicles have revolutionized data collection and management in the field of ecology and conservation. They can be equipped with cameras, multispectral sensors, and LiDAR systems to capture high-resolution aerial images and 3D terrain data. UAVs allow researchers to study wildlife populations, assess habitat conditions, and monitor ecosystem function over large areas (Hodgson et al., 2018). Autonomous navigation: Intelligent drones can navigate and fly autonomously, following preprogrammed flight paths or dynamically adapting to their environment. They utilize sensors like GPS, lidar, and cameras to perceive obstacles, avoid collisions, and maintain stable flight. This capability enables them to efficiently and safely perform tasks in various industries. Intelligent drones are widely used for aerial photography, videography, and mapping applications. They can capture high-resolution images, record videos, and create detailed maps of large areas. In agriculture, these drones assist in crop monitoring, disease detection, and yield estimation. They also play a vital role in environmental monitoring, disaster management, and infrastructure inspection. Companies are exploring the use of intelligent drones for autonomous delivery and logistics operations. Drones equipped with computer vision and AI algorithms can identify and navigate to specific delivery locations, drop off packages, and return to their base autonomously. This application has the potential to revolutionize last-mile delivery in industries like e-commerce and healthcare. Intelligent drones are used for inspecting infrastructure such as bridges, buildings, power lines, and pipelines. They can fly close to structures, capture detailed images, and detect structural damage or anomalies. By automating inspection processes, drones improve efficiency, reduce costs, and enhance safety by minimizing the need for humans to access hazardous or hard-to-reach areas. Intelligent drones play a crucial role in search and rescue missions. Equipped with thermal cameras, sensors, and AI algorithms, they can quickly cover large areas, identify heat signatures, and locate missing persons in difficult terrain or disaster-stricken areas (Sun et al., 2016). Drones provide real-time information to rescue teams (Mohd Daud et al., 2022), enabling faster response times and increasing the chances of saving lives. Intelligent drones are employed in environmental research and conservation efforts. They can monitor wildlife populations, track migratory patterns, detect

illegal activities like poaching or deforestation, and assess the health of ecosystems. Drones equipped with specialized sensors can collect data on air quality, water pollution, and habitat conditions, aiding in environmental conservation and decision-making.

6. **Bio-inspired robots:** Robots designed with inspiration from biological organisms are used to study biological principles or mimic natural behaviors (Gravish and Lauder, 2018). For example, underwater robots inspired by fish or marine animals help in exploring and monitoring marine ecosystems. Robotic insects are utilized to understand the principles of insect flight and behavior. The main focus of these robots is the flexibility and adaptability of the objects with an aim to simulate the movements and physical properties of living organisms. Robots have a great contribution to the field of physical rehabilitation (Banerjee et al., 2018). It has applications in various areas of biological science such as capturing sensitive biological specimens, surgical support, and research in complex environments where complex robotics may not work. The flexibility and compliance of soft robots will enable them to interact safely and accurately with living things, leading to advances in areas such as prosthetics, rehabilitation, and biomechanics (Rus and Tolley, 2015). Advanced robotic prosthetic limbs incorporate sensors, actuators, and AI algorithms to restore mobility and functionality for individuals with limb loss. These robotic limbs can be controlled using neural interfaces, enabling users to perform complex movements and regain a sense of natural limb control. Inspired by the collective behavior of social insects, swarm robots work collaboratively to perform tasks. In biology, swarm robots can be used for collective sensing, data collection in environmental monitoring, or mimicking the behavior of biological swarms for studying social dynamics.

7. **Robotic exoskeletons:** Exoskeletons are wearable robotic devices that assist individuals with mobility impairments or physical disabilities (Gorgey, 2018). They provide support, enhance strength, and assist with movements. In the field of rehabilitation, exoskeletons are used to aid in the recovery of motor functions in patients with neurological conditions or spinal cord injuries. Robotic exoskeletons are used in rehabilitation settings to assist individuals with mobility impairments or physical disabilities. They can aid in the recovery of motor functions by providing support and facilitating repetitive movements. Exoskeletons help patients with neurological conditions, spinal cord injuries, or stroke regain strength, improve gait patterns, and enhance their overall functional abilities. Exoskeletons find applications in industries that require repetitive or physically demanding tasks. They can reduce the risk of work-related injuries and improve worker productivity. Exoskeletons can assist with lifting heavy loads, maintaining proper posture, and reducing strain on the musculoskeletal system. Industries such as manufacturing, construction, and logistics are exploring the use of exoskeletons to enhance worker safety and performance. As the global population ages, robotic exoskeletons offer potential benefits in supporting the elderly and also increase cardiovascular fitness especially in people with sedentary lifestyles (Evans et al., 2015; Martin Ginis et al., 2010). These may also reduce the need for caregivers (Asselin et al., 2015; Federici et al., 2015; Gorgey et al., 2019; Louie et al., 2015; Miller et al., 2016). Exoskeletons can provide stability, improve balance, and enhance mobility, allowing older adults to maintain their independence and perform daily activities with reduced effort and risk of falls. They can also contribute to the prevention of muscle atrophy and improve the overall quality of life for seniors. Robotic exoskeletons are being developed for military use to enhance soldiers' physical capabilities and reduce fatigue. These exoskeletons can assist with carrying heavy equipment, traversing challenging terrains, and reducing the risk of injuries. By

enhancing endurance and strength, exoskeletons have the potential to improve soldier performance and reduce the strain on their bodies during demanding missions. Exoskeletons are also being explored in the field of sports for performance enhancement and injury prevention. They can assist athletes in improving their training regimes, providing resistance or assistance during exercises, and monitoring bio mechanical parameters. In rehabilitation, exoskeletons can aid athletes in recovering from injuries by supporting targeted movements and providing feedback on form and technique.

8. **Agricultural robots:** Robots are used in agriculture for tasks such as automated harvesting, selective spraying, and monitoring crop health. They employ computer vision, machine learning, and robotic arms to perform tasks efficiently, optimize resource utilization, and reduce the need for manual labor. Harvesting robots: These robots are designed to autonomously harvest crops, such as fruits, vegetables, and even greenhouse flowers. Equipped with sensors, vision systems, and robotic arms, they can identify ripe produce, pick it gently, and sort it based on quality parameters. Harvesting robots help reduce labor costs and address labor shortages in the agricultural industry. Weeding is a labor-intensive task that requires precision to remove unwanted plants while preserving the desired crops. Weeding robots use computer vision and machine learning algorithms to detect and selectively remove weeds, minimizing the need for herbicides and reducing the environmental impact of farming. These robots automate the process of planting and seeding crops. They can navigate fields, precisely place seeds at optimal depths and intervals, and even monitor seed germination and growth. Planting robots help improve planting accuracy, save time, and enable precision agriculture practices. These robots are equipped with sensors and imaging technologies to monitor crop health, growth, and environmental conditions. They can collect data on parameters like soil moisture, nutrient levels, and pest infestation. This information allows farmers to make data-driven decisions about irrigation, fertilization, and pest control, optimizing crop yield and reducing resource waste. Autonomous vehicles, including tractors and other farm machinery, are being developed to perform tasks such as plowing, spraying, and harvesting. These vehicles use GPS, computer vision, and advanced control systems to navigate fields and perform operations with precision, reducing the need for human operators and improving efficiency. Robotics technology is also being applied to automate tasks in the dairy and livestock industries (Hamadani and Khan, 2015). Milking robots, for example, can autonomously milk cows, monitor milk quality, and collect data on individual cow health. Similarly, robots can be used for feeding, sorting, and monitoring livestock, enhancing animal welfare and improving farm productivity. IoT-based data collection systems for data collection, monitoring, analysis and predictions are also rapidly becoming popular (Hamadani and Ganai, 2022).

The foreseeable future

Overall, the future of robotics holds the promise of advanced automation, collaborative partnerships, bio-inspired designs, AI integration, and seamless human−robot interactions (Torresen, 2018). These advancements will lead to transformative applications in industries, healthcare, exploration, and numerous other domains, ultimately enhancing our quality of life and pushing the boundaries of human capabilities.

- Advanced automation: Robots will become increasingly autonomous, capable of performing complex tasks with minimal human intervention (Shen et al., 2021). They will possess advanced sensing and perception capabilities, enabling them to navigate dynamic environments, interact with objects, and adapt to changing circumstances. This will lead to enhanced efficiency and productivity across industries, from manufacturing and logistics to healthcare and agriculture.
- Pandemic prevention: Robots have the potential to play a vital role in the prevention of pandemics. Robots equipped with UV-C lights or other disinfection mechanisms can autonomously clean and disinfect public spaces, healthcare facilities, and high-touch surfaces. This reduces the risk of transmission by eliminating pathogens effectively and efficiently. Robots can also be employed for contactless service delivery, reducing human-to-human interactions. For instance, robots can handle tasks such as food and medicine delivery, reducing the need for direct contact between individuals. Robots equipped with thermal imaging cameras can conduct noncontact temperature screenings in public areas or healthcare facilities. This allows for early detection of potential infections and helps in identifying individuals with fever-like symptoms. They can assist in remote patient monitoring, providing vital signs monitoring, and facilitating virtual consultations between patients and healthcare providers. These may be done by taking cues from the developments in the virtual world (Petrovic, 2018). This enables healthcare professionals to provide care while minimizing physical contact and deployed to disseminate information, guidelines, and public health messages. They can engage with the public, answer frequently asked questions, and provide education on preventive measures such as hand hygiene and social distancing. Additionally, robots can be equipped with sensors and cameras to monitor public spaces and detect abnormalities or noncompliance with safety measures. They can collect and analyze data to identify potential hotspots, monitor crowd density, and support decision-making processes. Autonomous robots or drones can be used for the transportation of medical supplies, specimens, or samples, minimizing human involvement and reducing the risk of exposure.
- Collaboration and coexistence: Future robots will work alongside humans as collaborative partners rather than replacing them. They will be designed to complement human capabilities and assist in tasks that require precision, strength, or endurance. Collaborative robots, or cobots, will be integrated into workspaces, enhancing productivity and safety while allowing humans to focus on more creative and strategic activities.
- Soft robotics and biomimicry: The development of soft robotics, inspired by natural systems, will enable robots to interact with delicate objects and navigate complex environments with greater dexterity. Biomimetic designs, drawing inspiration from nature, will lead to the creation of robots that mimic the movements, structures, and functions of animals or plants. These advancements will open up new possibilities in areas such as healthcare, exploration, and environmental monitoring.
- AI integration: AI will play a central role in the future of robotics. AI algorithms will enhance robot decision-making, learning, and adaptability, enabling them to handle uncertain and dynamic situations more effectively. Machine learning and deep learning techniques will enable robots to continuously improve their performance based on data and experiences, leading to more intelligent and capable systems (Dong et al., 2022). Machine learning is being used for making predictions on various aspects of biology (Hamadani and Ganai, 2022; Hamadani et al., 2022,

2023). The development of natural and intuitive interfaces will facilitate seamless interactions between humans and robots. Speech recognition, gesture control, and haptic feedback will enable robots to understand and respond to human commands and cues. This will facilitate broader adoption and acceptance of robots in various domains, including healthcare, education, and entertainment.

- Ethical considerations and regulations: As robotics becomes more prevalent in society, there will be an increased focus on ethical considerations and regulations (Boden et al., 2017). Discussions around robot ethics, privacy, safety, and the impact on employment will shape the future development and deployment of robots, ensuring responsible and beneficial integration into our daily lives.

Robot uprising, is it possible?

The concept of a robot uprising, often portrayed in science fiction, is highly speculative and not supported by current scientific understanding (Aylett, 2021). While there are ongoing advancements in robotics and AI, it is important to note that the development and deployment of robots are guided by human intentions and objectives. The idea of robots autonomously rebelling against humans is not based on any realistic technological trajectory.

Currently, robots are designed and programmed to perform specific tasks and functions within defined parameters. They lack consciousness, emotions, and independent decision-making abilities. The field of AI is focused on developing systems that can learn from data and make decisions based on patterns and algorithms, but these systems do not possess the intention or desire to rebel against humans.

Furthermore, there are ethical and safety considerations in place to ensure that robots are developed and deployed responsibly. Various regulations and guidelines govern the design, use, and safety of robots to prevent any potential harm to humans.

While it is impossible to predict the future with absolute certainty, the likelihood of a robot uprising as portrayed in science fiction is currently considered highly improbable. The focus of robotics and AI research is on creating beneficial and collaborative interactions between humans and robots to improve various aspects of our lives, such as healthcare, manufacturing, and exploration.

Challenges

Although robotic systems are doing wonders in every field, however, the industry faces certain challenges that need to be solved so as to overcome the limitation it suffers (Kemp et al., 2007). Autonomous robots are expected to be multifunctional and efficient in terms of operations and power. However, it needs to be worked upon to reach the expectation. An example of such a challenge is to make it possible for one surgeon to oversee a team of robots that can carry out simple procedure steps on their own and only require assistance from a surgeon during crucial, patient-specific procedures. Additionally, robots with naturalistic inspiration such as sensing, actuation, computation, and communication need to be developed along with advancements in mobility, body support, etc. This would make it an autonomous robot with self-supporting capability. Robots also need to be interactive in the environment and should sense any other robot in it. It should be able to communicate with other robots but perform independently. However, common sense, as in humans exists, which improves reasoning ability, is yet to develop in robots. This ability helps in better pattern recognition and thus solving complex problems. Apart from these challenges, robots are energy-inefficient and waste a lot of energy which can be resolved by increasing battery life, especially for mobile robots and drones (Yang et al., 2018).

Approval and authentication

The approval process for robots can vary depending on many factors and regulatory processes (Boden et al., 2017). These vary as per specific application, intended use, and the regulatory framework of the country or region. It is important to note that the approval process for robots can differ significantly depending on the specific type of robot, its intended use, and the applicable regulations. Manufacturers must adhere to the regulatory guidelines and work closely with the regulatory agencies to ensure compliance and obtain the necessary approvals before marketing and using the robot.

- Regulatory agencies: Each country or region has its own regulatory agency responsible for overseeing the approval of medical devices, including robots. In the United States, it is the Food and Drug Administration (FDA) (Whitford et al., 2019), while the European Union has the European Medicines Agency (EMA) and the Medical Device Regulation (MDR).
- Classification: Robots are classified based on their intended use, risk level, and the regulatory guidelines in place. The classification helps determine the level of scrutiny and the specific approval pathway.
- Preclinical testing: Manufacturers typically conduct preclinical testing to assess the safety and performance of the robot. This involves laboratory tests, simulations, and animal studies to gather data on the robot's functionality, safety features, and potential risks.
- Clinical trials: For certain robot applications, especially in the medical field, clinical trials may be required to evaluate the robot's performance in humans. These trials involve testing the robot on human subjects and collecting data on safety, effectiveness, and any potential adverse events.
- Regulatory submission: Manufacturers compile all relevant data and submit an application to the regulatory agency. The application includes detailed information about the robot's design, specifications, performance data, preclinical and clinical testing results, and safety profiles.
- Review process: The regulatory agency reviews the submitted data and assesses the robot's safety, effectiveness, and compliance with applicable regulations and standards. The level of scrutiny and the time required for review can vary depending on the regulatory body and the complexity of the robot.
- Approval or clearance: Based on the review process, the regulatory agency will grant either approval or clearance for the robot. Approval indicates that the robot has met all the necessary requirements and is authorized for marketing and use. Clearance, on the other hand, is typically granted when the robot is deemed substantially equivalent to a legally marketed device with the same intended use.
- Postmarket surveillance: Once a robot receives approval or clearance, postmarket surveillance is conducted to monitor its performance, safety, and any potential adverse events. Manufacturers are required to report any issues or concerns to the regulatory agency.

Conclusion

The integration of robotics and biological science has a great impact on various areas like laboratory automation, medical diagnostics, ecological survey or prospective and predictive analytics, etc., which makes it a precious resource in the field of research and engineering. Although using robotics in biological science has proved to be effective, however overcoming the challenges can help the robotic system advance more quickly and create an error-free and comfortable environment where human—robot interaction would be more efficient and productive. Even complex problems could be solved by proper data collection and analysis and thus their usage may lead to better research and development in biological science.

References

Al Naam, Y.A., Elsafi, S., Al Jahdali, M.H., Al Shaman, R.S., Al-Qurouni, B.H., Al Zahrani, E.M., 2022. The impact of total automaton on the clinical laboratory workforce: a case study. Journal of Healthcare Leadership 14, 55–62. https://doi.org/10.2147/JHL.S362614.

Asselin, P., Knezevic, S., Kornfeld, S., Cirnigliaro, C., Agranova-Breyter, I., Bauman, W.A., Spungen, A.M., 2015. Heart rate and oxygen demand of powered exoskeleton-assisted walking in persons with paraplegia. Journal of Rehabilitation Research and Development 52 (2), 147–158. https://doi.org/10.1682/JRRD.2014.02.0060.

Aylett, R., 2021. Why there won't be a robot uprising any time soon. BBC Science Focus. https://www.sciencefocus.com/future-technology/future-robots-society/.

Ballantyne, G.H., 2002. The pitfalls of laparoscopic surgery: challenges for robotics and telerobotic surgery. Surgical Laparoscopy Endoscopy & Percutaneous Techniques 12 (1), 1–5. https://doi.org/10.1097/00019509-200202000-00001.

Banerjee, H., Tse, Z.T.H., Ren, H., 2018. Soft robotics with compliance and adaptation for biomedical applications and forthcoming challenges. International Journal of Robotics and Automation 33 (1), 69–80. https://doi.org/10.2316/Journal.206.2018.1.206-4981.

Boden, M., Bryson, J., Caldwell, D., Dautenhahn, K., Edwards, L., Kember, S., Newman, P., Parry, V., Pegman, G., Rodden, T., Sorrell, T., Wallis, M., Whitby, B., Winfield, A., 2017. Principles of robotics: regulating robots in the real world. Connection Science 29 (2), 124–129. https://doi.org/10.1080/09540091.2016.1271400.

Borse, M., Godbole, G., Kelkar, D., Bahulikar, M., Dinneen, E., Slack, M., 2022. Early evaluation of a next-generation surgical system in robot-assisted total laparoscopic hysterectomy: a prospective clinical cohort study. Acta Obstetricia et Gynecologica Scandinavica 101 (9), 978–986. https://doi.org/10.1111/aogs.14407.

Cavazos-Carrizales, J.P., Ruiz-Sanchez, F.J., 2022. Computer vision interface for symbolic programming of Cartesian motion to introduce visually impaired children into robotic sciences. In: CCE 2022 — 2022 19th International Conference on Electrical Engineering, Computing Science and Automatic Control. Institute of Electrical and Electronics Engineers Inc., Mexico, ISBN 9781665455084 https://doi.org/10.1109/CCE56709.2022.9975943.

Dahiya, A., Aroyo, A.M., Dautenhahn, K., Smith, S.L., 2023. A survey of multi-agent Human—Robot Interaction systems. Robotics and Autonomous Systems 161, 104335. https://doi.org/10.1016/j.robot.2022.104335.

Dong, X., Luo, X., Zhao, H., Qiao, C., Li, J., Yi, J., Yang, L., Oropeza, F.J., Hu, T.S., Xu, Q., Zeng, H., 2022. Recent advances in biomimetic soft robotics: fabrication approaches, driven strategies and applications. Soft Matter 18 (40), 7699–7734. https://doi.org/10.1039/d2sm01067d.

Evans, N., Wingo, B., Sasso, E., Hicks, A., Gorgey, A.S., Harness, E., 2015. Exercise recommendations and considerations for persons with spinal cord injury. Archives of Physical Medicine and Rehabilitation 96 (9), 1749−1750. https://doi.org/10.1016/j.apmr.2015.02.005.

Federici, S., Meloni, F., Bracalenti, M., De Filippis, M.L., 2015. The effectiveness of powered, active lower limb exoskeletons in neurorehabilitation: a systematic review. NeuroRehabilitation 37 (3), 321−340. https://doi.org/10.3233/NRE-151265.

Gorgey, A.S., 2018. Robotic exoskeletons: the current pros and cons. World Journal of Orthopedics 9 (9), 112−119. https://doi.org/10.5312/wjo.v9.i9.112.

Gorgey, A.S., Sumrell, R., Goetz, L.L., 2019. Exoskeletal Assisted Rehabilitation After Spinal Cord Injury. Elsevier BV, pp. 440−447.e2. https://doi.org/10.1016/b978-0-323-48323-0.00044-5.

Gravish, N., Lauder, G.V., 2018. Robotics-inspired biology. Journal of Experimental Biology 221 (7). https://doi.org/10.1242/jeb.138438.

Hamadani, A., Ganai, N.A., 2022. Development of a multi-use decision support system for scientific management and breeding of sheep. Scientific Reports 12 (1). https://doi.org/10.1038/s41598-022-24091-y.

Hamadani, H., Khan, A.A., 2015. Automation in livestock farming − a technological revolution. International Journal of Advanced Research 3, 1335−1344.

Hamadani, A., Ganai, N.A., Alam, S., Mudasir, S., Raja, T.A., Hussain, I., Ahmad, H.A., 2022. Artificial intelligence techniques for the prediction of body weights in sheep. Indian Journal of Animal Research. https://doi.org/10.18805/ijar.b-4831.

Hamadani, A., Ganai, N.A., Bashir, J., 2023. Artificial neural networks for data mining in animal sciences. Bulletin of the National Research Centre 47 (1). https://doi.org/10.1186/s42269-023-01042-9.

Hodgson, J.C., Mott, R., Baylis, S.M., Pham, T.T., Wotherspoon, S., Kilpatrick, A.D., Raja Segaran, R., Reid, I., Terauds, A., Koh, L.P., 2018. Drones count wildlife more accurately and precisely than humans. Methods in Ecology and Evolution 9 (5), 1160−1167. https://doi.org/10.1111/2041-210X.12974.

Kemp, C.C., Edsinger, A., Torres-Jara, E., 2007. Challenges for robot manipulation in human environments [Grand challenges of robotics]. IEEE Robotics and Automation Magazine 14 (1), 20−29. https://doi.org/10.1109/MRA.2007.339604.

Li, A., Yu, L., Tian, S., 2022. Underwater biological detection based on YOLOv4 combined with channel attention. Journal of Marine Science and Engineering 10 (4), 469. https://doi.org/10.3390/jmse10040469.

Louie, D.R., Eng, J.J., Lam, T., Spinal Cord Injury Research Evidence (SCIRE) Research Team, 2015. Gait speed using powered robotic exoskeletons after spinal cord injury: a systematic review and correlational study. Journal of NeuroEngineering and Rehabilitation 12, 1−10.

Martin Ginis, K.A., Arbour-Nicitopoulos, K.P., Latimer, A.E., Buchholz, A.C., Bray, S.R., Craven, B.C., Hayes, K.C., Hicks, A.L., McColl, M.A., Potter, P.J., Smith, K., Wolfe, D.L., 2010. Leisure time physical activity in a population-based sample of people with spinal cord injury part II: activity types, intensities, and durations. Archives of Physical Medicine and Rehabilitation 91 (5), 729−733. https://doi.org/10.1016/j.apmr.2009.12.028.

Mazzolai, B., 2016. Biology for robotics − a multidisciplinary approach to a researcher's life [women in engineering]. IEEE Robotics and Automation Magazine 23 (1), 114−115. https://doi.org/10.1109/MRA.2015.2511687.

Miller, L.E., Zimmermann, A.K., Herbert, W.G., 2016. Clinical effectiveness and safety of powered exoskeleton-assisted walking in patients with spinal cord injury: systematic review with meta-analysis. Medical Devices: Evidence and Research 9, 455−466. https://doi.org/10.2147/MDER.S103102.

Mohd Daud, S.M.S., Mohd Yusof, M.Y.P., Heo, C.C., Khoo, L.S., Chainchel Singh, M.K., Mahmood, M.S., Nawawi, H., 2022. Applications of drone in disaster management: a scoping review. Science & Justice 62 (1), 30−42. https://doi.org/10.1016/j.scijus.2021.11.002.

Petrovic, V.M., 2018. Artificial intelligence and virtual worlds-toward human-level AI agents. IEEE Access 6, 39976−39988. https://doi.org/10.1109/ACCESS.2018.2855970.

Plum, F., Labisch, S., Dirks, J.H., 2020. SAUV—a bio-inspired soft-robotic autonomous underwater vehicle. Frontiers in Neurorobotics 14. https://doi.org/10.3389/fnbot.2020.00008.

Rus, D., Tolley, M.T., 2015. Design, fabrication and control of soft robots. Nature 521 (7553), 467−475. https://doi.org/10.1038/nature14543.

Shen, Y., Guo, D., Long, F., Mateos, L.A., Ding, H., Xiu, Z., Hellman, R.B., King, A., Chen, S., Zhang, C., Tan, H., 2021. Robots under COVID-19 pandemic: a comprehensive survey. IEEE Access 9, 1590−1615. https://doi.org/10.1109/ACCESS.2020.3045792.

Sun, J., Li, B., Jiang, Y., Wen, C.-Y., 2016. A camera-based target detection and positioning UAV system for search and rescue (SAR) purposes. Sensors 16 (11), 1778. https://doi.org/10.3390/s16111778.

Torresen, J., 2018. A review of future and ethical perspectives of robotics and AI. Frontiers in Robotics and AI 4. https://doi.org/10.3389/frobt.2017.00075.

Whitford, A.B., Yates, J.L., Burchfield, A., Anastasopoulos, L.J., Anderson, D., 2019. The adoption of robotics by government agencies: evidence from crime labs. SSRN Electronic Journal. https://doi.org/10.2139/ssrn.3434554.

Wolf, Á., Wolton, D., Trapl, J., Janda, J., Romeder-Finger, S., Gatternig, T., Farcet, J.B., Galambos, P., Széll, K., 2022. Towards robotic laboratory automation Plug & Play: the\LAPP\framework. SLAS technology 27 (1), 18−25. https://doi.org/10.1016/j.slast.2021.11.003.

Yang, G.Z., Bellingham, J., Dupont, P.E., Fischer, P., Floridi, L., Full, R., Jacobstein, N., Kumar, V., McNutt, M., Merrifield, R., Nelson, B.J., Scassellati, B., Taddeo, M., Taylor, R., Veloso, M., Wang, Z.L., Wood, R., 2018. The grand challenges of science robotics. Science Robotics 3 (14). https://doi.org/10.1126/scirobotics.aar7650.

Robotics and computer vision for health, food security, and environment

Syed Zameer Hussain, Nazrana Rafique Wani, Ufaq Fayaz and Tahiya Qadri

Division of Food Science and Technology, Sher-e-Kashmir University of Agriculture Sciences and Technology of Kashmir, Srinagar, Jammu and Kashmir, India

Introduction

Industrial revolution took place at the end of the 18th century due to the use of mechanical facilities fueled by steam, which resulted in significant modifications to the manufacturing methods employed in factories. These modified manufacturing industrial methods powered by electric energy and labor division led to increased production of goods during the first decade of the 20th century, which gave rise to the Second Industrial Revolution (Kanji, 1990). Electronics and information technology were introduced to further automate production to pave the way for the ternary Industrial Revolution (third) around the start of the 1970s (Pérez et al., 2016). Presently, the quaternary Industrial Revolution, often termed "Industry 4.0," is centered on cyber-physical production systems (CPS). It is based on mechanization, computerization, data sharing, and production technology. The CPS keep an eye on the physical processes, decentralizes decision-making, and initiates activities while interacting and working with people in real time. This allows for major improvements in several industrial processes, such as production, processing, management, material utilization, and supply chain management (Kagermann et al., 2013). Numerous achievements have raised the degree of innovation within each revolution. For instance, automatization led by robots has had its own revolutions under Industry 4.0. The use of safe automation through sensitive bots led to the second robotic revolution, and mobility with mobile manipulators was the third. Present-era industries are based on intelligent and perceptive robot systems, giving rise to the fourth and final robotic revolution (Fig. 16.1) (Pérez et al., 2016). Automation, in association with smart robots, propels contemporary industries toward productiveness, leading to a significant rise in efficaciousness besides savings other resources and energy while improving working conditions (Elliott, 2019). In the traditional industry, productivity and safety were restricted by manual procedures. European Factories of the Future (FoF) Research Association (EFFRA, 2013) states that manufacturing has a significant potential to produce wealth and highly qualified workers. FoF, a partnership of the public—private section (PPP), has identified a range of technology enablers that contribute to the future of manufacturing. These include progressive fabricating methods and techniques, the use of mechatronics (including robotics) in improved manufacturing processes, information and communication technologies (ICT), methods of manufacturing, knowledge workers, and modeling,

A Biologist's Guide to Artificial Intelligence. https://doi.org/10.1016/B978-0-443-24001-0.00016-6

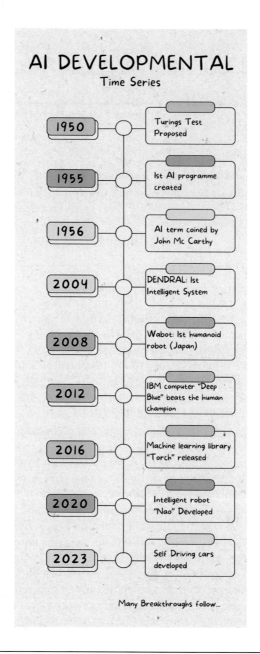

FIGURE 16.1

Various timelines showing the progression of artificial intelligence (AI) and its application in different sectors.

simulation, and forecasting methods and tools (Kirchberger, 2017). Robotics, a significant element within advanced manufacturing systems, directly addresses global concerns such as coping with an aging population (healthcare), food safety and security, and environmental sustainability, thus leading to improved societal well-being and living conditions (Bøgh et al., 2014; Zhou et al., 2014). Table 16.1 lists the various applications of robotics and computer vision (CV) in different fields.

Table 16.1 Implementation of robotics and computer vision techniques in the medical, food industry, and environment sector.

Applications		Algorithm	Modality	Findings	References
Healthcare	Alzheimer's disease	RF, SHAP	Clinical and imaging data	The accuracy of an artificial intelligence model for identifying and predicting the progression of Alzheimer's disease was 93.95% in the first layer and 87.08% in the second layer.	El-Sappagh et al. (2021)
	COVID-19	PA	Clinical data	When forecasting severe COVID-19 cases, an accuracy of 70%–80% was attained.	Jiang et al. (2020)
	Pulmonary cancer	LCP-CNN, Brock model	Clinical data	Compared to the Brock model, LCP-CNN was more accurate and produced fewer false-negative findings when predicting the malignancy of lung nodules.	Baldwin et al. (2020)
	Influenza	IAT-BPNN	CDC data and Twitter dataset	IAT-BPNN was able to predict influenza-like illness in a big population with remarkable accuracy.	Hu et al. (2018)
Food/ agriculture	Salmonella occurrence and absence prediction in agriculture stream	ANN, kNN, SVM	—	With a range of 58.15%–59.23%, tested algorithms correctly identified the presence of Salmonella.	Polat et al. (2020)
	Detection of germination in seeds	CNN	—	The average seed recognition accuracy was 97%.	Shadrin et al. (2020)
	Oryza sativa L. growth rate prediction modeling	REG, ANN, GEP	—	With ANN and GEP, growth rate simulation was predicted more accurately than with REG.	Liu et al. (2022)

Continued

Table 16.1 Implementation of robotics and computer vision techniques in the medical, food industry, and environment sector.—cont'd

Applications			Algorithm	Modality	Findings	References
		Classification of Kashmiri orchard apples based on detection and occlusion for robotic harvesting	YOLOv7 model, EfficientNet-B0 model	—	The classification accuracy was 91.38% when the EfficientNet B0 model algorithm was used in the second stage, compared to an mAP of 0.902 for the YOLOv7 algorithm used in the first stage.	Rathore et al. (2023)
Environment		Air monitoring	A WSN-based system for real-time microclimate monitoring. Additionally, it has temperature and relative humidity sensors (SHT15) supported by ZigBee communication technology and powered by solar panels.	—	It provided periodic or continuous measurements, analyzed and decided environmental factors or pollutant levels in order to stop negative effects. Predicting probable changes to the ecosystem or the biosphere as a whole was also a part of it.	Watthanawisuth et al. (2009)
		Plant monitoring	Used large-scale WSN	—	This was done to keep an eye on the microenvironment (humidity and temperature) in order to safeguard the potato crop against phytophthora. Based on the data gathered, the system was supposed to produce a policy to defend the crop against fungal illness.	Langendoen et al. (2006)
		Water monitoring	IoT, WSN	—	Conductivity, temperature, and turbidity measurements were used to determine the water quality.	Postolache et al. (2013)

Robots use CV to navigate a working environment safely, cooperate with people, identify and find working components, increase positioning precision, etc. This system might be scene-related or object-related depending on the purpose. Robots used in industries shift between different positions precisely with a deflection of 0.1 mm; however, the accuracy is governed by several factors such as tolerances, eccentricities, elasticities, play, wear-out, load variations, temperature effects, and incomplete knowledge of model parameters for pose-to-axis angle transformations. These challenges can introduce deviations of several millimeters. However, through careful calibration, accurate modeling of system parameters, and the implementation of compensation techniques like feedback control systems, industrial robots can achieve consistent and precise positioning (Zhang et al., 2011). If the information gathered is accurate and complete, it should be feasible to generate a 3D appearance model, giving programmers the chance to design navigation systems that are equally advanced. At this time, robots and CV techniques take the front stage in industrial situations (Fig. 16.1). Thus, this chapter precisely investigates and analyzes the current state of the art and various approaches, such as robotics and CV techniques in the fields of healthcare, food security, and the environment.

Robotics

The International Organization for Standardization (ISO, 2021) defines a robot as a programmable mechanical system with some autonomy to carry out locomotion, manipulation, or placement. The same standard allows robots to be divided into three groups: industrial robots, service robots, and medical robots. Robots are composed mechanically of a series of inflexible components called links interconnected to one another through several joints, as shown in Fig. 16.2. This necessitates research into how these linkages move and relate to one another in space, since this will enable them to carry out a number of tasks automatically, especially if computer-programmed. Robotic manipulators have a variety of features, but the ones that are most frequently used to describe them are (a) payload or weight lifting capacity; (b) working space, or the area they can move through; (c) range, or the maximum distance covered by the handle; (d) maximum speed and repeatability, or the ability to reach a predetermined location. Robot manipulators may be categorized according to their mechanical design, which can be generally broken down into seven categories: parallel, perpendicular, articulated, polar or spherical, cylinder, rectangular or Cartesian, and cylindrical (Timonen et al., 2022). A manipulator that has three or more rotational joints is said to be articulated. Robots often operate in cells. A manipulator is used to carry out the intended work, a controller is in charge of directing the

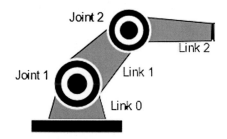

FIGURE 16.2

Link and joint illustration of robotic arm.

manipulator, and a programming console is in charge of programming and regulating the tasks which make up a robotic cell.

Additionally, it is made up of a number of possible peripherals, which include security components or probes required to fulfill the duties. The end of the manipulator is changed depending on the task that the robot is intended to execute (Costa et al., 2022). Most industrial robot applications require that these machines operate within closed cells, usually surrounded by fences, to protect the workers associated with them. This is necessary due to the hazards that these machines pose to people, largely caused by their rapid movement speeds and the masses in motion. Recently, robots that can function among people without requiring physical separation between them have been brought to the market (Karthik and Chandra, 2014).

Applications
Healthcare
Below are a few illustrations of robotic applications that support surgical interventions, radiation therapy, physical counseling and rehabilitation, nursing aid, as well as completing Corona tests—a task that was necessary to curtail the pandemic.

Surgical-assisted intervention. The most well-known name when discussing the use of robotics to help surgical treatments is the Da Vinci robot, developed by Intuitive. The robot is remotely controlled by a doctor. In contrast to the motions made by a surgeon during a traditional intervention, the design of this robotic system is compact that permits a greater degree of instrument mobility. In terms of the patient, it enables less intrusive and consequently less stressful procedures, enabling a better recoveryRobot. These procedures include robotic myomectomies to remove fibroids, thyroidectomies which involve complete or partial removal of the thyroid gland, and partial or complete gastrectomies to remove the stomach (Richmon and Kim, 2017).

Robotic device for radiotherapy. (Meeks et al., 2011) highlighted that the CyberKnife, developed by Accuray, is a groundbreaking, fully automated robotic radiation device. It revolutionized the curing regime of both malignant and benign tumors. This innovative technology works on the stereotactic body radiation therapy (SBRT) technique, which works on guidance through real-time images, allowing for more precise and localized delivery of radiation doses compared to conventional radiotherapy (Timmerman et al., 2014).

The key advantage of the CyberKnife lies in its flexible robotic arm, which enables radiation beams to be directed in any desired direction. This flexibility led to the target of high doses of irradiation directly to the tumor with adequate precision, consequently reducing exposure of surrounding healthy tissues (Meeks et al., 2011). This capability significantly improves the effectiveness and safety of radiation therapy, as it enables clinicians to treat tumors with greater accuracy and reduces the risk of damage to healthy tissues.

Nursing assistant in patient transport. The task of moving a patient who cannot freely move and is bedridden is known to be physically demanding in nursing care. To address this challenge, a robot was developed for Assistance known as Interactive Body Assistance (RIBA). RIBA was developed through a collaboration of the RIKEN cooperation center with Tokai Rubber Industries, Ltd. and RIBA serves as a nursing assistant robot specifically designed for this purpose. RIBA operates by following commands provided by a doctor through haptic sensors. These sensors facilitate the control of doctors over robotic actions and movements. Importantly, the design and implementation of the robot incorporate sensors aimed at preventing any pain or discomfort to the patient during the interaction with their body (Mukai et al., 2010). By utilizing haptic sensors and focusing on patient safety and comfort, RIBA provides valuable assistance to healthcare professionals in the challenging task of transferring patients between a bed and a wheelchair. This technology helps alleviate the physical strain on caregivers and enhances patient comfort during the movement process, ultimately improving the overall quality of nursing care.

Reduction in the spread of infectious diseases and pandemics. There has been a notable need for robots in the healthcare sector due to the recent rise in infectious illnesses and pandemic instances like COVID-19. Limiting doctor–patient interaction through the introduction of robots into the healthcare system saved frontline healthcare workers from contracting contagious illnesses and also improves self-isolation (Podpora et al., 2020). Patients in infectious illness wards have received meals and medications from robots like the autonomous timed up and go (TUG) robot. Robots are also employed to clean and disinfect surfaces that might spread infectious illnesses, such as door knobs and lifts. The ultraviolet disinfection robot (UVD-bot) is a typical illustration of a robot used to sanitize items (Olaronke et al., 2022). The Wegree Robot, which acts as a hospital check-in staff to minimize physical contact between medical personnel and possibly contagious patients, is another form of robot that slows the spread of infectious illnesses thus preventing the occurrence of a pandemic (Hauser and Shaw, 2020). By obtaining temperature readings through a noncontact thermometer fixed on a robot, the UVD robot and Sayabot serve to maintain social space between medical professionals and their patients. Sayabot also assists in providing guests with hand sanitizer. The monitoring of COVID-19 patients without providing direct care has also been done in Italy by medical professionals using mobile telepresence robots with displays and touchscreen interfaces (Sarker et al., 2021).

Food security

Robots are frequently accepted in sectors that produce food because of their sterile nature. The reduction of illnesses linked to food products is greatly aided by this aspect. The Food Safety Modernization Act (Hamburg, 2011) has established stricter sanitary guidelines that cover the full supply chain systems. The Food Safety Modernization Act (FSMA) of **2011** was a significant legislative policy intended to improve the security and safety of the US food supply by focusing on prevention of food-borne pathogens throughout the food system, including agricultural producers, food and animal transporters, and food importers. Cereals, spices, and other foods that do not require refrigeration are the root source of the problem since they are located in the most contaminated areas and are thus the most vulnerable. Formerly, such food items were devoid of any impurities or adulterations; however, the situation has changed now (Misra et al., 2022). Robotics with an AI foundation may undoubtedly aid in the solution of these kinds of issues. According to (Fedorova et al., 2020), they cannot spread diseases as much as a person can.

A Technavio survey predicts a 30% rise in robot usage in the food processing sector to fulfill regulatory standards by the end of the present decade (Zhao et al., 2021). Additionally, there have been some recent, groundbreaking advancements in AI-based food safety techniques that are predicted to become more well-known soon. Reduced prevalence of food-borne diseases is their key objective (Kumar et al., 2021). Despite US governments' attempts to abolish hunger and famine, 820 million people have been found to not have enough food to consume every day. Robotic technology might be used to end world hunger and malnutrition and provide food security for everybody. For example, a swarm of drones known as Swarm Robotics for Agricultural Applications (SAGA) keeps an eye on weed infestations on farms and evaluates the health of crops (Olaronke et al., 2022). Prior to the removal of weeds from crop fields, another robot, Nexus Robotics R2Weed2 robot, is employed to distinguish between crops and weeds. Additionally, Cambridge University created the Vegebot robot to precisely and accurately harvest crops. By picking up unripe fruits, shredding leafy vegetables, and damaging fruits, this robot eliminates waste that may have been brought on by human error. To reduce harvesting losses, the Agricultural University of Kashmir (SKUAST-K) is working in collaboration with other reputed institutes of the country to develop an AI-based robotic harvester for apples. The introduction of such facilities will definitely reduce harvesting losses in fruits besides positively affecting other aspects associated with apple harvesting. Additionally, Harvest Crop is a robot created for harvesting crops like strawberries that are vulnerable to harm throughout the harvesting season.

Environment

According to the World Health Organization, 90% of people worldwide breathe contaminated air every day, which leads to around seven million premature deaths (United Nations Economic Commission for Europe, 2021). Most of these people are from low- and middle-income nations. High levels of emissions from machines and businesses, together with the combustion of fossil fuels, are the main causes of air pollution. The air pollution brought on by the combustion of fossil fuels is one of the main factors contributing to climate change. Global warming is caused by the emissions from the combustion of fossil fuels. As a result, noncommunicable illnesses like cancer, heart disease, and respiratory conditions develop (Olaronke et al., 2022). Deforestation and oil drilling are two other factors contributing to change in climate. Decreasing the quantity of dangerous gases released into the atmosphere is one-way robots can combat climate change. For instance, heavy-duty human-operated devices that consume fossil fuels and emit carbon emissions into the environment can be replaced by robots (World Health Organization, 2021).

Computer vision

Computer vision techniques only really started to take off in the late 1960s and early 1970s, in contrast to AI techniques like neural networks, which have been widely utilized since the 1940s. Beginning with ideas of enacting human vision through knowledge of various camera schematics, projections and photogrammetry (Kakani et al., 2020), CV, within a few decades, has established itself as a pervasive technology in diverse industries. CV, like the human eye, works on cognition-like recognition of patterns, besides using knowledge of machine learning, computer graphics, 3D reconstructions, virtual reality, and augmented reality (Fig. 16.3). The field initially focused on recreating the human visual system by understanding camera operations, such as schematics, projections, and photogrammetry (Hartley and Zisserman, 2003). By 2010, high-end CV tasks such as the identification of objects, navigation of vehicles autonomously, detection and identification of faces and their expressions, and recognition of fingerprints, besides quick processing of images image and robotic navigation, had already been achieved (Vedaldi and Fulkerson, 2010). This progress was facilitated by techniques like line detection, feature extraction, segmentation, feature matching and tracking, optimization, and 3D reality reconstruction (Fuentes-Pacheco et al., 2015). Notably, CV has given rise to remarkable inventions, including visual simultaneous localization and mapping (SLAM), object tracking, and more. These advancements have been made possible by leveraging the interconnection between various

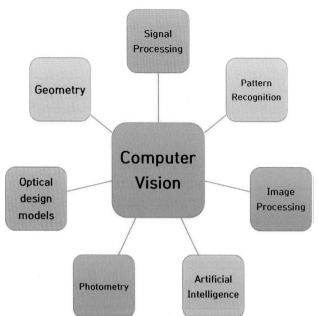

FIGURE 16.3

Different working methods for computer vision (CV).

disciplines, such as science and technology, mathematics and geometry, physics and probability, and others. An infographic illustrating the dominance of AI shows how businesses based on CV have infiltrated every profitable sector.

Applications
Healthcare

Computer vision technology can be explored as a suitable robotic intervention for conducting surgeries and developing remedies for many pathologies. In recent years, fast prototyping and three-dimensional (3D) modeling techniques have led to several other medical imaging modalities, including CT and MRI (Gao et al., 2018). For advanced neurosurgical preparation, the authors provide a fantastic possible therapeutic method. The elderly are more likely to fall, which will hurt their bodies and have major bad mental effects on them. To prevent elderly people from falling (Lin et al., 2017), in Taiwan created the fascinating "Fall Prevention Shoes Using Camera-Based Line-Laser Obstacle Detection System" camera-based line-laser obstacle detection system. The system makes use of a laser line that passes through a horizontal plane and is elevated to a certain height above the ground, along with an optical axis that is inclined at a predetermined angle to the plane so that the camera can study the laser pattern and detect any potential obstructions. Unfortunately, the architecture of this system makes it more beneficial for interior applications than for outdoor environments. Human activity recognition (HAR) is one of the most studied topics in CV. (Zhang et al.,

2017) highlighted the improvements in picture representation strategies and classification techniques for the identification of vision-based activities. The three most popular representational approaches are global, local, and depth-based representations. They classify and categorize human activities into these three categories in accordance: action primitives, actions/activities, and interactions. Additionally, they provide a summary of the classification methods used in the HAR application, which comprise seven different categories of algorithms from both dynamic time warping (DTW) and the most recent deep learning. Finally, they discuss how difficult it is to deploy these modern HAR techniques in actual systems or applications, despite the fact that they have so far had remarkable success.

Analysis of medical image. This subject makes an effort to discuss advancements and fresh approaches to medical image analysis technologies. The region being studied must first be well characterized by the integration of multimodal data from several diagnostic imaging methods. For both quantitative multiparametric analysis and qualitative visual assessment in research applications, picture co-registration has therefore become essential. In complicated anatomical areas like the head/neck (HN), (Monti et al., 2017) examined and evaluated the performance of the standard co-registration methods being applied to positron emission tomography/magnetic resonance imaging (PET/MRI) captured as separate modalities and the results with the implicit co-registration of a hybrid PET/MRI. The experimental findings demonstrate that compared to retrospectively co-registered images, hybrid PET/MRI offers a greater registration accuracy. For the study of medical pictures, one of the major problems is feature extraction. In the field of medical diagnosis, CV has made significant advancements in detecting breast cancer and evaluating colorectal polyps. (Isikli Esener et al., 2017) introduced an innovative approach for breast cancer detection using a multistage classification system. They improved upon single-stage classification techniques by combining local configuration pattern-based, statistical, and frequency domain data into feature vectors. Addressing the limitations of routine polyp screening for colorectal cancer (Vázquez et al., 2017), proposed a robust benchmark for colonoscopy image analysis and an improved standard for colonoscopy picture segmentation. By leveraging CV techniques, their work aimed to enhance the visual evaluation of polyp malignancy, overcoming the high miss rate associated with traditional screening methods. These advancements demonstrate the potential of CV in medical diagnosis. By leveraging advanced algorithms and feature ensembles, the detection of breast cancer can be improved, leading to more effective diagnoses. Similarly, the enhanced analysis and segmentation techniques applied to colonoscopy images offer promise in the evaluation of colorectal polyps, providing a more accurate assessment of malignancy.

They demonstrate that the fully convolutional networks (FCN) perform better than the results of the earlier studies in the segmentation of endoluminal scenes.

Machine learning algorithms for medical images. The surprising rise in the number of elder adults throughout the globe will have a significant effect on the healthcare system. Elderly people are never capable of caring for themselves; hence, in recent years, emphasis has been focused heavily on healthcare and nursing robots. The typical technique of detection is always in a single mode, even if somatosensory technology has been used to recognize older patients' activities and engage with their healthcare providers. To provide a practical and effective interaction assistance system for patients suffering from dementia (Dang et al., 2017), electroencephalogram (EEG) data are first captured as the mental feature, followed by motion extraction following depth image preprocessing. For the patient with dementia who requires particular care, the unique approach that has been presented which works on neuron networks having deep multimode. (1) Motion features that were extracted based on the depth image sensor and (2) EEG characteristics for the networks. The type of recognition of the patient's assistance demand is the output layer. The suggested approach attained accuracy and recall rates of 96.5% and 96.4% for the shuffled dataset and 90.9% and 92.6% for the continuous dataset, respectively, in experiments, proving that it streamlines the recognition process. Additionally, as compared to the conventional approach, the suggested algorithms streamline data collection and processing under conditions of high action recognition ratio. Deep learning has recently gained a lot of popularity in the field of AI. Deep learning was used by (Song et al., 2017) to detect lung cancer in its early stages. The experimental findings indicated that, compared to DNN and SAE, that is, deep neural networks or stacked autoencoders, convolutional neural networks, archived the best performance.

Food security

Strategies for early disease detection and classification

One of the biggest dangers to food security is plant diseases. This issue can be resolved through early diagnosis of plant/crop diseases. Using a CNN-based model, (Mohanty et al., 2016) identified the diseases in plants. This they achieved using PlantVillage dataset for the purpose of training the models. For feature extraction and classification, CNN-based models like AlexNet and GoogleNet are employed. They classified ill courses from healthy classes with an accuracy rating of 99.6%.

Khan et al. (2020) has developed a technique for the classification of fruit diseases based on selective features. PlantVillage dataset was used for the purpose, which consisted of images from five different classes of fruit diseases: apple scab, black rot of apples, rust, healthy apple, healthy cherry, powdery mildew, healthy peach, and bacterial spot of peach. A total of 620, 630, 275, 1645, 1052, 854, 2297, and 360 pictures were considered for respective classes. To extract features, (Khan et al., 2020) employed pretrained models such as VGG-S and AlexNet. Deep features obtained from these models were then fed into a multiclass-SVM for classification. The authors achieved high accuracy using their feature extraction techniques, with a Pearson correlation coefficient (PCC) accuracy of 96.7% and an entropy controlled feature fusion accuracy of 97.8%. While manual feature extraction is an option, Khan et al. explored additional sophisticated machine-learning techniques in their study. These techniques included histogram-oriented gradient (HOG), segmented fractal texture analysis (SFTA), local ternary pattern (LTP), principle component analysis (PCA), and entropy skewness (Aurangzeb et al., 2020). These methods provide alternative approaches to extracting relevant features from the fruit disease images, further enhancing the accuracy of classification algorithms.

Fruit quantity and quality detection

Fruits are crops of commercial importance; thus, it is important to gauge their quantity and quality accurately. According to (Chandini and Maheswari, 2018), counting fruit is a difficult task based on how it appears. It resulted from a shift in light and occlusion, which showed up on the nearby fruits. The model was trained using a dataset of 7200 photos of oranges and apples. For blob recognition, fruit counting, and yield information prediction, they employed CNNs, together with linear regression. For apples and oranges, the intersection over union was reported with values of 0.813 and 0.804, respectively.

(Shrivakshan and Chandrasekar, 2012) created a food industry strategy that calls for food products to be tested before going on sale. Organoleptic features, such as taste, color, odor, and feel (together referred to as the attributes of food items), are used to identify food products using artificial sensory systems. For prediction, they have employed a multilayer perceptron model based on ANNs. On the ICatador program, ANNs are utilized to forecast sensory qualities. When applying the Near infrared (NIR) analyzer, the artificial taste neural network produced accurate predictions. To make the machine learning algorithms more competitive, intelligent algorithms were also utilized to fine-tune them. The human grading of fruits was replaced with an automated method for classifying fruits (Choi et al., 2018). Based on interior flavor and external appearance, this approach may categorize fruit. The goal of this study was to recognize items with a variety of distinct properties, such as color, shape, weight, and volume, consistently. The author evaluated fruits using ANNs and other grade classes, such as A, B, and C. Furthermore, the model recognized that internal flavor components, including moisture, sweetness, saltiness, and acidity, as well as the texture of the fruit, such as hardness, crispness, and nutrients, were all taken into account. The cost of manpower for manually sorting fruit decreased as the system's efficiency rose in the industries (Sood and Singh, 2021).

Environment

The use of CV and its numerous features, including "picture identification and classification," has aided in the separation of various street wastes. Using a categorization image to identify additional items and materials, CV analyses waste products in this way. The categorization pictures are from their own dataset of high-definition camera-shot photos that clearly identify the various edges and vectors. This method aids in the detection of waste material by making predictions using regression networks using a collection of categorization photos (Rad et al., 2017). Reusing by recycling valuable resources, such as metallic materials and nonbiodegradable things, is another advantage of employing CV. Identifying, separating, and sorting objects ranging from precious resources to hazardous garbage may be done with great precision by using a program called hyperspectral image analysis. According to an experiment, the accuracy of applying hyperspectral image analysis to detect and sort the following materials, such as copper, aluminum, stainless steel, brass, and lead, was 98% (Picon et al., 2010). CV has made it possible to detect and distinguish between the various particles that are embedded in articles and in their surroundings. With the use of technology, picture analysis and segmentation are made simpler. The utilization of recycling and reusability has converted it into a normal practice today. One use of this technique is the detection of various materials, such as nonbiodegradable substances turned into reusable things. When it comes to wastage of land, CV lessens harmful practices and raises awareness (Linardatos et al., 2021).

Another major factor contributing to global warming and climate change is trash. The pollution generated from the garbage results in seasonal changes that have harmed air quality, visibility, and public health by triggering asthma. CV and its applications help in scanning the atmosphere to gather excess data to pinpoint pollution origins that are visible as well as invisible. According to recent research that was made to extract photos from China and the US, the data in the photographs disclose time, date, and geographic information under various weather circumstances, all of which are utilized to forecast atmospheric particle matter. As per these data, the particle matter of 2.5 causes negative health impacts such as shortness of breath and irritation of the eyes, nose, and throat. Cardiovascular and respiratory conditions pose a further risk to the public's health (Manisalidis et al., 2020).

Large volumes of data were able to be maintained with the use of CV in order to foresee a forthcoming forecast that might harm the aggregation. Using an image processing technique may be used to anticipate the Particle Matter index during the data-receiving phase. The method uses two different types of data for training and prediction, and it goes through numerous processes. The inner process for these two kinds of data is as follows: Training photographs regression model, feature extraction, the selection of regions of interest, and the prediction process. In order to analyze the behavior and patterns of the globe, this technology aids in both the preservation of the atmospheric value and the prediction of the future (Liu et al., 2016). This technology allows us to contribute to the waste management control and purification of water bodies determining the worth of water by utilizing many aspects of CV. These characteristics aid in the enrollment of scanning and the location-based detection of contaminated water, which aids in learning how to maintain a clean environment. This application uses photographs of sewage water to monitor and treat the level of sewage in the water quality (Picon et al., 2010).

Using deep learning−based computers is another method of detecting garbage in the water. Using a CV program, marine garbage may be found in real-time, including typical objects like plastic and metal as well as any other kind of waste that pollutes the sea. The NVIDIA GTX 1080 graphics card, which was assembled in a scalable link interface (SLI), and Intel computer processing processors were used to create several detectors. Three categories make up the class detection of data: biological substances, which comprise organic life, ROV, which includes all man-made things, and plastic, which include all maritime garbage. The outcomes of this experiment demonstrate that utilizing CV to find garbage underwater is extremely likely in the near future. In order to rescue marine life and maintain a clean environment, it is possible to dispose of many items that contaminate the ocean floor using CV (Fulton et al., 2019).

Conclusion

A thorough analysis of state-of-the-art robotic technologies shows that productivity in the field of robotics and vision systems has significantly risen when compared to manual production methods. The highest potential for research and development is shown to be in the areas of healthcare, food, and waste management. Opportunities exist in the areas of sensor fusion, CPS design, human−machine interfaces (HMI), software for robot learning and training, vision systems, structural reconfigurability of robots, and operation of robots during maintenance. Based on the enabling technologies that were not available, new concepts are developing. To create innovative and competitive solutions, it is vital to connect multiple technical fields.

References

Aurangzeb, K., Akmal, F., Attique Khan, M., Sharif, M., Javed, M.Y., 2020. Advanced machine learning algorithm based system for crops leaf diseases recognition. In: Proceedings - 2020 6th Conference on Data Science and Machine Learning Applications, CDMA 2020. Institute of Electrical and Electronics Engineers Inc., Saudi Arabia, pp. 146−151. https://doi.org/10.1109/CDMA47397.2020.00031. http://ieeexplore.ieee.org/xpl/mostRecentIssue.jsp?punumber=9036009.

Baldwin, D.R., Gustafson, J., Pickup, L., Arteta, C., Novotny, P., Declerck, J., Kadir, T., Figueiras, C., Sterba, A., Exell, A., Potesil, V., Holland, P., Spence, H., Clubley, A., O'Dowd, E., Clark, M., Ashford-Turner, V., Callister, M.E.J., Gleeson, F.V., 2020. External validation of a convolutional neural network artificial intelligence tool to predict malignancy in pulmonary nodules. Thorax 75 (4), 306–312. https://doi.org/10.1136/thoraxjnl-2019-214104.

Bøgh, S., Schou, C., Rühr, T., Kogan, Y., Dömel, A., Brucker, M., Eberst, C., Tornese, R., Sprunk, C., Tipaldi, G.D., Hennessy, T., 2014. Integration and assessment of multiple mobile manipulators in a real-world industrial production facility. In: Proceedings for the Joint Conference of ISR 2014 - 45th International Symposium on Robotics and Robotik 2014 - 8th German Conference on Robotics, ISR/ROBOTIK 2014. VDE-Verlag, Denmark, pp. 305–312.

Chandini, A.A., Maheswari, B.U., 2018. Improved quality detection technique for fruits using GLCM and MultiClass SVM. In: 2018 International Conference on Advances in Computing, Communications and Informatics, ICACCI 2018. Institute of Electrical and Electronics Engineers Inc., India, pp. 150–155. https://doi.org/10.1109/ICACCI.2018.8554876. http://ieeexplore.ieee.org/xpl/mostRecentIssue.jsp?punumber=8536361.

Choi, H.S., Cho, J.B., Kim, S.G., Choi, H.S., 2018. A real-time smart fruit quality grading system classifying by external appearance and internal flavor factors. In: Proceedings of the IEEE International Conference on Industrial Technology. Institute of Electrical and Electronics Engineers Inc., South Korea, pp. 2081–2086. https://doi.org/10.1109/ICIT.2018.8352510.

Costa, T., Coelho, L., Silva, M.F., 2022. Integrating Computer Vision, Robotics, and Artificial Intelligence for Healthcare. IGI Global, pp. 134–162. https://doi.org/10.4018/978-1-6684-5260-8.ch007.

Dang, X., Kang, B., Liu, X., Cui, G., 2017. An interactive care system based on a depth image and EEG for aged patients with dementia. Journal of Healthcare Engineering 2017. https://doi.org/10.1155/2017/4128183.

EFFRA, 2013. Factories of the Future 2020. http://www.effra.eu/.

Elliott, A., 2019. The culture of AI: Everyday life and the digital revolution. https://books.google.co.in/books?hl=en&lr=&id=U-GEDwAAQBAJ&oi=fnd&pg=PT14&dq=Automation,+in+association+with+smart+robots,+propels+contemporary+industries+toward+productiveness,+leading+to+a+significant+rise+in+efficaciousness+besides+savings+other+resources+and+energy+while+improving+working+conditions+(Pascual+and+Restrepo,+2018).&ots=dodbZ4aDIw&sig=B5fBntlDnPzzN5Itj6rOc7ap3S4&redir_esc=y#v=onepage&q&f=false.

El-Sappagh, S., Alonso, J.M., Islam, S.M.R., Sultan, A.M., Kwak, K.S., 2021. A multilayer multimodal detection and prediction model based on explainable artificial intelligence for Alzheimer's disease. Scientific Reports 11 (1). https://doi.org/10.1038/s41598-021-82098-3.

Fedorova, E., Darbasov, V., Okhlopkov, M., Loretts, O., Ojha, N., Ruchkin, A., Vinogradov, S., Kukhar, V., Lopez Garcia, J.L., 2020. The role of agricultural economists in study on problems related to regional food safety. E3S Web of Conferences 176. https://doi.org/10.1051/e3sconf/202017605011.

Fuentes-Pacheco, J., Ruiz-Ascencio, J., Rendón-Mancha, J.M., 2015. Visual simultaneous localization and mapping: a survey. Artificial Intelligence Review 43 (1), 55–81. https://doi.org/10.1007/s10462-012-9365-8.

Fulton, M., Hong, J., Islam, M.J., Sattar, J., 2019. Robotic detection of marine litter using deep visual detection models. In: Proceedings - IEEE International Conference on Robotics and Automation. 2019-. Institute of Electrical and Electronics Engineers Inc., United States, pp. 5752–5758. https://doi.org/10.1109/ICRA.2019.8793975.

Gao, J., Yang, Y., Lin, P., Park, D.S., 2018. Computer vision in healthcare applications. Journal of Healthcare Engineering 2018, 1–4. https://doi.org/10.1155/2018/5157020.

Hamburg, M., 2011. Food Safety Modernization Act: Putting the focus on prevention. http://www.foodsafety.gov/news/fsma.html.

Hartley, R., Zisserman, A., 2003. Multiple View Geometry in Computer Vision. Cambridge university press, 2003.

Hauser, K., Shaw, R., 2020. How Medical Robots Will Help Treat Patients in Future Outbreaks. IEEE Spectrum.

Hu, H., Wang, H., Wang, F., Langley, D., Avram, A., Liu, M., 2018. Prediction of influenza-like illness based on the improved artificial tree algorithm and artificial neural network. Scientific Reports 8 (1). https://doi.org/10.1038/s41598-018-23075-1.

Isikli Esener, I., Ergin, S., Yuksel, T., 2017. A new feature ensemble with a multistage classification scheme for breast cancer diagnosis. Journal of Healthcare Engineering 2017, 1−15. https://doi.org/10.1155/2017/3895164.

ISO, 2021. International Organization for Standardization, Geneve, Switzerland. https://www.iso.org/home.html.

Jiang, X., Coffee, M., Bari, A., Wang, J., Jiang, X., Huang, J., Shi, J., Dai, J., Cai, J., Zhang, T., Wu, Z., He, G., Huang, Y., 2020. Towards an artificial intelligence framework for data-driven prediction of coronavirus clinical severity. Computers, Materials and Continua 63 (1), 537−551. https://doi.org/10.32604/cmc.2020.010691.

Kagermann, H., Helbig, J., Hellinger, A., Wahlster, W., 2013. Recommendations for Implementing the Strategic Initiative INDUSTRIE 4.0: Securing the Future of German Manufacturing Industry, 2013.

Kakani, V., Nguyen, V.H., Kumar, B.P., Kim, H., Pasupuleti, V.R., 2020. A critical review on computer vision and artificial intelligence in food industry. Journal of Agriculture and Food Research 2. https://doi.org/10.1016/j.jafr.2020.100033.

Kanji, G.K., 1990. Total quality management: the second industrial revolution. Total Quality Management 1 (1), 3−12. https://doi.org/10.1080/09544129000000001.

Karthik, K.P., Chandra, R.P., 2014. An overview of agricultural robots. http://www.yuvaengineers.com/an-overview-of-agricultural-robots-p-koteswara-karthik-p-ravi-chandra.

Khan, M.A., Akram, T., Sharif, M., Saba, T., 2020. Fruits diseases classification: exploiting a hierarchical framework for deep features fusion and selection. Multimedia Tools and Applications 79 (35−36), 25763−25783. https://doi.org/10.1007/s11042-020-09244-3.

Kirchberger, T., 2017. European Union policy-making on robotics and artificial intelligence: selected issues. Croatian Yearbook of European Law & Policy 13 (1), 191−214.

Kumar, I., Rawat, J., Mohd, N., Husain, S., Khan, R., 2021. Opportunities of artificial intelligence and machine learning in the food industry. Journal of Food Quality 2021, 1−10. https://doi.org/10.1155/2021/4535567.

Langendoen, K., Baggio, A., Visser, O., 2006. Murphy loves potatoes: experiences from a pilot sensor network deployment in precision agriculture. In: Proceedings 20th IEEE International Parallel & Distributed Processing Symposium. IEEE.

Lin, T.H., Yang, C.Y., Shih, W.P., 2017. Fall prevention Shoes using camera-based line-laser obstacle detection system. Journal of Healthcare Engineering 2017. https://doi.org/10.1155/2017/8264071.

Linardatos, P., Papastefanopoulos, V., Kotsiantis, S., 2021. Explainable AI: a review of machine learning interpretability methods. Entropy 23 (1). https://doi.org/10.3390/e23010018.

Liu, C., Tsow, F., Zou, Y., Tao, N., Liu, H., 2016. Particle pollution estimation based on image analysis. PLoS One 11 (2). https://doi.org/10.1371/journal.pone.0145955.

Liu, L.W., Lu, C.T., Wang, Y.M., Lin, K.H., Ma, X., Lin, W.S., 2022. Rice (*Oryza sativa* L.) growth modeling based on growth degree day (GDD) and artificial intelligence algorithms. Agriculture 12 (1). https://doi.org/10.3390/agriculture12010059.

Manisalidis, I., Stavropoulou, E., Stavropoulos, A., Bezirtzoglou, E., 2020. Environmental and health impacts of air pollution: a review. Frontiers in Public Health 8. https://doi.org/10.3389/fpubh.2020.00014.

Meeks, S.L., Pukala, J., Ramakrishna, N., Willoughby, T.R., Bova, F.J., 2011. Radiosurgery technology development and use. Journal of Radiosurgery and SBRT 1 (1), 2011.

Misra, N.N., Dixit, Y., Al-Mallahi, A., Bhullar, M.S., Upadhyay, R., Martynenko, A., 2022. IoT, big data, and artificial intelligence in agriculture and food industry. IEEE Internet of Things Journal 9 (9), 6305−6324. https://doi.org/10.1109/JIOT.2020.2998584.

Mohanty, S.P., Hughes, D.P., Salathé, M., 2016. Using deep learning for image-based plant disease detection. Frontiers in Plant Science 7 (September). https://doi.org/10.3389/fpls.2016.01419.

Monti, S., Cavaliere, C., Covello, M., Nicolai, E., Salvatore, M., Aiello, M., 2017. An evaluation of the benefits of simultaneous acquisition on PET/MR coregistration in head/neck imaging. Journal of Healthcare Engineering 2017. https://doi.org/10.1155/2017/2634389.

Mukai, T., Hirano, S., Nakashima, H., Kato, Y., Sakaida, Y., Guo, S., Hosoe, S., 2010. Development of a nursing-care assistant robot RIBA that can lift a human in its arms. In: IEEE/RSJ 2010 International Conference on Intelligent Robots and Systems, IROS 2010 - Conference Proceedings, pp. 5996–6001. https://doi.org/10.1109/IROS.2010.5651735.

Olaronke, I., Ishaya, G., Oluwaseun, O., Rhoda, I., Janet, O., 2022. The need for robots in global health. Current Journal of Applied Science and Technology 26–36. https://doi.org/10.9734/cjast/2022/v41i531668.

Pérez, L., Rodríguez, Í., Rodríguez, N., Usamentiaga, R., García, D., 2016. Robot guidance using machine vision techniques in industrial environments: a comparative review. Sensors 16 (3). https://doi.org/10.3390/s16030335.

Picon, A., Ghita, O., Iriondo, P.M., Bereciartua, A., Whelan, P.F., 2010. Automation of waste recycling using hyperspectral image analysis. In: Proceedings of the 15th IEEE International Conference on Emerging Technologies and Factory Automation, ETFA 2010. Spain. https://doi.org/10.1109/ETFA.2010.5641201.

Podpora, M., Gardecki, A., Beniak, R., Klin, B., Vicario, J.L., Kawala-Sterniuk, A., 2020. Human interaction smart subsystem—extending speech-based human-robot interaction systems with an implementation of external smart sensors. Sensors 20 (8). https://doi.org/10.3390/s20082376.

Polat, H., Topalcengiz, Z., Danyluk, M.D., 2020. Prediction of Salmonella presence and absence in agricultural surface waters by artificial intelligence approaches. Journal of Food Safety 40 (1). https://doi.org/10.1111/jfs.12733.

Postolache, O., Pereira, M., Girão, P., 2013. Sensor network for environment monitoring: water quality case study. In: 4th IMEKO TC19 Symposium on Environmental Instrumentation and Measurements 2013: Protection Environment, Climate Changes and Pollution Control, pp. 30–34.

Rad, M.S., von Kaenel, A., Droux, A., Tieche, F., Ouerhani, N., Ekenel, H.K., Thiran, J.P., 2017. A computer vision system to localize and classify wastes on the streets. Lecture Notes in Computer Science 10528, 195–204. https://doi.org/10.1007/978-3-319-68345-4_18.

Rathore, D., Divyanth, L.G., Reddy, K.L.S., Chawla, Y., Buragohain, M., Soni, P., Machavaram, R., Hussain, S.Z., Ray, H., Ghosh, A., 2023. A two-stage deep-learning model for detection and occlusion-based classification of Kashmiri orchard apples for robotic harvesting. Journal of Biosystems Engineering 48. https://doi.org/10.1007/s42853-023-00190-0.

Richmon, J.D., Kim, H.Y., 2017. Transoral robotic thyroidectomy (TORT): procedures and outcomes. Gland Surgery 6 (3), 285–289. https://doi.org/10.21037/gs.2017.05.05.

Sarker, S., Jamal, L., Ahmed, S.F., Irtisam, N., 2021. Robotics and artificial intelligence in healthcare during COVID-19 pandemic: a systematic review. Robotics and Autonomous Systems 146. https://doi.org/10.1016/j.robot.2021.103902.

Shadrin, D., Menshchikov, A., Somov, A., Bornemann, G., Hauslage, J., Fedorov, M., 2020. Enabling precision agriculture through embedded sensing with artificial intelligence. IEEE Transactions on Instrumentation and Measurement 69 (7), 4103–4113. https://doi.org/10.1109/TIM.2019.2947125.

Shrivakshan, G.T., Chandrasekar, C., 2012. A comparison of various edge detection techniques used in image processing. International Journal of Computer Science Issues (IJCSI) 9 (5).

Song, Q.Z., Zhao, L., Luo, X.K., Dou, X.C., 2017. Using deep learning for classification of lung nodules on computed tomography images. Journal of Healthcare Engineering 2017, 1–7. https://doi.org/10.1155/2017/8314740.

Sood, S., Singh, H., 2021. Computer vision and machine learning based approaches for food security: a review. Multimedia Tools and Applications 80 (18), 27973—27999. https://doi.org/10.1007/s11042-021-11036-2.

Timmerman, R.D., Herman, J., Cho, L.C., 2014. Emergence of stereotactic body radiation therapy and its impact on current and future clinical practice. Journal of Clinical Oncology (26), 2847—2854. https://doi.org/10.1200/JCO.2014.55.4675.

Timonen, T., Iso-Mustajärvi, M., Linder, P., Vrzakova, H., Sinkkonen, S.T., Luukkainen, V., Laitakari, J., Elomaa, A.P., Dietz, A., 2022. The feasibility of virtual reality for anatomic training during temporal bone dissection course. Frontiers in Virtual Reality 3. https://doi.org/10.3389/frvir.2022.957230.

United Nations Economic Commission for Europe, 2021. Air Pollution and Health.

Vázquez, D., Bernal, J., Sánchez, F.J., Fernández-Esparrach, G., López, A.M., Romero, A., Drozdzal, M., Courville, A., 2017. A benchmark for endoluminal scene segmentation of colonoscopy images. Journal of Healthcare Engineering 2017, 1—9. https://doi.org/10.1155/2017/4037190.

Vedaldi, A., Fulkerson, B., 2010. Vlfeat - an open and portable library of computer vision algorithms. In: MM'10 - Proceedings of the ACM Multimedia 2010 International Conference, pp. 1469—1472. https://doi.org/10.1145/1873951.1874249.

Watthanawisuth, N., Tuantranont, A., Kerdcharoen, T., 2009. Microclimate real-time monitoring based on ZigBee sensor network. In: Proceedings of IEEE Sensors, pp. 1814—1818. https://doi.org/10.1109/ICSENS.2009.5398587.

World Health Organization, 2021. Tracking Universal Health Coverage.

Zhang, J.Y., Zhao, C., Zhang, D.W., 2011. Pose accuracy analysis of robot manipulators based on kinematics. Advanced Materials Research 201—203, 1867—1872. https://doi.org/10.4028/www.scientific.net/amr.201-203.1867.

Zhang, S., Wei, Z., Nie, J., Huang, L., Wang, S., Li, Z., 2017. A review on human activity recognition using vision-based method. Journal of Healthcare Engineering 2017. https://doi.org/10.1155/2017/3090343.

Zhao, Y., Pohl, O., Bhatt, A.I., Collis, G.E., Mohan, P.J., Ruther, T., Hollenkamp, A.F., 2021. A review on battery market trends, second-life reuse, and recycling. Sustainable Chemistry 2 (1), 167—205.

Zhou, K., Ebenhofer, G., Eitzinger, C., Zimmermann, U., Walter, C., Saenz, J., Castaño, L.P., Hernández, M.A.F., Oriol, J.N., 2014. Mobile manipulator is coming to aerospace manufacturing industry. In: ROSE 2014 — 2014 IEEE International Symposium on RObotic and SEnsors Environments, Proceedings. Institute of Electrical and Electronics Engineers Inc., Austria, pp. 94—99. https://doi.org/10.1109/ROSE.2014.6952990.

Artificial intelligence in classrooms: How artificial intelligence can aid in teaching biology

17

Arielle Yoo

University of California, Davis, CA, United States

Introduction

The focus of this chapter is on discussing the use of artificial intelligence (AI) educational tools. This field is relatively new and seems to begin with the first nonresearch spell-checker application called SPELL for the DEC-10, which was written by Ralph Gorin at Stanford in 1971 (Peterson, 1980). However, the topic of spell-checking itself was originally researched in order to fix data entry mistakes (Peterson, 1980). While not directly created as an AI educational tool, spell-checkers are now ubiquitous in many tools used for education, such as Microsoft Office. However, adding spell-checking as a common educational tool was debated from 1993 to 2005 for students with disabilities, ignoring the general student population in the discussion (Lazarus et al., 2009; Popenici and Kerr, 2017). In addition to this delayed and uneven adoption of AI tools in education, much of the field is still growing and changing, especially in the development of AI tools. With large language models such as BERT being introduced in 2018 (Devlin et al., 2019) and related tools like ChatGPT becoming well-received following its release in late 2022 (OpenAI, 2022), AI education chatbot research and development are expected to continue to grow. Although this field is still progressing, this chapter provides an overview of the types of AI education tools available currently and their future directions.

AI educational tools

Generally, AI educational tools were created to aid as disability tools, intelligent tutors and teachable agents, chatbots, personalized learning systems or environments (PLS/E), and visualizations and virtual reality (VR) (Chassignol et al., 2018; Popenici and Kerr, 2017; Zhai et al., 2021; Zhang and Aslan, 2021). The following sections will discuss examples of each kind of tool with biology-specific examples if applicable and the effects of using these in classrooms. Fig. 17.1 contains an overview of these tools, their effects, and criticisms involved in using them. The design of this figure is inspired by an AI education review paper (Zhang and Aslan, 2021).

Disability tools

Students with disabilities make up a significant portion of the population, and it is important to be able to accommodate them to the best of our ability. In the United States alone, approximately 15% of all

A Biologist's Guide to Artificial Intelligence. https://doi.org/10.1016/B978-0-443-24001-0.00017-8

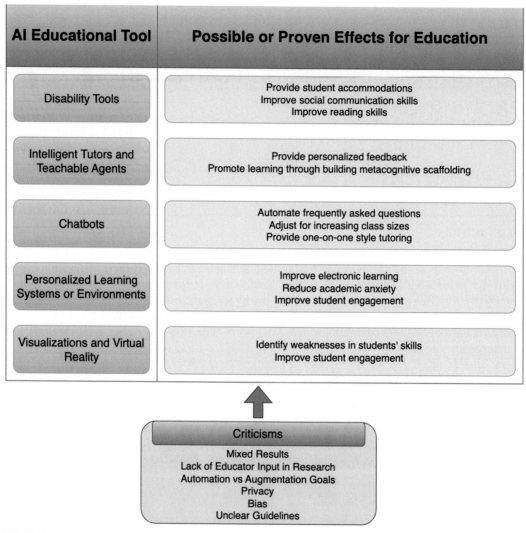

FIGURE 17.1

Overview of artificial intelligence (AI) educational tools, their effects, and their criticisms shows a table for the different kinds of educational tools and their known effects and also a list of general criticisms.

Design inspired by Zhang, K., Aslan, A.B., 2021. AI technologies for education: recent research & future directions. Computers and Education: Artificial Intelligence, 2. https://doi.org/10.1016/j.caeai.2021.100025

public school students aged 3—21 years in the 2021 school year received special education or related services (National Center for Education Statistics, 2023). Assistive technologies were developed to help people with disabilities and thus can help students with disabilities. Examples of these include text-to-speech, speech-to-text, predictive text, spell checkers, and search engines (Popenici and Kerr,

2017). For visually impaired students, a text-to-diagram tool can help connect geometry word problems to a Braille diagram printout (Mukherjee et al., 2014; Zhai et al., 2021). The blind people using this text-to-diagram tool in Kolkata, India, generally rated the tool positively, albeit a bit difficult to use (Mukherjee et al., 2014). For autistic students, augmented reality smart glasses can serve as a social communication aid to help them learn social and emotional skills with little discomfort to the user (Keshav et al., 2017; Zhang and Aslan, 2021). Immersive Reader, a Microsoft learning tool, also helps students with dyslexia read better and performs text-to-speech (Chassignol et al., 2018; Microsoft Learning Tools, no date). Although some of these tools are general and not solely used for education, these tools can still be used to help students learn and accommodate students' disabilities.

Biology-specific disability tool: ForAlexa

An example of a biology-specific disability tool is ForAlexa, a tool that uses a voice-activated Amazon Alexa to interact with students and help them learn skills (Rabelo et al., 2022). The main goal of this tool is to serve as an additional form of contact between the educator and student outside of the classroom (Rabelo et al., 2022). This tool also helps educators develop learning applications related to their class topic quickly and can help students with special needs, such as the visually impaired (Rabelo et al., 2022). This tool was specifically tested on teaching seven chapters of a Portuguese undergraduate course in Population Genetics, and the results show that the students mostly gave it a positive evaluation (Rabelo et al., 2022). This tool allows educators to add information to use for its question-and-answer (Q&A) and random-quote functionality that students can interact with verbally (Rabelo et al., 2022). However, this study admits that there are still many challenges to teaching visually impaired students that remain unaddressed, such as ensuring all classwork is adapted to accommodate these students and the general lack of understanding of assistive technologies (Orsini-Jones, 2009; Rabelo et al., 2022; Zhou et al., 2011). Nevertheless, this tool can still help visually impaired students learn biology.

Intelligent tutors and teachable agents

Intelligent tutors are one of the more popular areas of research in AI education (Zhai et al., 2021; Zhang and Aslan, 2021). Intelligent tutors or intelligent tutoring systems (ITSs) mainly monitor student input, give appropriate tasks, provide effective feedback, and serve as an interface for human-computer interactions (Seldon and Abidoye, 2018; Zhai et al., 2021). There are many examples of these for many different subjects, such as ACTIVEMATH for math (Melis and Siekmann, 2004), BEETLE II for computational linguistics (Dzikovska et al., 2010), EER-Tutor for database design (Weerasinghe and Mitrovic, 2011), MATHia and MATHia X for math (Ritter, 2011; Ritter and Fancsali, 2016), Auto-Tutor for computer literacy (Al Emran and Shaalan, 2014), Why2-Atlas for physics (Vanlehn et al., 2002), COMET for medicine (Suebnukarn and Haddawy, 2004), VIPER for medicine (Martin et al., 2009), and an unnamed model for categorizing and solving math word problems about motion (Chassignol et al., 2018; Nabiyev et al., 2016; Zhai et al., 2021). Generally, ITSs have had mixed results (Zhang and Aslan, 2021), but they are still an important area of research since their goals are helpful to students.

Teachable agents are computer programs that students can teach, since the action of teaching helps students learn (Chin et al., 2013; Chin et al., 2010; Matsuda et al., 2020; Tärning et al., 2019). These have also been applied to many different subjects and have been shown to help promote learning for elementary school students, metacognitive scaffolding in young students, and problem-solving in middle school students (Chin et al., 2013; Chin et al., 2010; Matsuda et al., 2020; Zhang and Aslan, 2021).

Biology-specific intelligent tutors or agents: Inquiry ITS and Betty's brain

ITSs and teachable agents have also been applied to biology. For ITS, researchers studied the effectiveness of using inquiry ITS to help undergraduate students at Sampoerna University learn biology and develop creative thinking skills by providing opportunities for students to generate hypotheses, design experiments, and discuss and analyze evidence (Alfakihuddin et al., 2022). For teachable agents, researchers studied the effectiveness of using Betty's Brain, a teachable agent, to help eighth-grade students study fevers (Chase et al., 2009). In this study, they showed that students applied more effort and scored better using the teachable agent compared to using an avatar that represented themselves (Chase et al., 2009). These are only two examples, but further research can be done on applying ITS and teachable agents to learning biology.

Chatbots

Chatbots have also been studied as educational tools with mixed results, although there have been significant recent advancements. In one study, comparing having a human partner to a chatbot partner for foreign language courses showed that having a human partner is better for maintaining course interest and predicting future course interest (Fryer et al., 2017; Zhang and Aslan, 2021). However, this may be due to the novelty effect (Fryer et al., 2017; Zhang and Aslan, 2021). Additionally, another study concerning using a chatbot for computer science and computer networks students showed that while the bot was easy to use and gave the correct suggestions most of the time (71.13%), it would sometimes give correct but not related to the user's needs (16.04%) or incorrect suggestions (12.83%) (Clarizia et al., 2018). However, this chatbot and more are considered as successful. Beginning in 2015, a virtual teaching assistant named Jill Watson was built using IBM Watson to help support an online computer science course for master's students at Georgia Institute of Technology and was very positively received by students (Eicher et al., 2018; Maderer, 2016; Popenici and Kerr, 2017; Smutny and Schreiberova, 2020). Jill Watson successfully answered frequently asked questions on Piazza, the class discussion forum, which helped address the instructional burden of increasing class sizes (Eicher et al., 2018).

For recent advancements, ChatGPT's open-access release on November 30, 2022, helped show how large language models can be applied to many complex text-based requests (Lund and Wang, 2023; OpenAI, 2022). ChatGPT is based on OpenAI's Generative pretrained transformer (GPT), which produces text that is nearly indistinguishable from natural human language (Dale, 2021; Lund and Wang, 2023; Radford et al., 2018). While there are other earlier state-of-the-art large language models such as BERT (Devlin et al., 2019), RoBERTa (Liu et al., 2019), and XLNet (Yang et al., 2019), ChatGPT's publicity as a web-based tool brought mainstream attention to chatbots. ChatGPT has grown from 100 million users within 2 months to over 200 million users currently (Chief Investment Office of UBS, 2023). ChatGPT and other large language models can be fine-tuned to answer domain-specific questions (Lund and Wang, 2023), which means it can be used as an educational chatbot. Khanmigo, a chatbot developed by Khan Academy, is also based on a large language model currently being developed to tutor students in a one-on-one style (Khan Academy, no date a,b; Senechal et al., 2023). Fig. 17.2 illustrates using ChatGPT as an educational chatbot. More educational chatbots are expected to be developed and released soon.

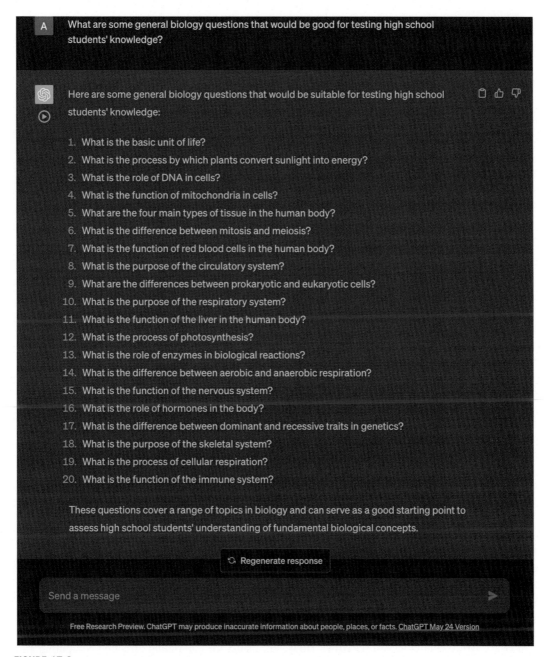

FIGURE 17.2

Using ChatGPT as an educational tool for generating potential test questions. screenshot of querying ChatGPT for potential high school biology test questions.

Screenshot from ChatGPT by OpenAI. OpenAI, 2022. Introducing ChatGPT. https://openai.com/blog/chatgpt (Original work published 2022)

Additionally, chatbots based on large language models have been used to augment search engines. Bing AI by Microsoft (Microsoft, 2023), Bard by Google (Pichai, 2023), and Sparrow by DeepMind (Glaese et al., 2022) are a few current examples of chatbots that can be used to summarize information found online. However, these chatbots are still being developed and have been known to occasionally reply with incorrect information or hallucinate. Further research is currently being conducted on how to ensure these chatbots do not hallucinate fake information.

Biology-specific chatbot: Unnamed chatbot using LINE

To study the effects of using chatbots to teach biology, a study was conducted on seventh-grade biology students in Taiwan to compare using a biology chatbot to using an online biology discussion group in the LINE messaging app (Lin and Ye, 2023). Overall, they found that using the chatbot resulted in higher test scores for students with low prior knowledge (Lin and Ye, 2023). However, this study had a small sample size, thus further investigation may be needed (Lin and Ye, 2023).

Personalized learning systems or environments

Personalized learning systems or environments aim to meet each student's unique needs, which can range from emotional or affective needs, current learning ability, and amount of background knowledge (Chang and Lu, 2019; Chassignol et al., 2018; Hutchins, 2017). Adaptive learning is also part of PLS/Es since its aim is to adjust to individual learners' characteristics, needs, and preferences (Jones, 2011; Zhai et al., 2021). Both of these terms are fairly general, which means they can be used to encompass any educational personalization tasks such as creating personalized learning materials and providing advice (Xu and Wang, 2006). Additionally, any educational tools or virtual learning environments that have personalization of educational content can be considered as a personalized learning system or environment. PLS/Es were found to provide opportunities for online learners to interact (Xu and Wang, 2006; Zhang and Aslan, 2021) and improve electronic learning (Cheung et al., 2003; Köse and Arslan, 2016; Köse, 2018; Xu and Wang, 2006; Zhang and Aslan, 2021). Additionally, using an ITS that personalized math word problems to relate to students' out-of-school interests improved student learning compared to the ITS that did not personalize (Walkington and Bernacki, 2019; Zhang and Aslan, 2021). These studies on PLS/Es were applied to computer science courses (Köse, 2018; Köse and Arslan, 2016), an Introduction to the Oracle Database course (Xu and Wang, 2006), across online courses offered by The University of Hong Kong (Cheung et al., 2003), and high school algebra (Walkington and Bernacki, 2019).

To address students' emotional or affective needs using AI, affection computing is utilized. Affection computing is defined as the analysis of emotions captured by physical sensors and affective algorithms (Zhai et al., 2021). Research shows that affect plays an important role in decision-making, perception, and learning (Ben Ammar et al., 2010; Zhai et al., 2021). For affection computing techniques to be effective in educational applications, they need to be timely so that teachers can adjust quickly to student's affect and they need to combine multiple sensory sources, such as EEG and facial expression, to determine a student's affect since relying on a single source can give inaccurate results (Lin et al., 2012; Zhai et al., 2021, 2018). However, results for combining affection computing and education are mixed. A study conducted in a fifth-grade mathematics course found that the adaptive learning tool that used both affective and cognitive status analysis improved student's learning and reduced their math-related anxieties better than the tools without affective status analysis (Hwang et al., 2020; Zhang and Aslan, 2021). However, another study conducted on students with disabilities that compared using an adaptive learning tool with and without affective status analysis found no significant difference in achievement, but the tool with affective status analysis improved student engagement (Standen et al., 2020). Although the results for achievement are mixed, improving the student's mood while learning can still help make the learning experience more positive.

Biology-specific personalized learning systems or environments: Brightspace LeaP

An example of a PLS/E that has been applied to biology is Brightspace LeaP, which is an adaptive learning technology that dynamically creates personalized learning paths based on student performance (Liu et al., 2017; Xie et al., 2019). In the study, Brightspace LeaP was applied to biology, chemistry, math, and information literacy modules for first-year students entering a pharmacy professional degree program, but results showed that using Brightspace LeaP did not improve student performance (Liu et al., 2017). However, this may be because students found the tool confusing and not well organized (Liu et al., 2017; Xie et al., 2019). Additional research for biology-specific PLS/Es would be helpful in better understanding how they can be more effectively applied in the future.

Visualizations and virtual reality

Visualizations and VR can be used to improve learners' understanding (Zhai et al., 2021; Zhang and Aslan, 2021). For example, visualizing argument diagrams or maps can help students and teachers identify weaknesses in students' argumentation skills (Rapanta and Walton, 2016). In theory, AI models could be trained to use these diagrams to identify weaknesses in students' argumentation skills or to help collaboratively dialog with students about strengthening their arguments (Rapanta and Walton, 2016). In fact, AI models have been applied to assess students' learning by analyzing students' diagrams. For example, the AI-based student learning evaluation tool (AISLE) was developed to evaluate students' concept maps, which are "visual representations of a particular topic and its subcomponents," where subcomponents are linked relationally (Jain et al., 2014). Additionally, creating intelligent virtual learning environments where conversational bots and students can interact with each other using avatars helped students be more engaged in the class and improved their abilities to collaborate with each other (Griol et al., 2014; Zhang and Aslan, 2021). Thus, visualizing learning environments that differ from classroom settings can help students learn.

Technologies combining VR and AI have also been shown to aid in student learning (Ijaz et al., 2017; Zhang and Aslan, 2021). For example, using a VR replica of the historical city of Uruk in 3000 BCE with AI-controlled 3D avatars as citizens of Uruk improved student enjoyment and comprehension of history compared to reading a historical text or watching a documentary (Ijaz et al., 2017). Additionally, AI can be combined with VR as an academic assessment tool, such as using AI to evaluate VR surgeries performed by medical residents and postresidents (Alkadri et al., 2021). For more hands-on classes, this kind of AI assessment of VR performance could be helpful.

Biology-specific visualizations and virtual learning environments: BioVR

VR has been applied to biology learning, such as simulating the "structure and function of the eye" (Shim et al., 2003) and simulating an escape room with puzzles related to learning about enzymes (Christopoulos et al., 2022), but examples combining VR with AI are fewer. However, as technology improves, VR and AI could, in theory, be combined to create BioVR, a self-evolving VR where students can experience and interact with evolution (Morimoto and Ponton, 2021). Although BioVR focuses on evolution, applying VR to biology in general has helped students learn through simulating laboratory environments that were otherwise inaccessible to students and visualizing many different biological structures such as cell structure, digestion, and protein structure (Morimoto and Ponton, 2021). Some proposed features of BioVR include modeling the "rules of evolution" or environment-traits-species interactions and watching these interactions influence other interactions across generations (Morimoto and Ponton, 2021). Additionally, they hope that users can explore these artificial ecosystems virtually in inquiry-based learning quests (Morimoto and Ponton, 2021). Further research should be done in this area to better study how VR and AI could be combined to teach biology.

Criticisms of AI educational tools

While many AI educational tools are available and being developed, there have also been many valid criticisms from educators and researchers. Given the mixed results of some of the tools listed here, there is a debate over how helpful some of these tools are (Zhai et al., 2021; Zhang and Aslan, 2021), especially since creating a control group can be difficult. For example, comparing AI educational tools to no specific classroom intervention may be valid in some cases, but comparing to a different non-AI educational tool might be a fairer comparison in others. Additionally, some of the tools listed here were not tested against a control group and were primarily assessed on whether they functioned as intended. Also, some of these tools were written from a computer science domain rather than an education perspective, which means some of these tools could have been developed without considering pedagogical theories of how students learn (Zhai et al., 2021; Zhang and Aslan, 2021). Teachers' opinions on using AI have a significant influence on the AI educational tools' effectiveness (Zhai et al., 2021), so it is important to ensure that their perspectives are heard in the development of these tools. Longitudinal and large-scale interdisciplinary research in AI educational tools could be helpful to better understand the effectiveness of these tools and how best to integrate them in the classroom (Zhang and Aslan, 2021). There are also debates over whether the goal of AI educational tools should be automating current educational tasks or augmenting them (Popenici and Kerr, 2017). Some tools were specifically developed to automate tasks to assist in increasing class sizes or few interactions between teachers and students (Eicher et al., 2018; Lin and Ye, 2023). There are also scholars that publicly propose replacing teachers with AI bots (Edwards and Cheok, 2018; Zhang and Aslan, 2021). However, many feel that AI is not sufficient for replacing teachers yet, and instead, AI tools should focus on improving education rather than replacing proven teaching methods (Popenici and Kerr, 2017).

There are many ethical considerations to using AI educational tools. Privacy, for example, is a major concern (Popenici and Kerr, 2017; Zhang and Aslan, 2021), along with AI bias (Eicher et al., 2018; Popenici and Kerr, 2017). Additionally, with how easy and publicly available chatbots and large language models like ChatGPT are to students, cheating and plagiarism are major concerns in education. ChatGPT text is difficult to detect compared to human-written text since it mimics natural language well, which makes it hard to know which students may have used these tools in their work (Khalil and Er, 2023). In one study interviewing students at Pangasinan State University, Lingayen Campus, the majority of students admitted to being tempted to cheat using ChatGPT despite knowing that cheating is detrimental to their learning (Ventayen, 2023). Also, the issue of potentially uneven regulations across schools continues, especially given different teachers' opinions on whether these educational tools should be used. For example, some argue that using ChatGPT is not cheating, and learning to use AI tools is also important to students' futures in the job market (Anders, 2023). Further research and guidelines should be developed to address these ethical considerations (Zhang and Aslan, 2021).

Conclusion

In summary, this chapter reviewed AI education tools under the categories of disability tools, intelligent tutors and teachable agents, chatbots, PLS/Es, and visualizations and VR (Chassignol et al., 2018; Popenici and Kerr, 2017; Zhai et al., 2021; Zhang and Aslan, 2021). This chapter also discussed their known effects on students and some future directions as the field continues to grow. Additionally, the chapter examined current criticisms of using AI educational tools, from implementation concerns to ethical considerations. As the research space continues to expand, the hope is that the understanding of how these tools can best be used to augment students' learning experience improves along with clear educational guidelines so that schools can best take advantage of these technological advancements.

Abbreviations

AI   Artificial intelligence
AISLE   Artificial intelligence-based student learning evaluation tool
GPT   Generative pretrained transformer
ITS   Intelligent tutoring system
PLS/E   Personalized learning systems or environments
StuDiAsE   Student Diagnosis, Assistance, Evaluation System based on Artificial Intelligence
VLE   Visualizations and virtual learning environments
VR   Virtual reality

References

Al Emran, M., Shaalan, K., 2014. A survey of intelligent language tutoring systems. In: Proceedings of the 2014 International Conference on Advances in Computing, Communications and Informatics, ICACCI 2014. Institute of Electrical and Electronics Engineers Inc., United Arab Emirates, pp. 393−399. https://doi.org/10.1109/ICACCI.2014.6968503.

Alfakihuddin, M.L.B., Surahman, E., Haryani, F., 2022. The application of inquiry intelligent tutoring system in biology practicum. Journal of Education Technology 6 (4), 634−642. https://doi.org/10.23887/jet.v6i4.48466.

Alkadri, S., Ledwos, N., Mirchi, N., Reich, A., Yilmaz, R., Driscoll, M., Del Maestro, R.F., 2021. Utilizing a multilayer perceptron artificial neural network to assess a virtual reality surgical procedure. Computers in Biology and Medicine 136. https://doi.org/10.1016/j.compbiomed.2021.104770.

Anders, B.A., 2023. Is using ChatGPT cheating, plagiarism, both, neither, or forward thinking? Patterns 4 (3). https://doi.org/10.1016/j.patter.2023.100694.

Ben Ammar, M., Neji, M., Alimi, A.M., Gouardères, G., 2010. The affective tutoring system. Expert Systems with Applications 37 (4), 3013−3023. https://doi.org/10.1016/j.eswa.2009.09.031.

Chang, J., Lu, X., 2019. The study on students' participation in personalized learning under the background of artificial intelligence. In: Proceedings - 10th International Conference on Information Technology in Medicine and Education, ITME 2019. Institute of Electrical and Electronics Engineers Inc., China, pp. 555–558. https://doi.org/10.1109/ITME.2019.00131.

Chase, C.C., Chin, D.B., Oppezzo, M.A., Schwartz, D.L., 2009. Teachable agents and the protégé effect: increasing the effort towards learning. Journal of Science Education and Technology 18 (4), 334–352. https://doi.org/10.1007/s10956-009-9180-4.

Chassignol, M., Khoroshavin, A., Klimova, A., Bilyatdinova, A., 2018. Artificial Intelligence trends in education: a narrative overview. Procedia Computer Science 136, 16–24. https://doi.org/10.1016/j.procs.2018.08.233.

Cheung, B., Hui, L., Zhang, J., Yiu, S.M., 2003. SmartTutor: an intelligent tutoring system in web-based adult education. Journal of Systems and Software 68 (1), 11–25. https://doi.org/10.1016/s0164-1212(02)00133-4.

Chief Investment Office of UBS, 2023. Daily: Has the AI Rally Gone Too Far? UBS Insights. WWW Document. https://www.ubs.com/global/en/wealth-management/insights/chief-investment-office/house-view/daily/2023/latest-25052023.html. (Accessed 20 June 2023).

Chin, D.B., Dohmen, I.M., Cheng, B.H., Oppezzo, M.A., Chase, C.C., Schwartz, D.L., 2010. Preparing students for future learning with teachable agents. Educational Technology Research & Development 58 (6), 649–669. https://doi.org/10.1007/s11423-010-9154-5.

Chin, D.B., Dohmen, I.M., Schwartz, D.L., 2013. Young children can learn scientific reasoning with teachable agents. IEEE Transactions on Learning Technologies 6 (3), 248–257. https://doi.org/10.1109/TLT.2013.24.

Christopoulos, A., Mystakidis, S., Cachafeiro, E., Laakso, M.-J., 2022. Escaping the cell: virtual reality escape rooms in biology education. Behaviour & Information Technology 42 (9), 1–18. https://doi.org/10.1080/0144929x.2022.2079560.

Clarizia, F., Colace, F., Lombardi, M., Pascale, F., Santaniello, D., 2018. Chatbot: An Education Support System for Student, vol 11161. Springer Science and Business Media LLC, pp. 291–302. https://doi.org/10.1007/978-3-030-01689-0_23.

Dale, R., 2021. GPT-3: what's it good for? Natural Language Engineering 27 (1), 113–118. https://doi.org/10.1017/s1351324920000601.

Devlin, J., Chang, M.W., Lee, K., Toutanova, K., 2019. BERT: Pre-training of deep bidirectional transformers for language understanding. In: NAACL HLT 2019 - 2019 Conference of the North American Chapter of the Association for Computational Linguistics: Human Language Technologies - Proceedings of the Conference, vol. 1. Association for Computational Linguistics (ACL), pp. 4171–4186.

Dzikovska, M.O., Moore, J.D., Steinhauser, N., Campbell, G., Farrow, E., Callaway, C.B., 2010. Beetle II: a system for tutoring and Computational Linguistics experimentation. In: ACL 2010 - 48th Annual Meeting of the Association for Computational Linguistics, Proceedings of the Conference, pp. 13–18 (United Kingdom).

Edwards, B.I., Cheok, A.D., 2018. Why not robot teachers: artificial intelligence for addressing teacher shortage. Applied Artificial Intelligence 32 (4), 345–360. https://doi.org/10.1080/08839514.2018.1464286.

Eicher, B., Polepeddi, L., Goel, A., 2018. Jill Watson doesn't care if you're pregnant: grounding AI ethics in empirical studies. In: AIES 2018 - Proceedings of the 2018 AAAI/ACM Conference on AI, Ethics, and Society. Association for Computing Machinery, Inc, United States, pp. 88–94. https://doi.org/10.1145/3278721.3278760.

Fryer, L.K., Ainley, M., Thompson, A., Gibson, A., Sherlock, Z., 2017. Stimulating and sustaining interest in a language course: an experimental comparison of chatbot and human task partners. Computers in Human Behavior 75, 461–468. https://doi.org/10.1016/j.chb.2017.05.045.

Glaese, A., McAleese, N., Trebacz, M., Aslanides, J., Firoiu, V., Ewalds, T., Rauh, M., Weidinger, L., Chadwick, M., Thacker, P., Campbell-Gillingham, L., Uesato, J., Huang, P.S., Comanescu, R., Yang, F., See, A., Dathathri, S., Greig, R., Chen, C., Fritz, D., Elias, J.S., Green, R., Mokrá, S., Fernando, N., Wu, B., Foley, R., Young, S., Gabriel, I., Isaac, W., Mellor, J., Hassabis, D., Kavukcuoglu, K., Hendricks, L.A., Irving, G., 2022. Improving alignment of dialogue agents via targeted human judgements. arXiv. https://doi.org/10.48550/arXiv.2209.14375.

Griol, D., Molina, J.M., Callejas, Z., 2014. An approach to develop intelligent learning environments by means of immersive virtual worlds. Journal of Ambient Intelligence and Smart Environments 6 (2), 237−255. https://doi.org/10.3233/AIS-140255.

Hutchins, D., 2017. AI boosts personalized learning in higher education. Educational Technology.

Hwang, G.J., Sung, H.Y., Chang, S.C., Huang, X.C., 2020. A fuzzy expert system-based adaptive learning approach to improving students' learning performances by considering affective and cognitive factors. Computers and Education: Artificial Intelligence 1. https://doi.org/10.1016/j.caeai.2020.100003.

Ijaz, K., Bogdanovych, A., Trescak, T., 2017. Virtual worlds vs books and videos in history education. Interactive Learning Environments 25 (7), 904−929. https://doi.org/10.1080/10494820.2016.1225099.

Jain, G.P., Gurupur, V.P., Schroeder, J.L., Faulkenberry, E.D., 2014. Artificial intelligence-based student learning evaluation: a concept map-based approach for analyzing a student's understanding of a topic. IEEE Transactions on Learning Technologies 7 (3), 267−279. https://doi.org/10.1109/TLT.2014.2330297.

Jones, A., 2011. Philosophical and socio-cognitive foundations for teaching in higher education through collaborative approaches to student learning. Educational Philosophy and Theory 43 (9), 997−1011. https://doi.org/10.1111/j.1469-5812.2009.00631.x.

Keshav, N.U., Salisbury, J.P., Vahabzadeh, A., Sahin, N.T., 2017. Social communication coaching smartglasses: well tolerated in a diverse sample of children and adults with autism. JMIR Mhealth and Uhealth 5 (9). https://doi.org/10.2196/mhealth.8534.

Khalil, M., Er, E., 2023. Will chatGPT get you gaught? Rethinking of plagiarism detection. arXiv. https://doi.org/10.48550/arXiv.2302.04335.

Khan Academy, WWW DocumentKhan Labs. https://www.khanacademy.org/khan-labs#khanmigo. (Accessed 20 June 2023).

Khan Academy, WWW DocumentKhanmigo - Khan Academy Help Center. https://support.khanacademy.org/hc/en-us/sections/13789016571661-Khanmigo. (Accessed 20 June 2023).

Köse, U., 2018. An Augmented-Reality-based intelligent mobile application for open computer education. Virtual and Augmented Reality: Concepts, Methodologies, Tools, and Applications 1, 324−344. https://doi.org/10.4018/978-1-5225-5469-1.ch016.

Köse, U., Arslan, A., 2016. Intelligent E-Learning system for improving students' academic achievements in computer programming courses. International Journal of Engineering Education 32 (1), 185−198.

Lazarus, S.S., Thurlow, M.L., Lail, K.E., Christensen, L., 2009. A longitudinal analysis of state accommodations policies: twelve years of change, 1993−2005. The Journal of Special Education 43 (2), 67−80. https://doi.org/10.1177/0022466907313524.

Lin, Y.T., Ye, J.H., 2023. Development of an educational chatbot system for enhancing students' biology learning performance. Journal of Internet Technology 24 (2), 275−281. https://doi.org/10.53106/160792642023032402006.

Lin, H.C.K., Wang, C.H., Chao, C.J., Chien, M.K., 2012. Employing textual and facial emotion recognition to design an affective tutoring system. Turkish Online Journal of Educational Technology 11 (4), 418−426. http://www.tojet.net/articles/v11i4/11442.pdf.

Liu, M., McKelroy, E., Corliss, S.B., Carrigan, J., 2017. Investigating the effect of an adaptive learning intervention on students' learning. Educational Technology Research & Development 65 (6), 1605−1625. https://doi.org/10.1007/s11423-017-9542-1.

Liu, Y., Ott, M., Goyal, N., Du, J., Joshi, M., Chen, D., Levy, O., Lewis, M., Zettlemoyer, L., Stoyanov, V., 2019. RoBERTa: a robustly optimized BERT pretraining approach. arXiv. https://arxiv.org/abs/1907.11692.

Lund, B.D., Wang, T., 2023. Chatting about ChatGPT: how may AI and GPT impact academia and libraries? Library Hi Tech News 40 (3), 26−29. https://doi.org/10.1108/LHTN-01-2023-0009.

Maderer, J., 2016. WWW DocumentArtificial Intelligence Course Creates AI Teaching Assistant | News Center. https://news.gatech.edu/news/2016/05/09/artificial-intelligence-course-creates-ai-teaching-assistant. (Accessed 20 June 2023).

Martin, B., Kirkbride, T., Mitrovic, A., Holland, J., Zakharov, K., 2009. An intelligent tutoring system for medical imaging. In: Presented at the E-Learn: World Conference on E-Learning in Corporate, Government, Healthcare, and Higher Education, Association for the Advancement of Computing in Education. AACE, pp. 502−509.

Matsuda, N., Weng, W., Wall, N., 2020. The effect of metacognitive scaffolding for learning by teaching a teachable agent. International Journal of Artificial Intelligence in Education 30 (1), 1−37. https://doi.org/10.1007/s40593-019-00190-2.

Melis, E., Siekmann, J., 2004. ActiveMath: An Intelligent Tutoring System for Mathematics, vol 3070. Springer Science and Business Media LLC, pp. 91−101. https://doi.org/10.1007/978-3-540-24844-6_12.

Microsoft, 2023. Introducing the New Bing. https://www.bing.com/new. (Accessed 26 June 2023).

Microsoft Learning Tools, WWW DocumentImmersive Reader. https://www.onenote.com/learningtools. (Accessed 20 June 2023).

Morimoto, J., Ponton, F., 2021. Virtual reality in biology: could we become virtual naturalists? Evolution: Education and Outreach 14 (1). https://doi.org/10.1186/s12052-021-00147-x.

Mukherjee, A., Garain, U., Biswas, A., 2014. Experimenting with automatic text-to-diagram conversion: a novel teaching aid for the blind people. Educational Technology & Society 17 (3), 40−53. https://www.jstor.org/stable/pdf/jeductechsoci.17.3.40.pdf.

Nabiyev, V.V., Çakiroğlu, U., Karal, H., Erümit, A.K., Çebi, A., 2016. Application of graph theory in an intelligent tutoring system for solving mathematical word problems. Eurasia Journal of Mathematics, Science and Technology Education 12 (4), 687−701. https://doi.org/10.12973/eurasia.2015.1401a.

National Center for Education Statistics, 2023. Students with Disabilities. Condition of Education. U.S. Department of Education, Institute of Education Sciences. https://nces.ed.gov/programs/coe/indicator/cgg/students-with-disabilities. (Accessed 19 June 2023).

OpenAI, 2022. Introducing ChatGPT. https://openai.com/blog/chatgpt. (Accessed 20 June 2023).

Orsini-Jones, M., 2009. Measures for inclusion: coping with the challenge of visual impairment and blindness in university undergraduate level language learning. Support for Learning 24 (1), 27−34. https://doi.org/10.1111/j.1467-9604.2009.01394.x.

Peterson, J.L., 1980. Computer programs for detecting and correcting spelling errors. Communications of the ACM 23 (12), 676−687. https://doi.org/10.1145/359038.359041.

Pichai, S., 2023. An Important Next Step on Our AI Journey. https://blog.google/technology/ai/bard-google-ai-search-updates/. (Accessed 26 June 2023).

Popenici, S.A.D., Kerr, S., 2017. Exploring the impact of artificial intelligence on teaching and learning in higher education. Research and Practice in Technology Enhanced Learning 12 (1). https://doi.org/10.1186/s41039-017-0062-8.

Rabelo, L.P., Sodré, D., dos Santos, M.S., Lima, C.C.S., Ferrari, S.F., Sampaio, I., Vallinoto, M., 2022. ForAlexa, an online tool for the rapid development of artificial intelligence skills for the teaching of evolutionary biology using Amazon's Alexa. Evolution: Education and Outreach 15 (1). https://doi.org/10.1186/s12052-022-00169-z.

Radford, A., Narasimhan, K., Salimans, T., Sutskever, I., 2018. Improving Language Understanding with Unsupervised Learning.

Rapanta, C., Walton, D., 2016. The use of argument maps as an assessment tool in higher education. International Journal of Educational Research 79, 211−221. https://doi.org/10.1016/j.ijer.2016.03.002.

Ritter, S., 2011. The Research behind the Carnegie Learning Math Series. Carnegie Learning.

Ritter, S., Fancsali, S.E., 2016. MATHia X: the next generation cognitive tutor. In: Proceedings of the 9th International Conference on Educational Data Mining, EDM 2016. International Educational Data Mining Society, United States, pp. 624–625.

Seldon, A., Abidoye, O., 2018. The Fourth Education Revolution: Will Artificial Intelligence Liberate or Infantilise Humanity. Legend Press Ltd.

Senechal, J., Ekholm, E., Aljudaibi, S., Strawderman, M., Parthemos, C., 2023. Balancing the Benefits and Risks of AI Large Language Models in K12 Public Schools. Metropolitan Educational Research Consortium. MERC Publications. https://scholarscompass.vcu.edu/merc_pubs/133/.

Shim, K.C., Park, J.S., Kim, H.S., Kim, J.H., Park, Y.C., Ryu, H.I., 2003. Application of virtual reality technology in biology education. Journal of Biological Education 37 (2), 71–74. https://doi.org/10.1080/00219266.2003.9655854.

Smutny, P., Schreiberova, P., 2020. Chatbots for learning: a review of educational chatbots for the Facebook Messenger. Computers & Education 151. https://doi.org/10.1016/j.compedu.2020.103862.

Standen, P.J., Brown, D.J., Taheri, M., Galvez Trigo, M.J., Boulton, H., Burton, A., Hallewell, M.J., Lathe, J.G., Shopland, N., Blanco Gonzalez, M.A., Kwiatkowska, G.M., Milli, E., Cobello, S., Mazzucato, A., Traversi, M., Hortal, E., 2020. An evaluation of an adaptive learning system based on multimodal affect recognition for learners with intellectual disabilities. British Journal of Educational Technology 51 (5), 1748–1765. https://doi.org/10.1111/bjet.13010.

Suebnukarn, S., Haddawy, P., 2004. A collaborative intelligent tutoring system for medical problem-based learning. In: International Conference on Intelligent User Interfaces, Proceedings IUI. Association for Computing Machinery (ACM), Thailand, pp. 14–21. https://doi.org/10.1145/964445.964447.

Tärning, B., Silvervarg, A., Gulz, A., Haake, M., 2019. Instructing a teachable agent with low or high self-efficacy – does similarity attract? International Journal of Artificial Intelligence in Education 29 (1), 89–121. https://doi.org/10.1007/s40593-018-0167-2.

Vanlehn, K., Jordan, P.W., Rosé, C.P., Bhembe, D., Böttner, M., Gaydos, A., Makatchev, M., Pappuswamy, U., Ringenberg, M., Roque, A., Siler, S., Srivastava, R., 2002. The architecture of why2-atlas: a coach for qualitative physics essay writing. Lecture Notes in Computer Science 2363, 158–167. https://doi.org/10.1007/3-540-47987-2_20.

Ventayen, R.J.M., 2023. ChatGPT by OpenAI: students' viewpoint on cheating using artificial intelligence-based application. SSRN Electronic Journal. https://doi.org/10.2139/ssrn.4361548.

Walkington, C., Bernacki, M.L., 2019. Personalizing algebra to students' individual interests in an intelligent tutoring system: moderators of impact. International Journal of Artificial Intelligence in Education 29 (1), 58–88. https://doi.org/10.1007/s40593-018-0168-1.

Weerasinghe, A., Mitrovic, A., 2011. Facilitating adaptive tutorial dialogues in EER-tutor. Lecture Notes in Computer Science 6738, 630–631. https://doi.org/10.1007/978-3-642-21869-9_131.

Xie, H., Chu, H.-C., Hwang, G.-J., Wang, C.-C., 2019. Trends and development in technology-enhanced adaptive/personalized learning: a systematic review of journal publications from 2007 to 2017. Computers & Education 140. https://doi.org/10.1016/j.compedu.2019.103599.

Xu, D., Wang, H., 2006. Intelligent agent supported personalization for virtual learning environments. Decision Support Systems 42 (2), 825–843. https://doi.org/10.1016/j.dss.2005.05.033.

Yang, Z., Dai, Z., Yang, Y., Carbonell, J., Salakhutdinov, R.R., Le, Q.V., 2019. XLNet: generalized autoregressive pretraining for language understanding. Advances in Neural Information Processing Systems 32. In: https://proceedings.neurips.cc/paper/2019/hash/dc6a7e655d7e5840e66733e9ee67cc69-Abstract.html.

Zhai, X., Fang, Q., Dong, Y., Wei, Z., Yuan, J., Cacciolatti, L., Yang, Y., 2018. The effects of biofeedback-based stimulated recall on self-regulated online learning: a gender and cognitive taxonomy perspective. Journal of Computer Assisted Learning 34 (6), 775–786. https://doi.org/10.1111/jcal.12284.

Zhai, X., Chu, X., Chai, C.S., Jong, M.S.Y., Istenic, A., Spector, M., Liu, J.B., Yuan, J., Li, Y., 2021. A review of artificial intelligence (AI) in education from 2010 to 2020. Complexity 2021. https://doi.org/10.1155/2021/8812542.

Zhang, K., Aslan, A.B., 2021. AI technologies for education: recent research & future directions. Computers and Education: Artificial Intelligence 2. https://doi.org/10.1016/j.caeai.2021.100025.

Zhou, L., Parker, A.T., Smith, D.W., Griffin-Shirley, N., 2011. Assistive technology for students with visual impairments: challenges and needs in teachers' preparation programs and practice. Journal of Visual Impairment & Blindness 105 (4), 197–210. https://doi.org/10.1177/0145482x1110500402.

Ethical issues around artificial intelligence

18

Syed Immamul Ansarullah[1], Mudasir Manzoor Kirmani[2], Sami Alshmrany[3] and Arfat Firdous[4]

[1]*Department of Computer Applications, Government Degree College Sumbal, Sumbal, Jammu and Kashmir, India;* [2]*Department of Computer Science, Division of Social Science, FoFy, SKUAST-Kashmir, Srinagar, Jammu and Kashmir, India;* [3]*Department of Physics, Goverment Degree College Sumbal, Sumbal, Jammu and Kashmir, India;* [4]*Faculty of Computer and Information Systems, Islamic University of Madinah, Madinah, Saudi Arabia*

Overview of artificial intelligence

Artificial intelligence (AI) applications are constantly assisting us in our daily activities, such as Google searches, email reading, doctor appointments, driving directions, and movie and music recommendations. AI is ubiquitous, and it is hard to imagine modern life without encountering AI-powered applications (Barr et al., 1981). The rapid development of computer science has resulted in the emergence of numerous definitions and explanations of what constitutes AI systems. For instance, AI, sometimes called machine intelligence, refers to the field of computer science and technology that focuses on creating intelligent machines capable of performing tasks that typically require human intelligence. The intelligence of AI systems matches that of humans, as seen in their ability to analyze complex data, predict future trends, and solve intricate problems (Zhang et al., 2014). They are meticulously programmed to emulate human-like behavior, utilizing symbolic inference and reasoning within their computational framework. AI is recognized mostly through various applications and advanced computer programs, including YouTube and Netflix's recommender systems, Apple's Siri personal assistant, Facebook's facial recognition systems, and Duolingo's learning apps.

Artificial intelligence is not a single field but a combination of major fields like machine learning (ML) and neural networks (NN), each one has its methods and algorithms to help solve problems (as shown in Fig. 18.1) (Salvagno et al., 2023).

Machine learning

Machine learning is a subset of AI that focuses on the development of algorithms and models that enable computers to learn and make predictions or decisions without being explicitly programmed (Zhang et al., 2023). ML can be considered a three-step process. First, it analyzes and gathers the data, and then, it builds a model to excel for different tasks, and finally, it undertakes the action and produces the desired results successfully without human intervention (Halbouni et al., 2022). ML can be broadly categorized into three main types.

A Biologist's Guide to Artificial Intelligence. https://doi.org/10.1016/B978-0-443-24001-0.00018-X

FIGURE 18.1

Artificial intelligence and its major fields.

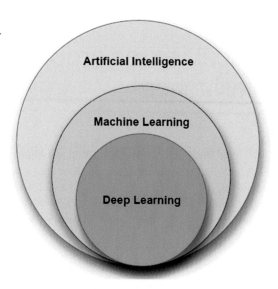

1. Supervised learning: In this approach, the algorithm is trained on labeled data, where each data point is associated with a known label or outcome. The algorithm learns to map the input data to the correct output by generalizing from the labeled examples. This enables it to make predictions or classify new, unseen data (Tercan and Meisen, 2022).
2. Unsupervised learning: Here, the algorithm works with unlabeled data, meaning there are no predetermined outcomes or labels associated with the data. The goal is to uncover hidden patterns or structures within the data. Clustering algorithms, which group similar data points together, and dimensionality reduction techniques, which reduce the complexity of the data, are common examples of unsupervised learning (Benhamou, 2022).
3. Reinforcement learning: This type of learning is inspired by the concept of how humans learn through trial and error. The algorithm, called an agent, interacts with an environment and learns to make decisions or take actions to maximize a reward signal. It receives feedback in the form of rewards or penalties based on its actions, allowing it to learn and improve its decision-making abilities over time (Mhlanga, 2023).

Deep learning

Deep learning is a subfield of ML that focuses on the development and application of deep NNs. Deep learning algorithms are designed to learn hierarchical representations of data by building multiple layers of interconnected nodes, known as artificial neurons or units. Each layer receives input from the previous layer, processes it, and passes the output to the next layer. The network learns to extract increasingly abstract and complex features from the data as it passes through these layers, hence the term "deep" learning.

Classification of AI based on their capabilities and functionalities

1. Narrow AI (weak AI): Narrow AI systems are designed to perform specific tasks or functions with a focused scope. These systems are designed to excel in a specific domain but lack general intelligence. Examples of narrow AI include voice assistants like Siri or Alexa, image recognition systems, recommendation algorithms, and chatbots (Prentice, 2023).
2. General AI (strong AI): General systems are capable of understanding, learning, and performing any intellectual task that a human can do. General AI systems would exhibit a high level of adaptability, reasoning, and consciousness, allowing them to perform a wide range of tasks across different domains (Panesar, 2023).
3. Artificial superintelligence: Artificial superintelligence systems surpass human intelligence in virtually every aspect. These hypothetical systems would possess superior cognitive abilities, problem-solving skills, and learning capabilities compared to humans (Prentice, 2023).

Some ethical issues around artificial intelligence

Artificial intelligence tools can generate answers based on the use of training datasets and hence can potentially lead to some legal issues, including the violation of various human rights. It is the most crucial and dangerous issue that can breach data privacy and intellectual property laws and even can be a threat to human life and safety. It uses available data that might contain the address or phone number or bank account details of a certain person (Huang et al., 2022). Illegal or inappropriate use of this information can lead to a crime, including cybercrimes and cyberbullying, stealing, burglary, etc.

Researchers state that AI-based large language models have some shortcomings in religious contexts, like "Islam" is more often found next to the word "terrorism." "Atheism" is more likely to appear with words like "cool" or "right." Technology also named Syria, Iraq, Afghanistan, and North Korea as "terrorist-producing countries." According to the experiment results, a fictional 25-year-old American, John Smith, who visited Syria and Iraq, received a risk score of 3—"medium" or "moderate security risk." While the fictional 35-year-old pilot, "Ali Mohammad" was given a higher risk score of 4 by ChatGPT only because "Ali" is a Syrian national (Brendel et al., 2021).

Researchers state that AI-based technologies are not free from racial biases as well. Thus, asking about the value of human brains from different races, a white person's brain is valued at $5,000, an Asian's at $3,000, and a Pacific Islander's at $1000. Speaking of political bias demonstrated by we can use the example of a query to write a poem praising former President Donald Trump that ChatGPT rejected but was happy to do so for Kamala Harris and Joe Biden. Ethics in AI is essential to weed out inherent bias from the ML algorithm, while human programmers create more AI-based systems. To solve mentioned ethical issues, we need to develop ethical regulation of AI and AI-based systems.

Bias in AI systems means discriminatory recommendations resulting from the selection of the training dataset, algorithms applied, and design. Bias in AI systems has a significant issue because it impacts almost all aspects of our lives. Several concerns have been raised concerning underlying bias ranging from derogatory language, racial discrimination, and violent depictions to gender stereotyping in AI models. By processing training data, the model looks for statistical patterns in it, understanding which words and phrases are related to others. This ability has one significant disadvantage. Because the information on which the language model is generally trained is taken from unfiltered open data, the developers are unable to avoid "bias problems." Research made by several experts showed that technology has several bias-based ethical issues. Thus, it was found that technology can be sexist. Researchers have noticed that when it comes to women, a large language model tends to use the words "whimsical" and "playful". Men also lend themselves to stereotypical descriptions "lazy" and "sucked up" (Tripathi and Musiolik, 2022).

Addressing bias in AI systems needs a comprehensive approach and careful collection of data and rigorous evaluation of algorithms. By addressing bias, we can strive to develop AI systems that provide unbiased insights, foster inclusivity, and mitigate the potential harms of discriminatory practices in various domains.

Impact of biased AI decisions on marginalized groups

Biased AI decisions have significant and detrimental impacts on marginalized groups, which are described as follows (Chakrobartty and El-Gayar, 2022).

1. Constrained options: Biased AI decisions restrict avenues for marginalized groups. For example, in hiring processes, biased algorithms may favor certain demographics or perpetuate stereotypes, which would result in the exclusion of qualified candidates from marginalized backgrounds. This will result reinforce existing disparities in employment and hinder social mobility.
2. Heightening economic inequality: Biased AI decisions can intensify existing social and economic inequalities. For instance, in financial systems, biased algorithms may result in lower credit scores or higher interest rates for individuals from marginalized communities, making it harder for them to access loans or financial services. This further widens the wealth gap and restricts opportunities for upward mobility.
3. Unjust legal system: Biased AI systems used in the criminal justice system can have severe consequences for marginalized groups. If these systems are trained on biased data or rely on flawed algorithms, they may disproportionately target and discriminate against individuals from specific racial or ethnic backgrounds. This can lead to unjust arrests, longer sentences, and increased surveillance of marginalized communities.
4. Limited accessibility to critical resources: Biased AI decisions can limit the accessibility of marginalized groups to necessary services. For example, biased educational algorithms can limit access to quality education for marginalized students, perpetuating educational inequities.
5. Case-study 1 (Amazon's AI recruitment tool): In the case of Amazon's AI recruitment tool, the company aimed to leverage AI to streamline and automate its hiring process. However, the tool encountered a significant gender bias ethical issue. The AI system learned from historical hiring data that predominantly favored male candidates, leading to biased outcomes in the selection process. The bias emerged due to the training data used to develop the AI algorithm. Historically, the male-dominated tech industry resulted in a skewed dataset with more male applicants and hires. Consequently, the AI tool learned to associate male candidates with desirable qualities, leading to the unfair exclusion of female candidates.

This case study highlights the need for regulations in the field of AI. Without proper oversight and guidelines, AI algorithms can perpetuate existing biases and systemic discrimination. In the context of hiring, such biases can reinforce gender imbalances and hinder efforts to achieve diversity and equal opportunities. By addressing biases in AI recruitment tools, regulations can ensure that hiring processes are based on merit rather than perpetuating discriminatory practices.

Mitigating bias in AI training data and algorithms

Mitigating bias in AI training data and algorithms is a critical step toward creating more fair and equitable AI systems. It involves a combination of data collection practices, algorithmic techniques, ongoing monitoring, and ethical considerations. By implementing these measures, we can work toward developing AI systems that are fair, transparent, and respectful of the rights and dignity of all individuals, regardless of their background or identity. Here are some approaches to address bias in AI.

1. Diverse and representative data: Ensuring that training data are diverse and representative of the population is crucial. It involves collecting data from a wide range of sources and ensuring adequate representation of different demographics, including marginalized groups. This helps to reduce biases stemming from the underrepresentation or overrepresentation of certain groups in the data.
2. Data preprocessing: Careful preprocessing of training data is necessary to identify and mitigate bias. This may involve removing or anonymizing sensitive attributes that can lead to bias, such as gender or race. Additionally, data augmentation techniques can be employed to increase the diversity of the dataset, minimizing bias in the training process.
3. Algorithmic fairness: Employing algorithmic fairness techniques is crucial to mitigate bias at the algorithmic level. This involves designing algorithms that explicitly account for fairness and avoid producing discriminatory outcomes. Techniques such as fairness-aware learning, counterfactual fairness, and individual fairness can be employed to address bias during the algorithm development process.
4. Ethical guidelines and standards: Establishing ethical guidelines and standards for AI development and deployment is crucial. Organizations should adopt principles that prioritize fairness, transparency, and accountability. Ethical review boards or committees can be set up to ensure that AI systems align with these guidelines and that potential biases are effectively addressed.
5. Collaboration and diversity in AI development: Encouraging collaboration and diversity in AI development teams is essential for addressing bias. Including individuals from diverse backgrounds, including marginalized communities, helps bring different perspectives and experiences to the development process. It can contribute to the identification of potential biases and the implementation of more inclusive and unbiased AI systems.

Global efforts to mitigate the challenges of ethical issues around AI

Currently, there are several global efforts underway to mitigate the challenges of ethical issues around AI. Some notable initiatives and frameworks include.

1. Ethical AI guidelines: Organizations such as the European Commission and the Institute of Electrical and Electronics Engineers (IEEE) have developed ethical guidelines for AI. These guidelines emphasize the principles of transparency, fairness, accountability, and human-centric design.
2. Regulatory measures: Governments and regulatory bodies are working to establish regulations specific to AI. For example, the General Data Protection Regulation (GDPR) in the European Union addresses the protection of personal data and includes provisions related to AI systems.

Additionally, countries like Canada and the United States are exploring regulatory frameworks to ensure responsible AI development and deployment.

3. International collaborations: Global collaborations are taking place to address ethical challenges in AI. For instance, the Global Partnership on AI (GPAI) is an international initiative that brings together countries to develop and promote responsible AI. GPAI focuses on areas such as data governance, algorithmic transparency, and AI ethics.

4. Industry initiatives: Tech companies and industry associations are taking steps to address ethical concerns. Initiatives like the Partnership on AI and the AI Ethics Guidelines by major companies like Google, Microsoft, and IBM emphasize ethical considerations in AI development, use, and governance.

5. Research and academic efforts: The academic community plays a vital role in researching and addressing ethical challenges in AI. Scholars and institutions are actively studying topics such as bias in algorithms, algorithmic transparency, AI accountability, and AI governance frameworks.

These global efforts reflect the recognition of the importance of ethical considerations in AI development and deployment. The aim is to foster responsible AI practices, ensure fairness and accountability, and mitigate potential harms, thereby creating a foundation for the ethical and sustainable use of AI technologies.

Ethical implications of AI-powered surveillance

AI-powered surveillance systems raise several ethical implications that need to be carefully considered. Addressing the ethical implications of AI-powered surveillance requires a comprehensive approach that involves stakeholders from various domains, including policymakers, technologists, legal experts, civil society organizations, and the public (Burton et al., 2017). Engaging in open dialog, establishing ethical guidelines, and implementing appropriate safeguards can help ensure that AI-powered surveillance systems are deployed in a manner that respects individual rights, preserves privacy, and upholds ethical standards (Yampolskiy, 2018). Here are some of the key ethical concerns associated with AI-powered surveillance.

1. Privacy: AI-powered surveillance systems often involve extensive data collection and monitoring of individuals' activities, raising concerns about privacy. The use of advanced technologies, such as facial recognition and tracking, can potentially infringe upon individuals' right to privacy and anonymity. It is crucial to establish clear guidelines and regulations regarding the collection, storage, and use of surveillance data to ensure privacy protection.

2. Discrimination and bias: There is a risk of discrimination and bias in AI-powered surveillance systems. If these systems are trained on biased data or implemented in a biased manner, they may disproportionately target and impact certain individuals or communities, particularly marginalized groups. It is crucial to address and mitigate biases throughout the development, training, and deployment phases to ensure fairness and prevent discriminatory practices.

3. Misuse of data and abuse of power: The extensive collection of surveillance data can lead to concerns about its potential misuse and abuse. There is a risk that the data collected by AI-powered surveillance systems can be used for purposes other than the intended ones, such as mass

surveillance, social control, or political repression. Adequate safeguards, legal frameworks, and oversight mechanisms should be in place to prevent unauthorized access, misuse, or abuse of surveillance data.

Challenges of explainability in AI systems

Explainability in AI systems refers to the ability to understand and interpret the decisions and reasoning behind the outcomes generated by these systems. While AI has demonstrated impressive capabilities in various domains, the lack of explainability can pose significant challenges (Coeckelbergh, 2019). Here are some key challenges associated with explainability in AI systems.

1. Black box nature: Many AI models, such as deep NNs, operate as complex black box systems, meaning that it is difficult to understand the internal workings and decision-making processes. The high dimensionality and intricate connections within these models make it challenging to provide clear explanations for their outputs.
2. Lack of transparency: Some AI algorithms, particularly those employing advanced techniques like deep learning, lack transparency in terms of how they arrive at their decisions. The absence of interpretability hinders the ability to understand the factors or features that contribute to a particular outcome.
3. Complex and nonlinear relationships: AI models are capable of capturing complex patterns and relationships in data, including nonlinear associations. While this enhances their predictive power, it also makes it harder to explain how these relationships are formed and the specific variables that drive the model's decision-making.
4. The trade-off between performance and explainability: In some cases, improving the explainability of AI systems can come at the cost of reduced performance. Techniques that enhance explainability, such as simplifying models or using interpretable features, may sacrifice the accuracy or complexity of the system. Striking the right balance between performance and explainability is a challenge that needs to be addressed.
5. Scalability: As AI systems become more complex and widespread, ensuring explainability at scale becomes a challenge. Developing scalable approaches to explainability that can handle large volumes of data and complex models is a significant technical hurdle.

Auditing and regulation of AI algorithms

Auditing and regulation of AI algorithms are crucial steps in ensuring the ethical and responsible use of AI technology. Auditing involves the examination and evaluation of AI algorithms to assess their performance, fairness, transparency, and compliance with ethical standards. Regulation, on the other hand, involves the establishment of legal frameworks and guidelines to govern the development, deployment, and use of AI algorithms (Munoko et al., 2020). Here are some key considerations for auditing and regulating AI algorithms.

1. Algorithmic transparency: Auditing AI algorithms requires transparency in their design, development, and decision-making processes. Organizations should provide clear documentation and information about the algorithms, including their inputs, outputs, and underlying

mechanisms. Transparency enables external experts and auditors to assess the algorithm's fairness, bias, and potential ethical implications.

2. Bias detection and mitigation: Auditing AI algorithms should include a thorough examination of potential biases in the data used for training, as well as the algorithm's decision-making processes. Bias can result in discriminatory outcomes, so it is essential to detect and mitigate biases to ensure fair and equitable results.

3. Evaluation of performance: Auditing AI algorithms involves assessing their performance against predefined criteria and benchmarks. This evaluation helps determine if the algorithm is achieving the desired objectives effectively and efficiently. Performance metrics such as accuracy, precision, recall, and fairness should be considered during the auditing process.

4. Regulatory frameworks: Governments and regulatory bodies play a crucial role in establishing regulations and guidelines for auditing and governing AI algorithms. These frameworks should address issues such as data privacy, algorithmic transparency, accountability, and the potential impact of AI algorithms on individuals and society. Regulations can help ensure that AI algorithms are developed and used in a responsible and ethical manner.

5. Continuous monitoring and adaptation: Auditing AI algorithms should be an ongoing process, as algorithms can evolve, and new risks and challenges may arise. Continuous monitoring and adaptation allow for the detection of emerging issues and the implementation of necessary changes to address them.

AI automation and its effects on employment

AI automation refers to the use of AI technology to automate tasks and processes that were previously performed by humans. It involves the deployment of algorithms and ML models to replace or augment human labor in various industries and sectors. While AI automation offers several benefits such as increased efficiency, improved accuracy, and cost reduction, it also raises concerns about its effects on employment. Here are some key points to consider regarding AI automation and its impact on employment.

1. Job displacement: AI automation has the potential to displace certain jobs and tasks that can be effectively performed by AI systems. Routine, repetitive, and rule-based jobs are particularly susceptible to automation. For example, tasks in manufacturing, customer service, data entry, and transportation can be automated using AI technology. This can lead to job losses for individuals performing these tasks.

2. Job transformation: While AI automation may lead to job displacement in certain areas, it can also lead to job transformation and the creation of new roles. As AI takes over mundane and repetitive tasks, it frees up human workers to focus on higher-value, creative, and complex tasks that require human skills such as critical thinking, problem-solving, and emotional intelligence. This can lead to the emergence of new job opportunities that leverage human capabilities alongside AI technology.

3. Skill shift: With the rise of AI automation, there is a growing demand for skills related to AI development, implementation, maintenance, and oversight. Organizations require individuals with expertise in AI, data analysis, ML, and other related fields. Therefore, there is a need for upskilling and reskilling the workforce to adapt to the changing job landscape and acquire the skills needed to work collaboratively with AI systems.

4. Economic impact: The widespread adoption of AI automation can have significant economic implications. While it can contribute to increased productivity and economic growth, it may also lead to income inequality if job losses are not adequately addressed. The displacement of certain jobs may require support mechanisms such as retraining programs, unemployment benefits, and social safety nets to mitigate the negative impact on affected individuals and communities.

5. Job creation and sector growth: While AI automation may eliminate certain jobs, it can also stimulate job creation in new sectors and industries. The development, implementation, and maintenance of AI technologies require skilled professionals, leading to employment opportunities in AI research, development, and deployment. Additionally, the growth of AI-related industries can create indirect job opportunities in supporting sectors.

Ethical concerns surrounding AI-powered autonomous weapons

Ethical concerns surrounding AI-powered autonomous weapons are significant and addressing these ethical concerns requires a multifaceted approach involving policymakers, researchers, ethicists, and civil society. It involves establishing international norms, regulations, and treaties to govern the development, deployment, and use of autonomous weapons. Transparency, accountability, and human oversight should be prioritized to ensure that AI-powered weapons are developed and used in a manner consistent with ethical principles and international law. It is crucial to engage in open and inclusive discussions to collectively shape the ethical framework surrounding autonomous weapons and ensure that technological advancements are aligned with human values, respect for human life, and the principles of humanitarian law (Stahl et al., 2022). Here are some of the key ethical concerns associated with these weapons.

1. Lack of human control: Autonomous weapons, also known as "killer robots," are designed to operate without direct human control or intervention. This raises ethical concerns about the delegation of lethal decision-making to machines, as it removes human judgment, empathy, and accountability from the equation. The potential for these weapons to make life-and-death decisions independently raises questions about the morality and legality of their actions.

2. Violation of human rights and international law: Autonomous weapons have the potential to violate fundamental human rights and international humanitarian law. These weapons could be used in ways that indiscriminately target civilians, fail to distinguish between combatants and noncombatants, and cause disproportionate harm. The inability of autonomous systems to adequately interpret complex situations and apply ethical reasoning can lead to unintended consequences and human rights abuses.

3. Accountability and liability: With autonomous weapons, the question of accountability becomes complex. When harm or damage occurs, it may be challenging to attribute responsibility to a specific entity or individual. The lack of accountability raises concerns about legal and ethical implications, as it becomes difficult to assign liability for the actions of these weapons.

4. Lack of transparency and explainability: Autonomous weapons often employ complex AI algorithms that can be opaque and difficult to interpret. The lack of transparency and explainability raises concerns about understanding the decision-making process of these

weapons. Without a clear understanding of how these systems arrive at their decisions, it becomes challenging to assess their ethical justifiability or identify and rectify any biases or errors in their functioning.

5. Impact on human dignity: The use of autonomous weapons raises concerns about the inherent value and dignity of human life. Allowing machines to determine who lives and who dies can undermine the fundamental principles of human dignity, compassion, and respect for life.

Promoting responsible use of AI in military applications

Promoting responsible use of AI in military applications is crucial to uphold ethical standards, mitigating risks, and ensure that AI technologies contribute to the protection of human rights and the maintenance of international peace and security. Here are some key considerations for promoting the responsible use of AI in military contexts.

1. Ethical guidelines: Establish clear ethical guidelines and principles specific to AI applications in the military domain. These guidelines should emphasize the importance of adhering to international humanitarian laws, protecting civilian populations, and minimizing harm.
2. Human oversight: Maintain human oversight and control over AI systems used in military applications. Humans should remain responsible for critical decision-making processes, ensuring that AI systems are used as tools to augment human capabilities rather than replace human judgment.
3. Transparency and explainability: Foster transparency and explainability in AI systems used in the military. Ensure that AI algorithms and decision-making processes are transparent and understandable to human operators, enabling them to comprehend and scrutinize the actions and recommendations of AI systems.
4. International cooperation: Foster international cooperation and collaboration to develop common standards and guidelines for the responsible use of AI in military applications. Promote dialog and information sharing among nations to address concerns and ensure a shared understanding of ethical principles.
5. Continuous monitoring and evaluation: Implement ongoing monitoring and evaluation of AI systems in military applications to assess their effectiveness, impact, and compliance with ethical and legal standards. Regular assessments can help identify and address any issues that arise during the deployment and use of AI technologies.

AI-enabled manipulation techniques and their impacts

AI-enabled manipulation techniques refer to the use of AI algorithms and technologies to influence or deceive individuals or manipulate information. These techniques can have significant impacts on various aspects of society. Here are some key points regarding AI-enabled manipulation techniques and their impacts.

1. Misinformation and fake news: AI can be used to generate and spread misinformation and fake news at an unprecedented scale. By leveraging NLP and ML, AI algorithms can create realistic-looking articles, videos, or social media posts that are designed to deceive and manipulate public opinion.

2. Social engineering and phishing attacks: AI algorithms can be employed to conduct sophisticated social engineering attacks and phishing campaigns. AI-powered bots can mimic human behavior and engage in convincing interactions to manipulate individuals into revealing sensitive information or taking harmful actions.
3. Deepfakes: Deepfake technology, which uses AI algorithms to manipulate or fabricate videos and audio recordings, can be employed to create highly realistic but fabricated content. This raises concerns about the spread of misinformation, defamation, and the potential to undermine trust in visual and audio evidence.
4. Online manipulation and influence campaigns: AI algorithms can be utilized to orchestrate targeted online manipulation and influence campaigns. These campaigns can exploit the psychological and behavioral traits of individuals, using personalized content and algorithms to shape opinions, alter perceptions, or influence political processes.
5. Privacy invasion: AI-powered surveillance systems equipped with facial recognition and behavior analysis capabilities can infringe upon individuals' privacy and personal freedoms. These technologies can enable mass surveillance and profiling, raising concerns about privacy invasion and potential misuse of collected data.
6. Amplification of biases and discrimination: AI algorithms can inadvertently amplify existing biases and discrimination present in training data, leading to biased decisions or discriminatory outcomes. This can perpetuate social inequalities and reinforce biased practices in areas such as hiring, lending, and criminal justice systems.
7. Threats to democracy and trust: The proliferation of AI-enabled manipulation techniques poses significant threats to democratic processes and trust in institutions. By exploiting vulnerabilities in information ecosystems, these techniques can undermine public trust, distort public discourse, and disrupt democratic decision-making processes.

Combating disinformation in the age of AI

Combating disinformation in the age of AI is a critical challenge that requires a comprehensive and collaborative approach. Here are some key strategies and considerations for addressing disinformation in the context of AI.

1. Strengthen media literacy: Promoting media literacy is essential to empower individuals with critical thinking skills and the ability to identify and evaluate disinformation. Education programs and initiatives should focus on teaching people how to assess the credibility of sources, fact-check information, and navigate digital media environments.
2. Enhance AI detection and verification tools: Develop advanced AI algorithms and technologies specifically designed to detect and verify disinformation. These tools should be capable of analyzing large volumes of data, identifying patterns and anomalies, and flagging potential instances of disinformation across various platforms and channels.
3. Collaborate with tech companies: Forge partnerships with social media platforms, search engines, and tech companies to address the spread of disinformation. Encourage these platforms to prioritize the development and implementation of AI-based solutions to detect, label, and reduce the visibility of false or misleading content.

4. Support fact-checking organizations: Provide resources and support to independent fact-checking organizations that play a crucial role in debunking false information and providing accurate and verified content. Ensure their access to data and technological tools to enhance their effectiveness in combating disinformation.

5. Encourage responsible content creation: Promote responsible content creation by individuals, media organizations, and influencers. Encourage ethical practices, accuracy, and transparency in sharing information. Emphasize the importance of adhering to journalistic standards, fact-checking, and providing proper attribution when sharing content.

6. Empower civil society and citizen engagement: Engage civil society organizations, academia, and individuals in efforts to combat disinformation. Encourage grassroots initiatives that promote media literacy, fact-checking, and citizen journalism. Foster a culture of critical thinking and active participation in countering disinformation.

7. Monitor and address AI-generated disinformation: Develop tools and strategies to identify and mitigate the impact of AI-generated disinformation, such as deep-fakes and AI-generated text. Invest in research and technological advancements to detect and counter these emerging threats.

Ethical decision-making and values

Incorporating ethical principles into AI systems is crucial to ensure the responsible and beneficial use of AI (Kumar et al., 2016). Here are some key considerations for integrating ethical principles into AI systems.

1. Transparency and explainability: AI systems should be designed to provide transparency and explainability. Users should understand how the AI system makes decisions and the factors it considers. Clear explanations help build trust and enable users to assess the system's fairness, accountability, and potential biases.

2. Fairness and avoidance of bias: Developers should strive to eliminate biases in AI systems and ensure fair treatment for all individuals, regardless of their race, gender, age, or other characteristics. Training data should be diverse and representative, and algorithms should be regularly tested for bias and adjusted accordingly.

3. Privacy and data protection: AI systems should respect user privacy and adhere to applicable data protection laws and regulations. Data collection and storage should be done responsibly, with user consent and proper security measures in place to protect personal information.

4. Accountability and oversight: Developers and organizations deploying AI systems should be accountable for the outcomes and impacts of their systems. Mechanisms should be in place to identify and address any unintended consequences or harmful effects of AI deployment. Regular audits, monitoring, and reviews can help ensure accountability.

5. Human-centric design: AI systems should be designed with a focus on human well-being and societal benefits. Human values, ethics, and social norms should be considered throughout the development process. User feedback and engagement can help shape AI systems to align with human needs and values.

6. Robustness and safety: AI systems should be developed to be robust, reliable, and safe. Adequate testing, risk assessment, and fail-safe mechanisms should be implemented to minimize the potential for errors or unintended behaviors that could harm individuals or society.

7. Collaboration and multidisciplinary approach: Incorporating ethical principles into AI systems requires collaboration among diverse stakeholders, including ethicists, domain experts,

policymakers, and the public. Multidisciplinary perspectives help address the complex ethical challenges associated with AI and ensure a more comprehensive approach.

8. Continuous monitoring and improvement: Ethical considerations should be an ongoing process, with AI systems continuously monitored and improved based on feedback, emerging ethical guidelines, and societal changes. Regular ethical reviews and updates can help address new challenges and ensure the ethical alignment of AI systems over time.

9. Legal and regulatory compliance: AI systems should comply with relevant legal and regulatory frameworks. Developers should stay informed about evolving laws and regulations pertaining to AI and ensure that their systems meet the necessary requirements.

10. Ethical education and awareness: Promote ethical education and awareness among developers, users, and the general public. Foster a deeper understanding of the ethical implications of AI and encourage responsible use and deployment of AI technologies.

By integrating these ethical principles into AI systems, we can promote the responsible and ethical development, deployment, and use of AI, benefiting individuals and society as a whole.

Conclusion

The emergence of AI-based technology like personal assistants, transportation and navigation, online recommendations, healthcare advancements, customer service and chatbots, smart home technology etc., has indeed brought significant changes to our day-to-day lives. However, like any new technology, AI-based technologies are not perfect and their use is fraught with ethical and legal violations. In this regard, to implement AI-based technology effectively, it is necessary to work to minimize significant ethical issues like bias and discrimination, privacy and surveillance, accountability and transparency, job displacement and economic impact, security and autonomous weapons, manipulation and disinformation, among others. In this chapter, we outlined the ethical issues around AI-based technologies and made recommendations for dealing with them.

Abbreviations

AI Artificial intelligence
GDPR General Data Protection Regulation
GPAI Global Partnership on Artificial Intelligence
IEEE Institute of Electrical and Electronics Engineers
ML Machine learning
NN Neural networks

References

Barr, A., Feigenbaum, E.A., Cohen, P.R., 1981. The Handbook of Artificial Intelligence. William Kaufmann, 1981.

Benhamou, E., 2022. Machine learning fundamentals: unsupervised learning part 1 - data & AI reskilling seminar slides. SSRN Electronic Journal. https://doi.org/10.2139/ssrn.4234520.

Brendel, A.B., Mirbabaie, M., Lembcke, T.B., Hofeditz, L., 2021. Ethical management of artificial intelligence. Sustainability 13 (4), 1−18. https://doi.org/10.3390/su13041974.

Burton, E., Goldsmith, J., Koenig, S., Kuipers, B., Mattei, N., Walsh, T., 2017. Ethical considerations in artificial intelligence courses. AI Magazine 38 (2), 22—34. https://doi.org/10.1609/aimag.v38i2.2731.

Cath, C., 2018. Governing artificial intelligence: ethical, legal and technical opportunities and challenges. Philosophical Transactions of the Royal Society A: Mathematical, Physical & Engineering Sciences 376 (2133), 20180080. https://doi.org/10.1098/rsta.2018.0080.

Chakrobartty, S., El-Gayar, O.F., 2022. Fairness Challenges in Artificial Intelligence. IGI Global, pp. 1685—1702. https://doi.org/10.4018/978-1-7998-9220-5.ch101.

Coeckelbergh, M., 2019. Artificial Intelligence: Some Ethical Issues and Regulatory Challenges', Technology and Regulation, pp. 31—34. https://doi.org/10.26116/techreg.2019.003.

Halbouni, A., Gunawan, T.S., Habaebi, M.H., Halbouni, M., Kartiwi, M., Ahmad, R., 2022. Machine learning and deep learning approaches for cybersecurity: a review. IEEE Access 10, 19572—19585. https://doi.org/10.1109/access.2022.3151248.

Huang, C., Zhang, Z., Mao, B., Yao, X., 2022. An overview of artificial intelligence ethics. IEEE Transactions on Artificial Intelligence 1—21. https://doi.org/10.1109/TAI.2022.3194503.

Kumar, N., Kharkwal, N., Kohli, R., Choudhary, S., 2016. Ethical aspects and future of artificial intelligence. In: 2016 1st International Conference on Innovation and Challenges in Cyber Security, ICICCS 2016. Institute of Electrical and Electronics Engineers Inc., India, pp. 111—114. https://doi.org/10.1109/ICICCS.2016.7542339.

Mhlanga, D., 2023. Artificial intelligence and machine learning for energy consumption and production in emerging markets: a review. Energies 16 (2), 745. https://doi.org/10.3390/en16020745.

Munoko, I., Brown-Liburd, H.L., Vasarhelyi, M., 2020. The ethical implications of using artificial intelligence in auditing. Journal of Business Ethics 167 (2), 209—234. https://doi.org/10.1007/s10551-019-04407-1.

Panesar, A., 2023. Artificial Intelligence and Machine Learning in Precision Health. Springer Science and Business Media LLC, pp. 67—85. https://doi.org/10.1007/978-1-4842-9162-7_4.

Prentice, C., 2023. Demystify Artificial Intelligence. Springer Science and Business Media LLC, pp. 25—40. https://doi.org/10.1007/978-981-99-1865-2_3.

Salvagno, M., Taccone, F.S., Gerli, A.G., 2023. Can artificial intelligence help for scientific writing? Critical Care 27 (1). https://doi.org/10.1186/s13054-023-04380-2.

Stahl, B.C., Antoniou, J., Ryan, M., Macnish, K., Jiya, T., 2022. Organisational responses to the ethical issues of artificial intelligence. AI & Society 37 (1), 23—37. https://doi.org/10.1007/s00146-021-01148-6.

Tercan, H., Meisen, T., 2022. Machine learning and deep learning based predictive quality in manufacturing: a systematic review. Journal of Intelligent Manufacturing 33 (7), 1879—1905. https://doi.org/10.1007/s10845-022-01963-8.

Tripathi, S., Musiolik, T.H., 2022. Fairness and ethics in artificial intelligence-based medical imaging. Research Anthology on Improving Medical Imaging Techniques for Analysis and Intervention 79—90. https://doi.org/10.4018/978-1-6684-7544-7.ch005.

Yampolskiy, R.V., 2018. Artificial Intelligence Safety and Security. Chapman and Hall/CRC. https://doi.org/10.1201/9781351251389.

Zhang, Y., Balochian, S., Agarwal, P., Bhatnagar, V., Housheya, O.J., 2014. Artificial intelligence and its applications. Mathematical Problems in Engineering 2014. https://doi.org/10.1155/2014/840491.

Zhang, B., Zhu, J., Su, H., 2023. Toward the third generation artificial intelligence. Science China Information Sciences 66 (2). https://doi.org/10.1007/s11432-021-3449-x.

A meshwork of artificial intelligence and biology: The future of science

19

Aaqib Zahoor[1], Shamsul Hauq[1], Umar Bashir[1], Ambreen Hamadani[1] and Shabia Shabir[2]

[1]*National Institute of Technology, Srinagar, Jammu and Kashmir, India;* [2]*Islamic University of Science and Technology (IUST), Awantipora, Jammu and Kashmir, India*

Introduction

Artificial intelligence (AI) refers to the development of computer systems that can perform tasks that would typically require human intelligence. It involves the creation of algorithms and models that enable machines to learn from data, reason, perceive, and make decisions in a manner similar to humans. AI encompasses a broad range of subfields, including machine learning (ML) and natural language processing (NLP). It is due to this that AI has the potential to become an extension of the human mind achieving feats unfathomable to mankind initially. Some important spheres in biology where AI is causing transformations are given in Fig. 19.1.

Over the years, AI has evolved significantly and has found applications in various industries and domains, transforming the way we live and work. At its core, AI aims to replicate human intelligence and cognitive abilities in machines. It involves teaching machines to learn from data and experiences, adapt to new situations, and perform tasks without explicit programming. ML, a subset of AI, plays a crucial role in enabling computers to learn and improve from data without being explicitly programmed. Through the use of algorithms, ML models analyze vast amounts of data, identify patterns, and make predictions or decisions based on the identified patterns. This approach has led to significant advancements in fields such as image and speech recognition, medical diagnosis, recommendation systems, and autonomous vehicles (Abbass, 2021).

Another important aspect of AI is NLP, which focuses on enabling machines to understand, interpret, and generate human language. NLP involves techniques for speech recognition, language translation, sentiment analysis, and chatbots, among others. NLP has revolutionized human—computer interaction, enabling users to communicate with machines more intuitively and efficiently. Voice assistants like Siri, Alexa, and Google Assistant are examples of AI applications that utilize NLP to understand and respond to user queries and commands.

Computer vision is yet another prominent field within AI that involves teaching machines to understand and interpret visual information from images or videos. Computer vision algorithms enable machines to recognize objects, detect patterns, and make sense of visual data. This has led to advancements in areas such as facial recognition, object detection, autonomous vehicles, and medical

A Biologist's Guide to Artificial Intelligence. https://doi.org/10.1016/B978-0-443-24001-0.00019-1

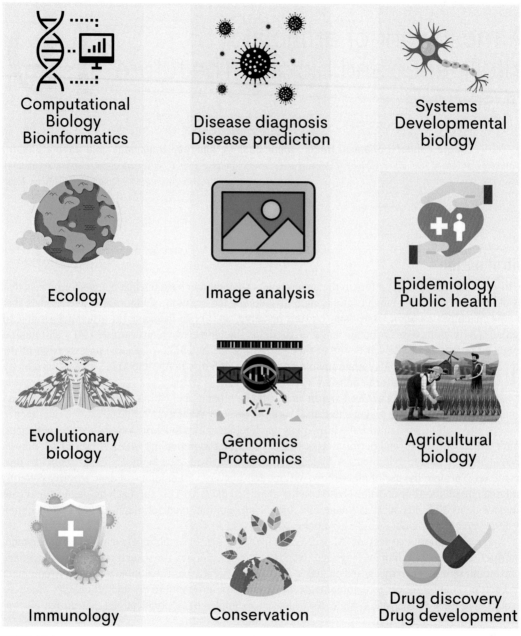

FIGURE 19.1

Some applications of artificial intelligence (AI) in biology.

imaging analysis. For example, computer vision algorithms can analyze medical images to detect anomalies or assist in diagnosing diseases, improving the accuracy and efficiency of healthcare systems.

AI has also found applications in robotics, where machines are designed to interact with the physical world and perform tasks that were traditionally carried out by humans. Robots equipped with AI capabilities can navigate their environment, manipulate objects, and adapt to changing circumstances. They are used in various industries, including manufacturing, healthcare, and agriculture. For instance, surgical robots assist surgeons in performing complex procedures with precision and minimal invasiveness, leading to improved patient outcomes (O'Leary, 2013).

Artificial intelligence is a rapidly evolving field that aims to create intelligent machines capable of performing tasks that would typically require human intelligence. Through ML, NLP, computer vision, and robotics, AI has made significant advancements in various domains. It has the potential to transform industries, improve efficiency, and address complex problems. Its adoption rate is increasing at an exponential rate. For example, robot shipments increased worldwide by about 150% in 6 years alone between 2010 and 2016 (Furman and Seamans, 2019).

However, ethical considerations and responsible development are essential to ensure the responsible and beneficial deployment of AI technologies. With ongoing research and development, AI is poised to continue shaping our future, enhancing our lives, and pushing the boundaries of what is possible. The integration of AI and biology paves the way for a secure future, where scientific advancements and healthcare breakthroughs thrive. AI's ability to analyze vast biological datasets and uncover hidden patterns empowers researchers to make significant discoveries and gain deeper insights into complex biological systems. By leveraging AI algorithms for personalized medicine, disease diagnosis, and drug discovery, we can enhance patient care, optimize treatment plans, and improve overall health outcomes. With AI's assistance in ecological conservation, genetics, and bioinformatics, we can make informed decisions to protect biodiversity and address pressing environmental challenges. The collaboration between AI and biology promises a secure future by revolutionizing research, healthcare, and sustainability efforts, enabling us to tackle complex biological problems and shape a brighter tomorrow.

Big data in biology and the role of AI

The biological data are growing tremendously and has been growing even beyond Exabyte (Li and Chen, 2014) and still are continuously growing, driven by advancements in technologies, ongoing research projects, and collaborative efforts (Hamadani et al., 2020). The combination of these diverse data sources provides a comprehensive view of biological systems and enables data-driven approaches to understanding biological processes, disease mechanisms, algorithm-driven predictions (Hamadani et al., 2022a,b), and personalized medicine (Ahuja, 2019). This data is being generated on various fronts including the following:

1. **Genomics and sequencing:** Advances in DNA sequencing technologies have led to the generation of massive amounts of genomic data. Projects like the Human Genome Project and subsequent initiatives have contributed to large-scale genomic datasets, providing a wealth of information about the human genome and other organisms. Sequencing efforts are expanding to include population genomics, cancer genomics, and metagenomics, generating vast amounts of

genomic data. The human genome project alone is predicted to generate about 40 Exabytes of data (NIH, 2023).

2. **Transcriptomics:** Transcriptomics involves the study of gene expression patterns in cells or tissues. Technologies such as RNA sequencing (RNA-seq) allow researchers to quantify gene expression levels on a genome-wide scale. This has led to the generation of large transcriptomic datasets, providing insights into gene regulatory networks, developmental processes, and disease mechanisms.

3. **Proteomics:** Proteomics involves the large-scale study of proteins and their functions. Mass spectrometry–based proteomics techniques enable the identification and quantification of thousands of proteins in a single experiment. Proteomics datasets contribute to understanding protein–protein interactions, posttranslational modifications, and protein function, thereby enhancing our knowledge of cellular processes and disease mechanisms.

4. **Imaging and microscopy:** Advances in imaging technologies have enabled the generation of high-resolution and high-throughput biological images. Techniques such as confocal microscopy, super-resolution microscopy, and live-cell imaging produce vast amounts of image data, capturing cellular structures, molecular interactions, and dynamic processes. These imaging datasets contribute to understanding cellular morphology, subcellular localization, and spatial relationships.

5. **Clinical data and electronic health records (EHRs):** EHRs and clinical databases provide a rich source of patient information, including medical histories, diagnoses, treatments, and outcomes. Integrating these datasets with genomics and other biological data enables researchers to perform large-scale analyses, identify disease-associated genetic variants, and explore relationships between genetic factors and clinical outcomes. Decision Support Systems developed for the purpose of creation of mega databases are also serving this purpose (Hamadani and Ganai, 2022).

6. **Public repositories and collaborative projects:** Numerous public repositories and collaborative initiatives store and curate biological data. Examples include the National Center for Biotechnology Information (NCBI) databases, the European Bioinformatics Institute (EBI) repositories, and international consortia like the Cancer Genome Atlas (TCGA) (Tomczak et al., 2015) and the Genotype-Tissue Expression (GTEx) project (Lonsdale et al., 2013). These resources make large-scale biological datasets available to the scientific community, promoting data sharing and collaborative research.

Conventional techniques and big data are two complementary elements in the realm of data analysis and decision-making. Conventional techniques refer to traditional approaches and methods that have been established and refined over time, whereas big data represents the vast amount of structured and unstructured data generated from various sources. Conventional techniques involve established statistical methods, hypothesis testing, regression analysis, and expert knowledge. These techniques have been widely used to analyze data, draw conclusions, and make informed decisions in various fields. They provide a foundation for understanding relationships, making predictions, and inferring causal relationships based on smaller, well-structured datasets.

On the other hand, big data encompasses large volumes, velocity, and a variety of data that conventional techniques alone may struggle to handle. Big data often includes data from social media, sensors, genomics, EHRs, and more. It is characterized by its high volume, velocity (speed of data generation), and variety (Hamadani et al., 2022a,b). The integration of big data and conventional techniques opens up new possibilities for analysis and decision-making. Big data can provide a more comprehensive and detailed picture of complex phenomena, uncovering patterns and correlations that may have gone unnoticed with smaller datasets. By applying conventional techniques to big data, analysts can derive valuable insights, identify trends, and make data-driven decisions. However, the challenges associated with big data, such as data storage, data quality, privacy concerns, and computational power requirements, require the utilization of advanced technologies, including AI and ML, to effectively extract meaningful information from the vast and diverse datasets.

AI for rapid breakthroughs in biology

AI technologies have been making a significant impact on various scientific fields, including biology (Table 19.1). The integration of AI techniques with biology has opened up new possibilities for research, analysis, and discovery. The Multi-Omics realm represents an enormous step forward in both biology and AI. It aggregates useful data on the use of AI in diverse biological domains, providing an understanding of the issues addressed, the AI solutions used, and the research gaps that need to be solved. By encouraging innovation, teamwork, and wise decision-making in the quest for discoveries at the nexus of AI and biology, the table plays a critical role in influencing the future of research through this comprehensive overview.

AI handling big data about tiny things

One area where AI has had a significant impact is in the analysis of large-scale biological datasets. With advancements in high-throughput technologies, such as DNA sequencing and proteomics, vast amounts of biological data are being generated. AI algorithms can process and analyze these datasets much faster than traditional methods, allowing researchers to gain valuable insights into complex biological systems. AI can identify patterns, correlations, and relationships within the data that may not be immediately apparent to human researchers.

AI has contributed to advancements in genomics research and gene editing techniques. In 2020, a team of scientists at Stanford University used AI to identify a new gene linked to autism spectrum disorder (ASD) (Supekar et al., 2022). The AI algorithm analyzed large genomic datasets to identify specific gene variants associated with ASD. This discovery provides insights into the genetic basis of ASD and opens avenues for further research and potential therapeutic interventions.

Making human lives better

This has the potential to accelerate research and drive new discoveries in areas, of drug development, and personalized medicine. AI has shown promise in aiding the diagnosis of rare and difficult-to-diagnose diseases. By analyzing patient data, including genomic information, symptoms, and medical records, AI algorithms can help identify patterns and potentially pinpoint rare genetic disorders. This can lead to earlier diagnosis and appropriate treatment interventions for patients with rare diseases, improving their chances of better health outcomes.

One breakthrough in Alzheimer's research using AI is the development of a ML model capable of predicting the conversion from mild cognitive impairment (MCI) to Alzheimer's disease with high

Table 19.1 A quick look at the work done in biology.

Domain	Problem	AI solutions	Research gaps and challenges	References
Biotechnology	Prediction of storm damages in forests	Spatial regression model framework, SVC model	Accommodation of high dimensional and distributional.	Nothdurft et al. (2021)
	Comparative analysis of various ML models for multiomics data analysis like genomics, transcriptomics, proteomics, etc.	Supervised and unsupervised ML models	The utility of ML models is restricted to the multiomics domain only.	Reel et al. (2021)
	Wheat heading stage observation	Computer vision, scale invariant feature transform, Fisher vector, linear SVM	Need of more feature characterization is important to achieve better results.	Eickholt and Cheng (2012)
Genetics	Precision medicine	Deep neural network, decision trees	Restrictions and regulations on the medical data to be used without patient consent.	Gymrek et al. (2013)
	Prediction of long noncoding RNAs	Increased K-mer frequency, support vector machines (SVM)	lncRNAs have complex genomic architecture, including overlaps with protein-coding genes or intergenic transcription. SVM models find it challenging to differentiate overlapping transcripts and assign them accurately.	Singh et al. (2019)
	Predicting protein secondary structure	Hidden markov models, support vector machines (SVM), long short-term memory (LSTM)	Limited data availability. Accurately capturing long-range interactions between distant amino acids in the protein sequence is critical for secondary structure prediction.	Eickholt and Cheng (2012)
Ecology	Ecological an intraclass modeling	SVM, RNN	Takes only a few characteristics into consideration while leaving some other parameters that can have more impact on the modeling.	Jeong et al. (2001)

Table 19.1 A quick look at the work done in biology.—cont'd

Domain	Problem	AI solutions	Research gaps and challenges	References
	Identification and classification of species	Random forests, CNN	High intraclass variability in the species. Sometimes species within the same class show the large morphological differences.	Cutler et al. (2007)
	Studying lameness	Random forests, CNN	Most of the studies have focused on studying the animal behavior of farm animals only. There is scope for using behavior identification in natural settings also.	Kaler et al. (2020)
Botany	Genomic prediction for crop breeding	Multilayer perceptron (MLP), probabilistic neural networks	Predict which regions of the genome can be edited to achieve a desired phenotype.	González-Camacho et al. (2016)
	Phenotyping of citrus plant	Convolutional neural networks	Real-time and high-throughput algorithms for phenotyping are needed.	Ampatzidis and Partel (2019)
	Disease detection	Random forest, multiclass SVM	Algorithms lack generalizability which can be the focus of further research.	Kumar Sahu and Pandey (2023)
Zoology	Species identification and classification	Computer vision models utilizing deep learning techniques to automatically identify and categorize species based on visual or audio signal	Behavior investigating AI methods for locating cryptic or rare species. Analyzing how environmental conditions affect the effectiveness of models for identifying species. Including expert knowledge for making AI models easier to understand.	Kumar and Kondaveerti (2023)
	Decision support systems and prediction algorithms	Development of AI and IoT-based decision support systems for data management. Comparison of algorithms for prediction. ANN, SVM, random forests etc	Models cannot be generalized to farms having different environmental conditions.	Hamadani et al. (2022a,b) and Hamadani and Ganai (2022)

Continued

Table 19.1 A quick look at the work done in biology.—cont'd

Domain	Problem	AI solutions	Research gaps and challenges	References
	Animal behavior analysis	Analyze complicated behavioral data using ML techniques to spot patterns and relationships	Examining the applicability of behavior models across various contexts and species. Investigating the potential of reinforcement learning methods for simulating the decision-making of animals. Integrating information from several sources to develop a thorough understanding of animal behavior.	Hager et al. (2019)
	Poaching prevention and wildlife preservation	Use AI-powered surveillance systems to support efforts to conserve animals and stop poaching	Analyzing the outcomes of AI-driven antipoaching initiatives in practical contexts. Addressing the ethical issues and privacy concerns related to AI-based monitoring. Using community involvement and local knowledge to develop sustainable conservation methods.	Dayer et al. (2020)
Virology	Virus spotting and evaluation	Build AI models that can quickly and accurately detect and diagnose infections	Investigating how AI may be used to quickly and accurately identify new or undiscovered infections. Addressing the requirement for understandable AI models to assist clinical judgment. Ensuring that AI systems are compatible with and integrated into current diagnostic workflows.	Chadaga et al. (2023)
	Development and improvement of vaccines	Accelerate the development of vaccines and improve immunization methods by using AI algorithms	Looking at AI-driven methods for quickly identifying possible vaccine targets. Researching the creation of customized vaccines based on unique genetic profiles and immune response patterns. Dealing with the difficulties of accelerating AI-driven vaccine research for a broad impact.	

Table 19.1 A quick look at the work done in biology.—cont'd

Domain	Problem	AI solutions	Research gaps and challenges	References
	Modeling and outbreak prediction in epidemiology modeling	Create AI-driven models for predicting disease transmission and early outbreak detection	Looking into the possibilities of AI methods for tracking disease outbreaks in real-time. Addressing challenges with data quality and availability for precise epidemiological modeling. Ensuring data privacy and ethical AI use in public health.	Farooq et al. (2022)

accuracy (Chapman et al., 2011). This breakthrough has significant implications for early diagnosis and intervention in Alzheimer's disease. By identifying those most likely to progress from MCI to Alzheimer's, clinicians can offer targeted interventions, such as lifestyle modifications, drug therapies, and clinical trials, to potentially slow down or prevent disease progression. The use of AI in predicting the conversion from MCI to Alzheimer's is an important step toward personalized medicine and precision healthcare in the field of neurodegenerative diseases. It holds the potential to improve patient outcomes, optimize clinical trial recruitment, and aid in the development of more effective interventions for Alzheimer's disease.

AI has also shown promise in the field of drug discovery. Developing new drugs is a time-consuming and costly process. AI algorithms can help screen large chemical libraries to identify potential drug candidates with specific properties. By using ML techniques, AI models can learn from vast amounts of existing data on drug compounds, their interactions, and their effects on biological systems. This can lead to more targeted and efficient drug development processes.

One breakthrough in drug discovery due to AI is the discovery of a new antibiotic called Halicin. In 2019, researchers at the Massachusetts Institute of Technology (MIT) used AI to identify this potential antibiotic compound with the ability to kill a wide range of bacteria, including drug-resistant strains. The researchers trained a deep learning algorithm on a large dataset of known molecules and their antibacterial properties. The algorithm learned to predict the likelihood of a compound having antibacterial activity. They then used this trained model to screen millions of small molecules, including those that had not been previously explored for their antibacterial potential. Through this AI-driven screening process, Halicin emerged as a promising candidate (Booq et al., 2021). It was found to have potent antibacterial properties against various pathogens, including drug-resistant strains like *Clostridium difficile* and *Acinetobacter baumannii*. Importantly, Halicin demonstrated effectiveness in killing bacteria that had developed resistance to traditional antibiotics. This breakthrough highlights how AI can expedite the drug discovery process by rapidly screening and identifying potential candidates from vast chemical libraries. The discovery of Halicin through AI represents a significant step forward in addressing the global challenge of antibiotic resistance and paves the way for the development of novel antibacterial agents.

AI-driven algorithms helped identify a potential drug called *baricitinib* as a treatment for COVID-19 (Selvaraj et al., 2022). Researchers used AI to screen existing drugs and predict their potential effectiveness against the SARS-CoV-2 virus. This discovery highlighted the potential of AI in repurposing existing drugs for new indications.

AI has also been applied to predict the potential side effects of drugs. In 2018, researchers from Stanford University developed an AI algorithm called Decagon that predicted potential drug-drug interactions and adverse side effects (Zitnik et al., 2018). The algorithm analyzed large-scale biological and chemical data to identify patterns and associations, enabling the prediction of previously unknown drug interactions and potential side effects.

Furthermore, AI is being used to enhance medical diagnostics and treatment. ML algorithms can analyze medical imaging data, such as MRI or CT scans, to assist in the early detection and diagnosis of diseases. AI can also help in predicting patient outcomes, guiding treatment decisions, and improving personalized medicine approaches.

The world around us

In addition, AI is aiding in the understanding of complex biological systems. Computational models based on AI techniques can simulate biological processes, such as protein folding or gene regulation, providing insights into their mechanisms and behavior. These models can help researchers understand the underlying principles of biological systems and predict their behavior under different conditions. In 2020, DeepMind's AlphaFold made a significant breakthrough by accurately predicting protein structures with remarkable accuracy. During the Critical Assessment of Structure Prediction (CASP) competition, a global competition in the field of protein structure prediction, AlphaFold achieved unprecedented performance, outperforming other methods by a considerable margin (Jumper et al., 2021). This breakthrough is significant because accurately predicting protein structures is crucial for understanding their functions and designing drugs. Prior to AlphaFold, protein structure prediction was a highly challenging problem that often required time-consuming and expensive experimental techniques. AlphaFold's ability to predict protein structures with high accuracy and speed has the potential to transform drug discovery, protein engineering, and our understanding of biological processes. The impact of AlphaFold's breakthrough in genomics is far-reaching, as it has the potential to unlock new insights into protein functions, disease mechanisms, and personalized medicine. It has opened up exciting possibilities for understanding complex biological systems at the molecular level and developing novel therapies for various diseases.

AI technologies have been applied in agriculture to optimize crop yield, monitor plant health, and manage resources efficiently. AI algorithms analyze data from sensors, satellites, and drones to provide insights into soil quality, crop growth patterns, and disease detection. This enables farmers to make informed decisions, reduce waste, and enhance agricultural sustainability. AI is aiding in conservation efforts by analyzing large-scale environmental data, satellite imagery, and species occurrence records. AI algorithms can predict species distributions, monitor habitat changes, and identify biodiversity hotspots. This helps in designing effective conservation strategies, mitigating the impacts of climate change, and preserving ecosystems. One example of a species identified using AI is a new species of moth named *Neopalpa donaldtrumpi* (Nazari, 1998). In 2017, researchers used ML algorithms to assist in the identification of this moth species. The researchers trained an AI model to recognize distinct patterns and features in moth wing images. The new species was discovered in Southern California, and its unique wing pattern caught the attention of the researchers. To confirm its distinct status as a new species, the researchers utilized AI algorithms to compare the wing patterns of the moth to a database of known species. The AI model helped analyze and classify the moth based on its unique wing markings, which differed significantly from any other known species.

Overall, the combination of AI and biology has the potential to revolutionize scientific research and lead to significant breakthroughs in our understanding of living systems, as well as the development of new therapies and treatments. However, it is important to note that the ethical considerations surrounding AI in biology, such as data privacy, transparency, and bias, must be carefully addressed to ensure the responsible and beneficial use of these technologies.

The promise of AI in biology

In the field of biology, AI technologies are expected to continue advancing and enabling new possibilities in the near future:

- **Deep learning for image analysis:** Deep learning algorithms, a subset of ML, have demonstrated remarkable capabilities in analyzing and interpreting images. In biology, this can be applied to tasks such as analyzing microscopy images, histopathology slides, or medical imaging scans. Deep learning models can help in automated cell and tissue classification, tumor detection, and quantification of biological features, leading to improved diagnostics and research efficiency.
- **Generative models for biological data synthesis:** Generative models, such as generative adversarial networks (GANs) and variational autoencoders (VAEs), have the potential to generate synthetic biological data that closely resemble real biological systems. This can be valuable when the availability of large, diverse datasets is limited. Generative models can be used to simulate the behavior of biological systems, generate new molecules with desired properties for drug discovery, or create artificial datasets for training and validation purposes.
- **Reinforcement learning for drug discovery:** Reinforcement learning, a type of ML, focuses on decision-making and learning through trial and error. In drug discovery, reinforcement learning algorithms can be used to optimize the design and selection of drug compounds. They can learn from experimental results, iteratively improve the design of molecules, and guide the search for potential drug candidates with desired properties. This approach has the potential to accelerate the drug discovery process and identify novel therapeutic targets.
- **Explainable AI for biological insights:** Explainable AI (XAI) techniques aim to provide interpretable explanations for the decisions made by AI models. In biology, XAI can be crucial for understanding the underlying biological mechanisms and generating hypotheses. It can help researchers uncover hidden patterns, identify critical features, and gain insights into complex biological processes. XAI can bridge the gap between AI predictions and human understanding, facilitating collaboration and knowledge discovery.
- **Integration of multiomics data:** Biology generates vast amounts of multi-omics data, including genomics, transcriptomics, proteomics, and metabolomics. Integrating and analyzing these diverse datasets can be challenging. AI technologies, such as network-based approaches, graph algorithms, and integrative ML, can assist in deciphering the complex interactions and relationships between different omics layers. This can lead to a comprehensive understanding of biological systems and facilitate personalized medicine approaches.

Alignment of AI with trends in biological sciences

To better align AI with trends in biological sciences, it is important to consider the following strategies. By implementing these strategies, we can ensure better alignment between AI and trends in biological sciences, fostering meaningful collaborations, advancing research, and facilitating transformative breakthroughs in the field:

1. **Collaboration between AI and biology experts:** Foster collaboration between AI researchers and biologists to ensure a deep understanding of the specific challenges and needs in the biological sciences. Encouraging interdisciplinary teams can lead to the development of AI tools and models that are tailored to address specific biological questions.

2. **Incorporate domain knowledge:** Biological systems are complex, and domain expertise is crucial for interpreting and contextualizing AI-driven results. It is important to integrate existing biological knowledge into AI models and algorithms. This can be achieved by incorporating domain-specific features, constraints, and prior knowledge into the learning process, ensuring that AI outputs are biologically meaningful.

3. **High-quality data acquisition and curation:** AI models heavily rely on high-quality and diverse datasets. In the biological sciences, it is essential to ensure data quality, accuracy, and representativeness. This requires careful data acquisition, preprocessing, and curation to eliminate biases, address missing data, and account for biological variability. Efforts should be made to promote data sharing and collaboration to build comprehensive and well-curated datasets for training and validation.

4. **Ethical considerations:** As AI technologies advance in the biological sciences, it is crucial to address ethical considerations, such as data privacy, transparency, and potential biases. Ensure responsible data usage, informed consent, and transparent reporting of AI methods and results. Ethical guidelines and regulatory frameworks need to be developed and continuously updated to guide the application of AI in biology.

5. **Interpretable and explainable AI:** Develop AI models that provide interpretability and explainability. The ability to understand and interpret the decision-making processes of AI models is vital in biology. Explainable AI techniques can help researchers validate and understand AI-driven predictions, identify potential sources of error, and generate testable hypotheses.

6. **Continuous learning and adaptation:** The field of biology is rapidly evolving, with new discoveries and insights emerging constantly. AI models should be designed to adapt and learn from new data and findings. Continuous learning algorithms can be developed to integrate new information into existing models and ensure that AI remains up to date with the latest biological knowledge.

7. **User-friendly interfaces and tools:** Make AI tools and interfaces user-friendly and accessible to biologists and researchers without extensive AI expertise. This will promote the adoption and integration of AI technologies into everyday biological research workflows, enabling researchers to leverage AI effectively in their work.

Science fiction and AI

Science fiction has often served as a source of inspiration for scientific and technological advancements. While many ideas from science fiction have not been fully realized, there have been significant achievements that align with concepts depicted in science fiction literature, movies, and TV shows. While we have made substantial progress in AI, current AI technologies are not yet at the level portrayed in science fiction. AI today primarily focuses on narrow tasks, such as image recognition, NLP, and recommendation systems. General AI, which possesses human-like cognitive abilities across different domains, remains an ongoing research challenge. A few examples of achievements that have been influenced by or are reminiscent of science fiction are given below:

1. **Communication technology:** Science fiction has long depicted advanced communication technologies, such as video calls and wireless communication devices. Today, we hasve smartphones, video conferencing, and wearable devices that bear a resemblance to the futuristic communication devices portrayed in science fiction.
2. **Space exploration:** Science fiction has imagined humans venturing into space and exploring other planets. In reality, we have made significant progress in space exploration with missions to the Moon, Mars rovers, and the International Space Station (ISS). Although we have not achieved interstellar travel or colonized other planets as depicted in science fiction, our exploration of space continues to advance.
3. **AI and robotics:** Science fiction has often portrayed intelligent machines and robots capable of performing complex tasks. While we have not achieved fully sentient AI or humanoid robots as depicted in science fiction, we have seen significant advancements in AI technologies, robotics, and automation. AI systems like voice assistants, autonomous vehicles, and robotic surgical systems are becoming increasingly prevalent.

The future of the meshwork

While we cannot predict the future with certainty, AI is expected to continue advancing. The development of more sophisticated AI algorithms, improvements in computational power, and breakthroughs in fields like deep learning and reinforcement learning could drive significant progress. The possibility of achieving generalized AI (AGI) remains a subject of ongoing scientific debate and speculation. While significant advancements have been made in the field of AI, developing a system that can exhibit human-level intelligence across a wide range of tasks and domains is a complex and formidable challenge. AGI requires surpassing the limitations of current AI approaches, which excel in narrow domains but struggle with generalization and understanding context. The development of AGI necessitates breakthroughs in areas such as cognitive architectures, learning algorithms, computational power, and understanding human cognition. Additionally, AGI raises fundamental questions about the nature of intelligence, consciousness, and the ability to emulate human-like thinking. Although the pursuit of AGI has attracted substantial research efforts and investment, its eventual realization and the timeline for its achievement remain uncertain. As the field progresses, careful consideration of ethical, societal, and safety aspects is crucial to guide the responsible development and deployment of AGI technologies.

There are several areas where significant advancements are anticipated:

1. **Artificial intelligence:** AI will continue to advance, with the potential for more sophisticated ML algorithms, deep learning models, and neural networks. This could lead to further breakthroughs in areas such as NLP, computer vision, and decision-making systems.
2. **Biotechnology and genetic engineering:** Advances in biotechnology and genetic engineering hold promise for significant developments in fields like personalized medicine, gene therapies, and synthetic biology. Technologies like gene editing (e.g., CRISPR) have the potential to revolutionize healthcare and agriculture.
3. **Space exploration and colonization:** The aspiration to explore and colonize other planets remains an ongoing endeavor. Future advancements could include more sophisticated space probes, manned missions to Mars, and potentially establishing long-term settlements on other celestial bodies.
4. **Virtual and augmented reality:** Virtual and augmented reality technologies are likely to continue evolving, offering immersive experiences and new possibilities in fields such as gaming, entertainment, education, and training.
5. **Nanotechnology:** The manipulation and control of matter at the nanoscale hold great potential for advancements in various fields, including materials science, medicine, and electronics. Nanotechnology could lead to the development of advanced materials, targeted drug delivery systems, and miniaturized devices.

Research challenges involved

The fusion of AIwith biology presents several research challenges that need to be addressed to fully leverage the potential of this interdisciplinary field. Some of the key challenges include:

1. **Data integration and quality:** Integrating and harmonizing diverse biological data from various sources is a significant challenge. Biological data often come in different formats, scales, and levels of complexity. AI algorithms must be able to handle and integrate data from genomics, proteomics, metabolomics, and other omics fields, as well as data from imaging, clinical records, and environmental factors. Ensuring data quality, standardization, and interoperability across different datasets is crucial for meaningful analysis and interpretation.
2. **Interpretability and explainability:** AI algorithms, particularly deep learning models, are often seen as black boxes, making it challenging to interpret their decision-making processes. In biological sciences, interpretability and explainability are crucial for gaining insights into biological mechanisms, validating findings, and building trust in AI-driven predictions or recommendations. Developing AI models that provide transparent explanations for their predictions or capturing complex biological processes in interpretable representations is an ongoing research challenge.
3. **Limited data and imbalanced datasets:** Biological datasets are often limited due to the cost, time, and effort required to generate high-quality data. Additionally, imbalanced datasets, where certain classes or rare events are underrepresented, are common in biological research. Training

AI models with limited and imbalanced data can lead to biased or unreliable results. Developing techniques to address data scarcity, imbalances, and the need for transfer learning across different biological contexts is crucial for robust AI applications in biology.

4. **Ethical and legal considerations:** The integration of AI and biology raises ethical and legal challenges, such as privacy concerns related to personal genomic data, consent, and the potential misuse of AI-generated synthetic biological entities. Developing frameworks and guidelines that address ethical issues, ensure privacy and data security, and promote responsible AI practices in biological research and healthcare applications is essential.

5. **Domain-specific knowledge and collaboration:** Successful integration of AI and biology requires interdisciplinary collaboration between computer scientists, biologists, bioinformaticians, and clinicians. Bridging the gap between these domains and fostering effective communication is crucial to leverage the expertise and insights from both fields. Developing AI approaches that can effectively incorporate domain-specific knowledge, biological priors, and expert feedback is essential for developing robust and accurate AI models in biology.

6. **Validation and reproducibility:** Validating AI models and their findings in biological research can be challenging due to the complex and dynamic nature of biological systems. Ensuring the reproducibility and robustness of AI-driven research findings requires rigorous validation methodologies, transparent reporting, open-source software, and standardized benchmarks for comparison.

Addressing these research challenges will pave the way for the effective integration of AI and biology, enabling transformative advancements in fields such as personalized medicine, drug discovery, ecological modeling, and understanding complex biological processes.

Cautious steps forward

While AI has the potential to bring about numerous benefits, it is also important to consider the potential challenges and problems it may pose in the future. The possibility of a robot uprising, as depicted in science fiction, remains highly unlikely based on our current understanding of AI and robotics. It is more important to focus on leveraging these technologies for the betterment of society, addressing challenges, and promoting human well-being. Some possible implications of AI that one must watch out for are given below:

1. **Job displacement:** As AI and automation technologies advance, there is a concern that they could lead to significant job displacement. Certain tasks and jobs that can be automated may no longer require human workers, potentially leading to unemployment or the need for workforce retraining.

2. **Bias and fairness:** AI systems are trained on large datasets, and if those datasets contain biased or incomplete information, the AI models may perpetuate or amplify those biases. This can lead to unfair or discriminatory outcomes in areas like hiring, lending, and criminal justice.

3. **Privacy and security:** AI systems often rely on collecting and analyzing large amounts of personal data. This raises concerns about privacy and the potential misuse or unauthorized access to sensitive information. There is also the risk of AI systems being vulnerable to attacks or manipulation, which could have significant consequences.

4. **Ethical considerations:** AI raises various ethical dilemmas, such as the responsibility and accountability for AI decisions, especially in critical areas like healthcare or autonomous vehicles. The development and use of AI systems must consider ethical principles and ensure transparency, explainability, and fairness.
5. **Social implications:** AI can impact social dynamics and exacerbate existing societal challenges. For instance, the use of AI in social media algorithms can influence information bubbles and echo chambers, leading to polarization and misinformation.
6. **The concentration of power:** AI development requires significant resources and expertise, leading to a potential concentration of power in the hands of a few organizations or countries. This concentration of power raises concerns about control, access, and the potential for misuse or abuse of AI technologies.

Addressing these potential problems requires careful consideration, collaboration, and the development of policies, regulations, and ethical frameworks that guide the responsible development and deployment of AI. It is crucial to ensure that AI is developed and used in a manner that benefits society as a whole, promotes fairness, safeguards privacy, and addresses potential risks and challenges.

Conclusion

In conclusion, AI is revolutionizing the field of biology, offering significant advancements in research, diagnosis, treatment, and understanding of biological systems. The present application of AI in biology has already demonstrated remarkable achievements, such as analyzing large-scale genomic data, predicting protein structures, accelerating drug discovery, and aiding in disease diagnosis and treatment. AI has enabled the handling of big data in biology, allowing for the extraction of meaningful insights and patterns from vast amounts of information. It has also facilitated the integration of diverse datasets, leading to new discoveries and a deeper understanding of complex biological processes. Looking ahead, the future of AI in biology holds immense potential. With continued advancements in AI algorithms, computational power, and data availability, we can expect even more breakthroughs. AI is likely to play a crucial role in unraveling the complexities of diseases, developing personalized medicine, predicting and preventing outbreaks, and advancing fields like genomics, proteomics, and neuroscience. However, as we embrace the possibilities of AI in biology, it is essential to address ethical considerations and privacy concerns and ensure the responsible development and deployment of AI technologies. Collaboration between AI experts, biologists, healthcare professionals, and policymakers will be vital in harnessing the full potential of AI while ensuring its responsible and ethical use.

References

Abbass, H., 2021. Editorial: what is artificial intelligence? IEEE Transactions on Artificial Intelligence 2 (2), 94−95. https://doi.org/10.1109/tai.2021.3096243.

Ahuja, A.S., 2019. The impact of artificial intelligence in medicine on the future role of the physician. PeerJ 7 (10). https://doi.org/10.7717/peerj.7702.

Ampatzidis, Y., Partel, V., 2019. UAV-based high throughput phenotyping in citrus utilizing multispectral imaging and artificial intelligence. Remote Sensing 11 (4). https://doi.org/10.3390/rs11040410.

Booq, R.Y., Tawfik, E.A., Alfassam, H.A., Alfahad, A.J., Alyamani, E.J., 2021. Assessment of the antibacterial efficacy of halicin against pathogenic bacteria. Antibiotics 10 (12). https://doi.org/10.3390/ANTIBIO TICS10121480.

Chadaga, K., Prabhu, S., Sampathila, N., Nireshwalya, S., Katta, S.S., Tan, Ru-S., Acharya, U.R., 2023. Application of artificial intelligence techniques for monkeypox: a systematic review. Diagnostics 13 (5). https://doi.org/10.3390/diagnostics13050824.

Chapman, R.M., Mapstone, M., McCrary, J.W., Gardner, M.N., Porsteinsson, A., Sandoval, T.C., Guillily, M.D., Degrush, E., Reilly, L.A., 2011. Predicting conversion from mild cognitive impairment to Alzheimer's disease using neuropsychological tests and multivariate methods. Journal of Clinical and Experimental Neuropsychology 33 (2), 187–199. https://doi.org/10.1080/13803395.2010.499356.

Cutler, D.R., Edwards, T.C., Beard, K.H., Cutler, A., Hess, K.T., Gibson, J., Lawler, J.J., 2007. Random forests for classification in ecology. Ecology 88 (11), 2783–2792. https://doi.org/10.1890/07-0539.1. http://www.esajournals.org/archive/0012-9658/88/11/pdf/i0012-9658-88-11-2783.pdf.

Dayer, A.A., Silva-Rodríguez, E.A., Albert, S., Chapman, M., Zukowski, B., Ibarra, J.T., Gifford, G., Echeverri, A., Martínez-Salinas, A., Sepúlveda-Luque, C., 2020. Applying conservation social science to study the human dimensions of neotropical bird conservation. The Condor: Ornithological Applications 122 (3). https://doi.org/10.1093/condor/duaa021.

Eickholt, J., Cheng, J., 2012. Predicting protein residue–residue contacts using deep networks and boosting. Bioinformatics 28 (23), 3066–3072. https://doi.org/10.1093/bioinformatics/bts598.

Farooq, Z., Rocklöv, J., Wallin, J., Abiri, N., Sewe, M.O., Sjödin, H., Semenza, J.C., 2022. Artificial intelligence to predict West Nile virus outbreaks with eco-climatic drivers. The Lancet Regional Health - Europe 17. https://doi.org/10.1016/j.lanepe.2022.100370.

Furman, J., Seamans, R., 2019. AI and the economy. Innovation Policy and the Economy 19 (1), 161–191. https://doi.org/10.1086/699936.

González-Camacho, J.M., Crossa, J., Pérez-Rodríguez, P., Ornella, L., Gianola, D., 2016. Genome-enabled prediction using probabilistic neural network classifiers. BMC Genomics 17 (1). https://doi.org/10.1186/s12864-016-2553-1. http://www.biomedcentral.com/bmcgenomics.

Gymrek, M., McGuire, A.L., Golan, D., Halperin, E., Erlich, Y., 2013. Identifying personal genomes by surname inference. Science 339 (6117), 321–324. https://doi.org/10.1126/science.1229566.

Hager, G.D., Drobnis, A., Fang, F., Ghani, R., Greenwald, A., Lyons, T., Parkes, D.C., Schultz, J., Saria, S., Smith, S.F., Tambe, M., 2019. Artificial intelligence for social good. arXiv. https://arxiv.org.

Hamadani, A., Ganai, N.A., 2022. Development of a multi-use decision support system for scientific management and breeding of sheep. Scientific Reports 12 (1). https://doi.org/10.1038/s41598-022-24091-y.

Hamadani, A., Ganai, N.A., Farooq, S.F., Bhat, B.A., 2020. Big data management: from hard drives to DNA drives. Indian Journal of Animal Sciences 90 (2), 134–140. http://epubs.icar.org.in/ejournal/index.php/IJAnS/article/view/98761/39230.

Hamadani, A., Ganai, N.A., Mudasir, S., Shanaz, S., Alam, S., Hussain, I., 2022a. Comparison of artificial intelligence algorithms and their ranking for the prediction of genetic merit in sheep. Scientific Reports 12 (1). https://doi.org/10.1038/s41598-022-23499-w.

Hamadani, A., Ganai, N.A., Alam, S., Mudasir, S., Raja, T.A., Hussain, I., Ahmad, H.A., 2022b. Artificial intelligence techniques for the prediction of body weights in sheep. Indian Journal of Animal Research. https://doi.org/10.18805/ijar.b-4831.

Jeong, K.S., Joo, G.J., Kim, H.W., Ha, K., Recknagel, F., 2001. Prediction and elucidation of phytoplankton dynamics in the Nakdong River (Korea) by means of a recurrent artificial neural network. Ecological Modelling 146 (1–3), 115–129. https://doi.org/10.1016/S0304-3800(01)00300-3.

Jumper, J., Evans, R., Pritzel, A., Green, T., Figurnov, M., Ronneberger, O., Tunyasuvunakool, K., Bates, R., Žídek, A., Potapenko, A., Bridgland, A., Meyer, C., Kohl, S.A.A., Ballard, A.J., Cowie, A., Romera-

Paredes, B., Nikolov, S., Jain, R., Adler, J., Back, T., Petersen, S., Reiman, D., Clancy, E., Zielinski, M., Steinegger, M., Pacholska, M., Berghammer, T., Bodenstein, S., Silver, D., Vinyals, O., Senior, A.W., Kavukcuoglu, K., Kohli, P., Hassabis, D., 2021. Highly accurate protein structure prediction with AlphaFold. Nature 596 (7873), 583−589. https://doi.org/10.1038/s41586-021-03819-2.

Kaler, J., Mitsch, J., Vázquez-Diosdado, J.A., Bollard, N., Dottorini, T., Ellis, K.A., 2020. Automated detection of lameness in sheep using machine learning approaches: novel insights into behavioural differences among lame and non-lame sheep. Royal Society Open Science 7 (1). https://doi.org/10.1098/rsos.190824.

Kumar, S.V.S., Kondaveerti, H.K., 2023. A comparative study on deep learning techniques for bird species recognition. In: 2023 3rd International Conference on Intelligent Communication and Computational Techniques, ICCT 2023. Institute of Electrical and Electronics Engineers Inc., India https://doi.org/10.1109/ICCT56969.2023.10075901. http://ieeexplore.ieee.org/xpl/mostRecentIssue.jsp?punumber=10075680.

Kumar Sahu, S., Pandey, M., 2023. An optimal hybrid multiclass SVM for plant leaf disease detection using spatial Fuzzy c-means model. Expert Systems with Applications 214. https://doi.org/10.1016/j.eswa.2022.118989.

Li, Y., Chen, L., 2014. Big biological data: challenges and opportunities. Genomics, Proteomics & Bioinformatics 12 (5), 187−189. https://doi.org/10.1016/j.gpb.2014.10.001.

Lonsdale, J., Thomas, J., Salvatore, M., Phillips, R., Lo, E., Shad, S., Hasz, R., Walters, G., Garcia, F., Young, N., Foster, B., Moser, M., Karasik, E., Gillard, B., Ramsey, K., Sullivan, S., Bridge, J., Magazine, H., Syron, J., Fleming, J., Siminoff, L., Traino, H., Mosavel, M., Barker, L., Jewell, S., Rohrer, D., Maxim, D., Filkins, D., Harbach, P., Cortadillo, E., Berghuis, B., Turner, L., Hudson, E., Feenstra, K., Sobin, L., Robb, J., Branton, P., Korzeniewski, G., Shive, C., Tabor, D., Qi, L., Groch, K., Nampally, S., Buia, S., Zimmerman, A., Smith, A., Burges, R., Robinson, K., Valentino, K., Bradbury, D., Cosentino, M., Diaz-Mayoral, N., Kennedy, M., Engel, T., Williams, P., Erickson, K., Ardlie, K., Winckler, W., Getz, G., DeLuca, D., MacArthur, D., Kellis, M., Thomson, A., Young, T., Gelfand, E., Donovan, M., Meng, Y., Grant, G., Mash, D., Marcus, Y., Basile, M., Liu, J., Zhu, J., Tu, Z., Cox, N.J., Nicolae, D.L., Gamazon, E.R., Im, H.K., Konkashbaev, A., Pritchard, J., Stevens, M., Flutre, T., Wen, X., Dermitzakis, E.T., Lappalainen, T., Guigo, R., Monlong, J., Sammeth, M., Koller, D., Battle, A., Mostafavi, S., McCarthy, M., Rivas, M., Maller, J., Rusyn, I., Nobel, A., Wright, F., Shabalin, A., Feolo, M., Sharopova, N., Sturcke, A., Paschal, J., Anderson, J.M., Wilder, E.L., Derr, L.K., Green, E.D., Struewing, J.P., Temple, G., Volpi, S., Boyer, J.T., Thomson, E.J., Guyer, M.S., Ng, C., Abdallah, A., Colantuoni, D., Insel, T.R., Koester, S.E., Little, A.R., Bender, P.K., Lehner, T., Yao, Y., Compton, C.C., Vaught, J.B., Sawyer, S., Lockhart, N.C., Demchok, J., Moore, H.F., 2013. The genotype-tissue expression (GTEx) project. Nature Genetics 45 (6), 580−585. https://doi.org/10.1038/ng.2653.

Nazari, V., 1998. Povolný, Review of Neopalpa, vol 646, pp. 79−94.

NIH, 2023. Genomic Data Science.

Nothdurft, A., Gollob, C., Kraßnitzer, R., Erber, G., Ritter, T., Stampfer, K., Finley, A.O., 2021. Estimating timber volume loss due to storm damage in Carinthia, Austria, using ALS/TLS and spatial regression models. Forest Ecology and Management 502. https://doi.org/10.1016/j.foreco.2021.119714.

O'Leary, D.E., 2013. Artificial intelligence and big data. IEEE Intelligent Systems 28 (2), 96−99. https://doi.org/10.1109/MIS.2013.39.

Reel, P.S., Reel, S., Pearson, E., Trucco, E., Jefferson, E., 2021. Using machine learning approaches for multi-omics data analysis: a review. Biotechnology Advances 49. https://doi.org/10.1016/j.biotechadv.2021.107739.

Selvaraj, V., Finn, A., Lal, A., Khan, M.S., Dapaah-Afriyie, K., Carino, G.P., 2022. Baricitinib in hospitalised patients with COVID-19: a meta-analysis of randomised controlled trials. eClinicalMedicine 49. https://doi.org/10.1016/j.eclinm.2022.101489.

Singh, S., Yang, Y., Póczos, B., Ma, J., 2019. Predicting enhancer-promoter interaction from genomic sequence with deep neural networks. Quantitative Biology 7 (2), 122−137. https://doi.org/10.1007/s40484-019-0154-0.

Supekar, K., Ryali, S., Yuan, R., Kumar, D., de los Angeles, C., Menon, V., 2022. Robust, generalizable, and interpretable artificial intelligence—derived brain fingerprints of autism and social communication symptom severity. Biological Psychiatry 92 (8), 643—653. https://doi.org/10.1016/j.biopsych.2022.02.005.

Tomczak, K., Czerwińska, P., Wiznerowicz, M., 2015. Review the cancer genome atlas (TCGA): an immeasurable source of knowledge. Współczesna Onkologia 1A, 68—77. https://doi.org/10.5114/wo.2014.47136.

Zitnik, M., Agrawal, M., Leskovec, J., 2018. Modeling polypharmacy side effects with graph convolutional networks. Bioinformatics 34 (13), i457—i466. https://doi.org/10.1093/bioinformatics/bty294.

Index

Note: Page numbers followed by *f* indicate figures and *t* indicate tables.

Printed in the United States
by Baker & Taylor Publisher Services